Geothermal energy: utilization and technology

Edited by
Mary H. Dickson and Mario Fanelli

CNR – Institute of Geosciences and Earth Resources

Routledge
Taylor & Francis Group

LONDON AND NEW YORK

earthscan
from Routledge

First published 2005 by Earthscan

Published 2022 by Routledge
2 Park Square, Milton Park, Abingdon, Oxon OX14 4RN
605 Third Avenue, New York, NY 10017

Routledge is an imprint of the Taylor & Francis Group, an informa business

Publisher's Note

The publisher has gone to great lengths to ensure the quality of this reprint
but points out that some imperfections in the original copies may be apparent.

ISBN 13: 978-1-138-99188-0 (pbk)
ISBN 13: 978-1-84407-184-5 (hbk)

Original title:
Geothermal Energy: Utilization and Technology
First published by the United Nations Educational, Scientific and Cultural
Organization (UNESCO), Paris, France
© UNESCO 2003
This edition has been published by arrangement with UNESCO

The designations employed and the presentation of material throughout this
publication do not imply the expression of any opinion whatsoever on the
part of UNESCO concerning the legal status of any country, territory, city or
area or of its authorities, or concerning the delimitation of its frontiers or
boundaries.

The authors are responsible for the choice and the presentation of the facts
contained in this book and for the opinions expressed therein, which are not
necessarily those of UNESCO and do not commit the Organization.

Typeset by S R Nova Pvt. Ltd, Bangalore, India
Cover design by Yvonne Booth

A catalogue record for this book is available from the British Library

Library of Congress Cataloging-in-Publication Data applied for

Geothermal energy: utilization and technology / edited by Mary H. Dickson and Mario Fanelli.
        p. cm.
Includes bibliographical references and index.
ISBN 1-84407-184-7
1. Geothermal engineering. 2. Geothermal resources. I. Dickson, Mary H. II. Fanellie. Mario.
TJ280.7.G443 2005
621.44-dc22

2005006659

Preface

UNESCO has actively promoted geothermal energy for over thirty years now, since the first UNESCO-sponsored geothermal training course in 1970, and the first UNESCO geothermal book in 1973. The progress achieved since then in the scientific and technological sectors and in terms of environmental awareness is immense, especially in the last ten years. For example, we are now able to construct more detailed and accurate models of geothermal systems; developments in exploration methodology and data interpretation have allowed us to cut back on the time and funds needed to investigate geothermal prospects; we are on the brink of creating artificial geothermal systems; we have increased the efficiency of geothermal power plants and direct-use plants, and have found innovative solutions to old problems; we have adopted heat pumps, which exploit the heat content of low-temperature resources, on a far greater scale and at a far faster rate than we could ever have imagined possible. In June 1992 in Rio, in December 1997 in Kyoto, and again in August 2002 in Johannesburg, at the World Summit on Sustainable Development, great emphasis was put on greenhouse gas emissions, and there were strong recommendations to reduce our consumption of fossil fuels, and to make better use of the renewable sources of energy, one of which is, of course, geothermal energy.

It is within this context that UNESCO is publishing *Geothermal Energy: Utilization and Technology*, as part of its new Renewable Energies series.

The book describes the various methods and technologies used to exploit the Earth's heat, *after* it has been extracted from the Earth's crust. It does *not* deal with the study of geothermal systems, the methods applied in geothermal research or the technologies used to extract geothermal resources, each of which would require an entire volume in order to be dealt with in any detail.

The reader of *Geothermal Energy: Utilization and Technology* will, ideally, have a scientific/technological background, and the technical reader will hopefully use it as a handbook. Other readers may find it a helpful guide to how geothermal resources can be utilized to our advantage and to the benefit of our environment. Finally, it should also prove an informative textbook for geothermal training courses.

The book is developed over nine chapters, the first of which provides an exhaustive overview of geothermal energy and the scientific and technological state of the art, as well as acting as a framework and reference point for the chapters that follow. Chapter 2 covers the generation of electricity, while Chapters 3 to 7 deals with the various non-electric uses of geothermal energy (district heating, space cooling, greenhouse heating, aquaculture and industrial applications). Chapter 8 discusses the environmental impact of geothermal energy, and, finally, Chapter 9 provides a complete review of the economic, financial and legal aspects of geothermal projects. The reader will find many stimulating case histories and practical examples throughout these chapters.

Geothermal Energy: Utilization and Technology will, we hope, serve in both the geothermal workplace and study room; the more it is used, the more we feel we will have contributed to sustainable development worldwide.

Mario Fanelli
Istituto di Geoscienze e Georisorse
CNR, Pisa

Contents

List of Figures   xi
List of Tables   xiv
List of Contributors   xvi
Notes on the Editors   xviii
Unit Conversion Tables   xix

1. Geothermal Background

Mary H. Dickson and Mario Fanelli

Aims   1
Objectives   1

Introduction   2
   Brief geothermal history   2
   Present status of geothermal utilization   3
Nature of geothermal resources   4
   The Earth's thermal engine   4
   Geothermal systems   8
Definition and classification of geothermal
   resources   10
Exploration   12
   Objectives of exploration   12
   Exploration methods   13
   Exploration programme   14
Utilization of geothermal resources   14
   Electricity generation   15
   Direct heat uses   16
   Economic considerations   19
Environmental impact   20
   Sources of pollution   21
Final considerations   22

References   24
Self-assessment questions   26
Answers   27

2. Electricity Generation

Roger B. Hudson

Aims   29
Objectives   29

Technical features of plant options   30
   Atmospheric exhaust conventional steam turbine   30
   Condensing exhaust conventional steam turbine   31
   Binary plant   36
   Biphase rotary separator turbo-alternator   38
Well-head generating units   41
   *Economic considerations regarding small geothermal
      plants*   42

Reference   44
Recommended literature   44

Appendix A
Thermodynamics of cycles   45
    Thermodynamics of flash process   45
    Thermodynamics of organic Rankine cycle   45
    Thermodynamics of biphase process   48

Appendix B
Principal manufacturers   49
    Manufacturers of conventional atmospheric exhaust
        or condensing steam turbine plants   49
    Manufacturers of geothermal binary plants   49
    Manufacturers of geothermal biphase rotary separator
        plants   49

Self-assessment questions   50
Answers   51

3. Space and District Heating

Einar Tjörvi Elíasson, Halldór Ármannsson,
Sverrir Thórhallsson, María J. Gunnarsdóttir,
Oddur B. Björnsson and Thorbjörn Karlsson

Aims   53
Objectives   53

Introduction   54
    Preamble   54
    Utilization of low-temperature geothermal
        resources   54
Resource considerations   54
    Resource development   54
    Temperature of fluid   55
    Available flow rates   55
    Chemistry of fluids   55
    Distance from potential market   57
Space heating (or cooling) needs   58
    Climate   58
    Population, population density   58
    Building types   59
    Techno-economic aspects   59
Hot water collection and transmission system   59
    Types of district heating systems   59
    Pipe systems   61

Equipment selection   64
    Down-hole pumps   64
    De-gassing tanks   65
    Heat exchangers   65
    Radiators   66
    Control equipment   66
    Heat pumps   67
Economic considerations   67
    Cost of drilling   67
    Cost of pipeline   67
    Capital investment   68
    Operating cost   69
    Cost of improving heat efficiency of buildings   69
    Cost of alternative thermal energy sources   69
Tariffs   70
    Sales policies and metering   70
Integrated uses   71
Environmental considerations   72
    Chemical pollution   72
    Thermal pollution   72
    Physical effects   72
    Social and economic considerations   73

Recommended literature   73
Self-assessment questions   74
Answers   77

4. Space Cooling

Kevin D. Rafferty

Aims   81
Objectives   81

Introduction   82
Air conditioning   82
    Lithium bromide/water cycle machines   82
    Performance   83
    Large tonnage equipment costs   84
    Small tonnage equipment   85
Commercial refrigeration   86
    Water/ammonia cycle machines   86
Absorption research   87
Materials   87
Conclusions   88

References   88
Self-assessment questions   89
Answers   90

5. Greenhouse Heating

Kiril Popovski

Aims   91
Objectives   91

Introduction   92
Energy aspects of protected crop cultivation   92
  Why protected crop cultivation   92
  Greenhouse climate   93
Characteristics of heat consumption   94
Technical solutions for geothermal greenhouse
  heating   95
  Factors influencing the choice of technological
    solution   95
  Hot-water transmission systems   95
  Combined uses   96
Geothermal greenhouse heating installations   98
  Classification   98
  Soil heating installations   100
  Soil–air heating installations   101
  Aerial pipe heating installations and convectors   102
  Fan-assisted convectors   102
  Other types of heating installation   102
Factors influencing the choice of heating
  installation   102
  Temperature profiles in the greenhouse   102
  Economic aspects   106
  Operating problems   107
  Environmental aspects   107
  Adaptation of technological solutions to local
    conditions   108
Final considerations   108

References   109
Self-assessment questions   111
Answers   112

6. Aquaculture

Aims   113
Objectives   113

Introduction to Geothermal Aquaculture

John W. Lund

Background   114
Examples of geothermal projects   114
General design and considerations   116
Additional information   117

References   117
Self-assessment questions   119
Answers   120

Aquaculture Technology

Kevin D. Rafferty

Introduction   121
Heat exchange processes   121
  Evaporative loss   121
  Convective loss   122
  Radiant loss   123
  Conductive loss   123
Reduction of heating requirements   124
  Surface cover   124
  Pond enclosure   124
  Thermal mass   125
Flow requirements   125

References   126
Self-assessment questions   127
Answers   128

7. Industrial Applications

Paul J. Lienau

Aims   129
Objectives   129

Introduction   130
Examples of industrial applications of
   geothermal energy   130
   Pulp, paper and wood processing   130
   Diatomite plant   132
   Vegetable dehydration   132
   Other industrial uses   132
Selected industrial applications   134
   Pulp and paper mill (Hornburg and Lindal, 1978)   134
   Drying lumber (VTN-CSL, 1977)   136
   Crop drying (Lienau, 1978)   138
   Vegetable and fruit dehydration (Lienau, 1978)   142
   Potato processing (Lienau, 1978)   145
   Heap leaching (Trexler et al., 1987, 1990)   148
   Waste-water treatment plant (Racine et al., 1981)   150

References   151
Self-assessment questions   153
Answers   154

8. Environmental Impacts and Mitigation

Kevin Brown and Jenny Webster-Brown

Aim   155
Objectives   155

Introduction   156
Physical impacts   156
   Land and the landscape   156
   Noise   156
   Natural geothermal features   157
   Heat-tolerant vegetation   157
   Hydrothermal eruptions   158
   Subsidence   158
   Induced seismicity   159
   Thermal effects of waste discharge   159
   Water usage   159
   Solid wastes   160
Impacts on air quality   160
   Composition of gas discharges   160
   Toxic and environmental effects   161

Impacts on water quality   162
   Composition of fluid discharges   163
   Toxicity and environmental effects   163
Social impacts   164
Workplace impacts   165
   Exposure to airborne contaminants   165
   Exposure to liquid contaminants   166
   Exposure to noise   167
   Exposure to heat   167
   OSH criteria and standards   167
Legislation and EIA   168
   Environmental impact assessment   168
   The consent application and award process   168
   Monitoring programmes   170
   The future   170

References   170
Recommended literature   171
Self-assessment questions   172
Answers   173

9. Economics and Financing

R. Gordon Bloomquist and George Knapp

Aims   175
Objectives   175

Introduction   176
Economic considerations   176
   Provision of fuel   176
   Project design and facility construction   178
   Revenue generation   185
Financing considerations   187
   Institutional framework   188
   Financing approaches and sources   191
   Contracts and risk allocation   194

References   200
Self-assessment questions   202
Answers   204

List of Figures

1.1 The Earth's crust, mantle and core. Top right: a section through the crust and the uppermost mantle 2

1.2 Schematic cross-section showing plate tectonic processes 7

1.3 World pattern of plates, oceanic ridges, oceanic trenches, subduction zones and geothermal fields. Arrows show the direction of movement of the plates towards the subduction zones. (1) Geothermal fields producing electricity; (2) mid-oceanic ridges crossed by transform faults (long transversal fractures); (3) subduction zones, where the subducting plate bends downwards and melts in the asthenosphere 7

1.4 Schematic representation of an ideal geothermal system 8

1.5 Model of a geothermal system. Curve 1 is the reference curve for the boiling point of pure water. Curve 2 shows the temperature profile along a typical circulation route from recharge at point A to discharge at point E 9

1.6 Diagram showing the different categories of geothermal resources. The vertical axis is the degree of economic feasibility; the horizontal axis is the degree of geological assurance 10

1.7 The Lindal diagram 14

1.8 Simplified flow diagram of the geothermal district heating system of Reykjavik 16

1.9 Simplified schemes of ground source heat pumps 17

1.10 Growth curves for some crops 18

1.11 Effect of temperature on growth or production of food animals 18

2.1 Atmospheric exhaust cycle simplified schematic 30

2.2 Sketch of Mitsubishi Modular-5/10 and MPT-2L mini-turbines 30

2.3 Atmospheric-exhaust (sea level) turbine steam consumption curves 31

2.4 Generalized capital cost for conventional steam turbine geothermal power development (1998) 31

2.5 Condensing cycle simplified schematic 32

2.6 Ansaldo 55 MW double-flow geothermal turbine 32

2.7 Condensing cycle process schematic 33

2.8 Condensing turbine steam consumption curves 35

2.9 Binary cycle simplified schematic 36

2.10 Isometric of biphase rotary separator 38

2.11 Topping system arrangement for biphase turbine with conventional condensing turbine 39

2.12 Ratio of net power of standard condensing biphase topping system to optimized single-flash condensing unit versus biphase inlet pressure (for constant mass flow) 40

2.13 Performance of condensing biphase topping system and optimized single-flash condensing unit versus fluid enthalpy 40

2.14 Temperature–entropy diagram of typical geothermal flashed-steam process and available isentropic heat drop 45

2.15 Characteristics of organic Rankine cycle using R114N compared with equivalent Rankine cycle using water 46

2.16   Temperature–thermal power diagram for heat exchange using water or R114 as secondary fluid   47

2.17   Temperature–entropy diagram for biphase rotary separator turbine system   48

3.1   Typical well drawdown curves   55

3.2   Typical reservoir drawdown curves   56

3.3   Heat loss in pipelines on the basis of 0 °C outdoor temperature   57

3.4   Typical duration curve   58

3.5   Main types of district heating system   60

3.6   Types of pipeline used for geothermal district heating systems   61

3.7   Effects of radiator size   64

3.8   Typical house heating system   65

3.9   Effect of water temperature on the size of radiators   66

3.10   Cost of low-temperature well drilling in Europe   68

3.11   Chloride variation in system 'a' with time   76

3.12   Chlorine variation in system 'b' with time   76

4.1   Diagram of two-shell lithium bromide cycle water chiller   82

4.2   Capacity of lithium bromide absorption chiller   83

4.3   Chiller and auxiliary equipment costs: electric and absorption   84

4.4   Simple payback on small absorption equipment compared to conventional rooftop equipment   85

4.5   Small tonnage absorption equipment performance   86

4.6   Required resource temperatures for water/ammonia absorption equipment   86

4.7   The COP for water/ammonia absorption equipment in refrigeration applications   87

4.8   Water/ammonia single and double effect regenerative cycle performance   87

5.1   Heat requirement fluctuations in a greenhouse; average conditions in January in a greenhouse in Gevgelia, Republic of Macedonia. E = solar radiation energy flux (Wh/m^2); T_a = outside air temperature (°C); T_{in} = optimal inside air temperature (°C); Q = greenhouse heat requirements (W)   94

5.2   Heat requirement fluctuations in a greenhouse over a typical year in Gevgelia, Republic of Macedonia. E = solar radiation energy flux (Wh/m^2); $T_{a.av}$ = monthly average outside air temperature (°C); Q = greenhouse heat requirements (W)   94

5.3   Percentage coverage of annual heat demand for greenhouse heating by alternative energy sources relative to Naaldwijk, Netherlands   94

5.4   Simple direct connection to the geothermal spring or well   95

5.5   Heat pump installation with simple connection to the well   96

5.6   Simplified flow diagram of the Bansko integrated geothermal system (Republic of Macedonia), consisting of greenhouse heating and different heat users of a hotel-spa complex   97

5.7   Technically improved geothermal installation in Srbobran (Yugoslavia), with extraction of the CH$_4$ from the geothermal water for electricity generation and direct heat use for peak demand in the greenhouse   98

5.8   Classification of low-temperature heating systems. Heating installations with natural air movement (natural convection): (a) aerial pipe heating; (b) bench heating; (c) low-position heating pipes for aerial heating; (d) soil heating. Heating installations with forced-air movement (forced convection): (e) lateral position; (f) aerial fan; (g) high-position ducts; (h) low-position ducts   99

5.9   Installation for heating the soil in greenhouses: (a) position of heating pipes; (b) temperature profile of the heated soil; (c) vertical temperature profile in the greenhouse   100

5.10   Installation for heating the air and soil in greenhouses (heating pipes placed on the soil surface). (a) Position of heating pipes; (b) different solutions for allocation of heating pipes (cultivation on benches, in pots on soil surface and in soil); (c) vertical temperature profile in greenhouse   101

5.11   Aerial heating installations made of smooth or finned steel pipes. (a) and (b) Position of aerial pipe heat exchangers for low-temperature heating fluids: (b1) along the plant rows; (b2) in the plant canopy; (b3) below the benches; (c) heat transfer coefficient for the greenhouse interior based on pipe diameter and temperature of the heating fluid; (d) vertical air temperature profile in a greenhouse heated by aerial pipe heating installations   103

5.12 (a) Position of the 'fan-jet' heating system in a greenhouse; (b) layout of a fan-assisted convector unit; (c) vertical temperature profiles in a greenhouse heated with fan convector units at different heights   104

5.13 Water/air heat exchanger in air distribution tube   104

5.14 Vertical temperature profiles in a greenhouse, depending on the type and location of the heating installation. (a) High aerial pipes; (b) high pipes; (c) low pipes; (d) high-positioned air-heaters; (e) 'fan-jet' system 2 m above the crop; (f) high-speed air heating; (g) convectors; (h) 'fan-jet' between the plants; (i) 'fan-jet' below the benches   105

5.15 Horizontal temperature distribution (°C) in a greenhouse heated by different types of heating installation, under the same outside climatic conditions   105

5.16 Vertical air temperatures in a greenhouse, heated by different types of heating installations   106

5.17 Heat cost of geothermal energy, heavy oil, coal and electricity, as a function of annual load factor. Annual load factor $L = (Q/P)\ 8{,}760 \times 3{,}600$, where Q is the heat consumption during the year (joules) and P is the heat capacity of the geothermal well (watts). Example of greenhouse heating in Bansko, Macedonia   106

5.18 Block diagram of the development of a geothermal greenhouse heating project   109

6.1   Optimum growing temperatures for selected animal and aquatic species   114

6.2   The geothermal aquaculture research project at Oregon Institute of Technology   115

7.1   Application temperature range for some industrial processes and agricultural applications   131

7.2   Pulp mills (Kraft process) process flow   134

7.3   Lumber drying process flow   136

7.4   Long-shaft, double-track, compartment kiln with alternately opposing internal fans   137

7.5   Location of fans and heat exchangers in kilns   139

7.6   Alfalfa drying and pelletizing process   140

7.7   Perforated false-floor system for bin drying of grain   140

7.8   Columnar grain dryer   141

7.9   Tunnel dryer, air flow pattern   142

7.10 Vegetable dehydration process flow   143

7.11 Multi-stage conveyor dryer using 110 °C geothermal fluid and 4 °C ambient air   145

7.12 Frozen french fry process flow   146

7.13 Potato processing flow diagram for geothermal conversion   147

7.14 Idealized thermally enhanced heap leach   148

7.15 Heap leach process flow   149

7.16 Soil temperature at a depth of 10 cm at Central Nevada Field Laboratory near Austin, Nev., (elevation 1,810 m)   149

7.17 Waste-water treatment process flow   150

8.1   Subsidence at the Wairakei geothermal field in New Zealand   159

8.2   The principal contaminants potentially occurring in gas, steam and fluid discharges, shown for a model, developed geothermal field   160

8.3   A sign warning of the potential danger posed by hydrogen sulfide in a well-head pit at Wairakei geothermal field in New Zealand   166

8.4   Flow diagram showing the consent application process under the Resource Management Act, New Zealand's principal environmental legislation   169

List of Tables

1.1 Installed geothermal generating capacities worldwide from 1995 to 2000 (from Huttrer, 2001) and at the beginning of 2003   4

1.2 Electric capacity from geothermal energy out of total electric capacity for some developing countries in 1998 (MW_e)   4

1.3 Non-electric uses of geothermal energy in the world (2000): installed thermal power (in MW_t) and energy use (in TJ/yr)   5

1.4 Classification of geothermal resources based on temperature (°C)   11

1.5 Energy and investment costs for electric energy production from renewables   20

1.6 Energy and investment costs for direct heat production from renewables   20

1.7 Probability and severity of potential environmental impact of direct-use projects   21

1.8 Geothermal potential worldwide   22

2.1 Expected manufacture and erection times for a typical well-head development   32

2.2 Manufacture and erection times for a typical development   36

2.3 Expected manufacture and erection times for modular binary and combined-cycle installations   38

2.4 Expected manufacture and erection periods for a typical biphase installation   41

2.5 Latent heat of vaporization and specific heat capacity   47

3.1 Maximum permissible levels of chemical concentration in freshwater in comparison with those typical in geothermal effluent   56

3.2 Common types of pipeline   62

3.3 Effect of insulation on typical house heating demand   70

3.4 Geothermal areas: short- and long-term capacities   75

3.5 Chemical analyses of fluids from geothermal areas (in mg/kg except for pH)   75

3.6 Advantages and disadvantages for the three areas considered   78

5.1 Geothermal greenhouses in the world   92

6.1 Crops that are good candidates for aquaculture   117

6.2 Temperature requirements and growth periods for selected aquaculture species   121

6.3 Summary of example of heat loss   124

6.4 Summary of example of heat loss using pool cover   124

6.5 Summary of example of heat loss using pond enclosure   125

7.1 Summary of geothermal applications   133

7.2 Comparison of pulp and paper process steam requirements   135

7.3 Typical kiln-drying schedules   138

7.4 Energy consumed in kiln-drying wood (Moore Dry Kiln Company, Oregon)   138

7.5   Minimum geothermal fluid temperatures for kiln drying at kiln inlet (Moore Dry Kiln Company, Oregon)   139

7.6   Product drying in a conveyor dryer   144

7.7   Conveyor dryer energy requirements   144

7.8   Potato processing temperature requirements   147

7.9   Waste-water treatment plant process temperatures   151

8.1   Contaminant concentrations in some representative steam and waste-bore-water discharges from developed geothermal fields. Concentration units are mg/kg, for all except mercury (μg/kg)   161

List of Contributors

Halldór Ármannsson
Chief Geochemist
ÍSOR – Icelandic GeoSurvey
Grensasvegur 9
108 Reykjavik
Iceland
E-mail address: h@isor.is
Web: www.isor.is

Oddur B. Björnsson
Chief Engineer
Fjarhitun hf. Consulting Engineering Company
Borgartún 17
105 Reykjavik
Iceland
E-mail address: oddur@fjarhitun.is

R. Gordon Bloomquist
Senior Scientist, Washington State University Energy Program
925 Plum St.
Town Square Building No. 4
PO Box 43165
Olympia, WA 98504-3165
USA
E-mail address: bloomquistr@energy.wsu.edu

Kevin L. Brown
Associate Professor, Geothermal Institute
University of Auckland
Private Bag 92019
Auckland
New Zealand
E-mail address: kl.brown@auckland.ac.nz

Mary H. Dickson
Associate Editor of Geothermics
Istituto di Geoscienze e Georisorse
National Research Council of Italy
Via G. Moruzzi, 1
56124 Pisa
Italy
E-mail address: marnell@igg.cnr.it

Einar Tjörvi Elíasson
CEO
KRETE Geothermal Consulting Ltd
Hjardarhaga 50
108 Reykjavik
Iceland
E-mail address: ete@krete.is and ete@os.is

Mario Fanelli
Consultant Geologist
Istituto di Geoscienze e Georisorse
National Research Council of Italy
Via G. Moruzzi, 1
56124 Pisa
Italy
E-mail address: fanelli@igg.cnr.it

María J. Gunnarsdóttir

Division Manager, Samorka (Federation of Icelandic Energy and Water Works)
Sudurlandsbraut 48
IS-108 Reykjavik
Iceland
E-mail address: mariaj@samorka.is

Roger B. Hudson

Engineering Group Manager, Power
PB Power, Asia Pacific
60 Cook St., Auckland
PO Box 3935 Auckland
New Zealand
E-mail address: HudsonR@pbworld.com

Thorbjörn Karlsson

Professor, University of Iceland
Reykjavik
Iceland
E-mail address: thorbj@verk.hi.is

George M. Knapp

Senior Partner, Energy Group
Squire, Sanders & Dempsey, L.L.P.
1201 Pennsylvania Avenue, NW
Washington, D.C. 20004
USA
E-mail address: gknapp@ssd.com

Paul J. Lienau

Director Emeritus, Geo-Heat Center
Oregon Institute of Technology
3201 Campus Dr.
Klamath Falls-OR 97601
USA
E-mail address: lienaupc@whidbey.net

John W. Lund

Professor of Civil Engineering
Director, Geo-Heat Center

Oregon Institute of Technology
3201 Campus Dr.
Klamath Falls-OR 97601
USA
E-mail address: lundj@oit.edu
Web: geoheat.oit.edu

Kiril Popovski

Director, International Summer School on Direct Applications of Geothermal Energy
Ul Dame Gruev br 1-III/16
Skopje 91000
F.Y.R. of Macedonia
E-mail address: isskiril@sonet.com.mk

Kevin D. Rafferty

Senior Mechanical Engineer/Associate Director
Geo-Heat Center
Oregon Institute of Technology
3201 Campus Dr.
Klamath Falls-OR 97601
USA
E-mail address: raffertk@oit.edu

Sverrir Thórhallsson

Head, Engineering Department
ÍSOR – Icelandic GeoSurvey
Grensasvegur 9
108 Reykjavik
Iceland
E-mail address: s@isor.is

Jenny Webster-Brown

Environmental Science
University of Auckland
Auckland
New Zealand
E-mail address: j.webster@auckland.ac.nz

Notes on the Editors

Mary Helen Dickson

Mary Helen Dickson obtained an M.A. in English and History from Glasgow University, Scotland. She began her career as a journalist with a Scottish daily newspaper. After moving to Italy, in 1972 she joined the staff of the International Institute for Geothermal Research of Pisa (National Research Council of Italy), which has now become a part of the much larger Istituto di Geoscienze e Georisorse (IGG).

Her main activity in the IGG is with the International School of Geothermics, organizing UNESCO-sponsored geothermal courses and congresses in Italy and abroad, and as Associate Editor of the international journal *Geothermics*.

She acted as consultant for geothermal energy for the UNITAR (United Nations Institute for Training and Research)/UNDP (United Nations Development Programme) Centre on Small Energy Resources; she also played an active role in the International Geothermal Association, as member of its Board of Directors and as Chairman of the Information Committee in the period 1998–2001, as well as serving on a number of other IGA Committees. She also collaborated in the organization of the World Geothermal Congress '95 (Italy), and the World Geothermal Congress 2000 (Japan).

She is the author of 45 publications.

Mario Fanelli

Mario Fanelli obtained a degree in Geology from Rome University. After a few years of studies in prehistory, in 1964 he began working in applied geology, as part of the newly created geothermal research programme of the National Research Council of Italy. He was a senior researcher with the International Institute for Geothermal Research in Pisa until 2001. He now collaborates with the Istituto di Geoscienze e Georisorse (CNR).

As Director of the International School of Geothermics he organized UNESCO-sponsored geothermal courses and congresses in Italy and abroad from 1970 to 2000. He was Secretary of the Organizing Committee of the first international geothermal symposium (United Nations Symposium on the Development and Utilization of Geothermal Resources, Pisa, 1970), and collaborated in the organization of the World Geothermal Congress '95 (Italy), and the World Geothermal Congress 2000 (Japan).

He headed the project 'Geothermal Gradient and Heat Flow in Italy' of the Commission of the European Communities, and from 1976 to 1981 was responsible for the Geothermal Data Bank (a joint project of the National Research Council of Italy, the Electricity Generating Authority of Italy and the US Department of Energy). He has also acted as consultant for geothermal energy for the UNITAR/UNDP Centre on Small Energy Resources.

He is the author of more than 70 publications.

Unit Conversion Tables

Class	To convert	into	Multiply by
Length	inch (in)	m	0.0254
	foot (ft)	m	0.3048
	mile	m	1,609
Velocity	ft/s	m/s	0.3048
	ft/min	m/s	0.0051
	miles per hour	m/s	0.447
	miles per hour	km/h	1.609
Area	ft^2	m^2	0.0929
	hectare (ha)	m^2	10,000
	hectare (ha)	km^2	0.01
Volume	in^3	m^3	1.639×10^{-5}
	ft^3	m^3	0.02832
	litre (l)	m^3	0.001
	gallon US (gal)	m^3	3.785×10^{-3}
	barrel (42 gal)	m^3	0.159
Mass	pound (lb)	kg	0.4536
	ton (US short)	kg	907
Density	lb/in^3	kg/m^3	2.768×10^4
	lb/ft^3	kg/m^3	16.02
	lb/gal (US)	kg/m^3	119.83
Force	pound-force (lb_f)	N (newton)	4.448
	kilogram-force (kg_f)	N	9.807

Class	To convert	into	Multiply by
Pressure	bar	Pa (pascal)	10^5
	millibar	Pa	100
	atm (760 mm Hg)	Pa	1.013×10^5
	mm Hg (0 °C)	Pa	133.3
	kg/cm²	Pa	9.807×10^4
	psi	Pa	6,895
	atm (760 mm Hg)	bar	1.013
	mm Hg (0 °C)	bar	1.333×10^{-3}
	kg/cm²	bar	0.9807
	psi	bar	0.06895
Energy	cal	J (joule)	4.184
	kcal	J	4,184
	British thermal unit (BTU)	J	1,055
	kWh	J	3.6×10^6
Power	lb-ft/s	W	1.356
	BTU/s	W	1,055
	BTU/h	W	0.2931
	HP (metric)	W	735.5
Enthalpy	BTU/lb	J/kg	2,326
	kcal/kg	J/kg	4,184
	BTU/lb	kJ/kg	2.326
	kcal/kg	kJ/kg	4.184
Volumetric Rate	ft³/s	m³/s	0.02832
	ft³/min	m³/s	4.719×10^{-4}
	gal/min	m³/s	6.309×10^{-5}
	gal/min	l/s	0.063
Mass Rate	metric ton/hour (t/h)	kg/s	0.2778
	short ton/hour (t/h)	kg/s	0.2519
	lb/h	kg/s	1.26×10^{-4}
Temperature	°C (Celsius)	K (kelvin)	K = °C + 273.1
	°F (Fahrenheit)	K	K = 5/9(°F + 459.7)
	°F	°C	°C = 5/9(°F - 32)
Thermal Conductivity	cal/(cm·s·°C)	W/mK	418.4
Permeability	darcy	m²	0.987×10^{-12}

Geothermal Background

Mary H. Dickson and Mario Fanelli
Istituto di Geoscienze e Georisorse, CNR, Pisa, Italy

AIMS

1. To define what is meant by the term *geothermal energy*, and the relationship between geothermal energy and geological phenomena of a planetary scale; to evaluate how much of this energy could be recovered and exploited by humankind. To identify the main physical mechanisms occurring in the shallowest parts of the Earth's crust, how these mechanisms can be harnessed to allow us to extract the Earth's heat, and the main methods of research that can be used.

2. To show how geothermal energy can be utilized in numerous applications, from electricity generation to a wide range of direct heat uses, and the benefits that can be gained from these uses.

3. To emphasize the relatively minor impact of geothermal energy on the environment, without underestimating the risks that effectively do exist.

4. Finally, to demonstrate the potential benefits – to a community and/or to a nation – of exploiting indigenous geothermal resources, with emphasis on the importance of first making careful assessments of each specific situation, especially as regards quality of the resource and socio-economic conditions, and also of defining the programme of action.

OBJECTIVES

When you have completed this chapter you should be able to:

1. define the nature of the Earth's heat, the phenomena related to the latter and how these phenomena influence the distribution of the geothermal areas in the world

2. define geothermal systems and explain how they function

3. define the main categories of geothermal energy

4. discuss the present status of the development of geothermal energy in the world

5. discuss the main research methods used in areas of potential geothermal interest

6. discuss the major forms of geothermal utilization

7. discuss the potential impact on the environment of geothermal energy.

1.1 INTRODUCTION

Heat is a form of energy, and *geothermal energy* is literally the heat contained within the Earth that generates geological phenomena on a planetary scale. Geothermal energy is often used nowadays, however, to indicate that part of the Earth's heat that can, or could, be recovered and exploited by humankind, and it is in this sense that we will use the term from now on.

1.1.1 Brief geothermal history

The presence of volcanoes, hot springs and other thermal phenomena must have led our ancestors to surmise that parts of the interior of the Earth were hot. However, it was not until a period between the sixteenth and seventeenth century, when the first mines were excavated to a few hundred metres below ground level, that humankind deduced, from simple physical sensations, that the Earth's temperature increased with depth.

The first measurements by thermometer were probably performed in 1740, in a mine near Belfort, in France (Bullard, 1965). By 1870 modern scientific methods were being used to study the thermal regime of the Earth, but it was not until the twentieth century, and the discovery of the role played by *radiogenic heat*, that we could fully comprehend such phenomena as heat balance and the Earth's thermal history. All modern thermal models of the Earth, in fact, must take into account the heat continually generated by the decay of the long-lived radioactive isotopes of uranium (U^{238}, U^{235}), thorium (Th^{232}) and potassium (K^{40}), which are present in the Earth (Lubimova, 1968).

Added to radiogenic heat, in uncertain proportions, are other potential sources of heat such as the primordial energy of planetary accretion. Realistic theories on these models were not available until the 1980s, when it was demonstrated that there was no equilibrium between the radiogenic heat generated in the Earth's interior and the heat dissipated into space from the Earth, and that our planet is slowly cooling down. To give some idea of the phenomenon involved and its scale, we will cite a *heat balance* from Stacey and Loper (1988), in which the total flow of heat from the Earth is estimated at 42×10^{12} W (conduction, convection and radiation). Of this figure, 8×10^{12} W comes from the crust, which represents only 2 per cent of the total volume of the Earth but is rich in radioactive isotopes, 32.3×10^{12} W comes from the mantle, which represents 82 per cent of the total volume of the Earth, and 1.7×10^{12} W comes from the core, which accounts for 16 per cent of the total volume and contains no radioactive isotopes. (See Figure 1.1 for a sketch of the inner structure of the Earth). Since the radiogenic heat of the mantle is estimated at 22×10^{12} W, the cooling rate of this part of the Earth is 10.3×10^{12} W.

In more recent estimates, based on a greater number of data, the total flow of heat from the Earth is about 6 per cent higher than the figure utilized by Stacey and Loper (1988). Even so, the cooling process is still very slow. The temperature of the mantle has decreased no more than 300 to 350 °C in 3 billion years, remaining at about 4,000 °C at its base. It seems probable that the total heat content of the Earth, reckoned above an assumed average surface temperature of 15 °C, is of the order of 12.6×10^{24} MJ, and that of the crust is of the order of 5.4×10^{21} MJ (Armstead, 1983). The thermal energy of the Earth is therefore immense, but only a fraction could be utilized by humankind. So far our utilization of this energy has

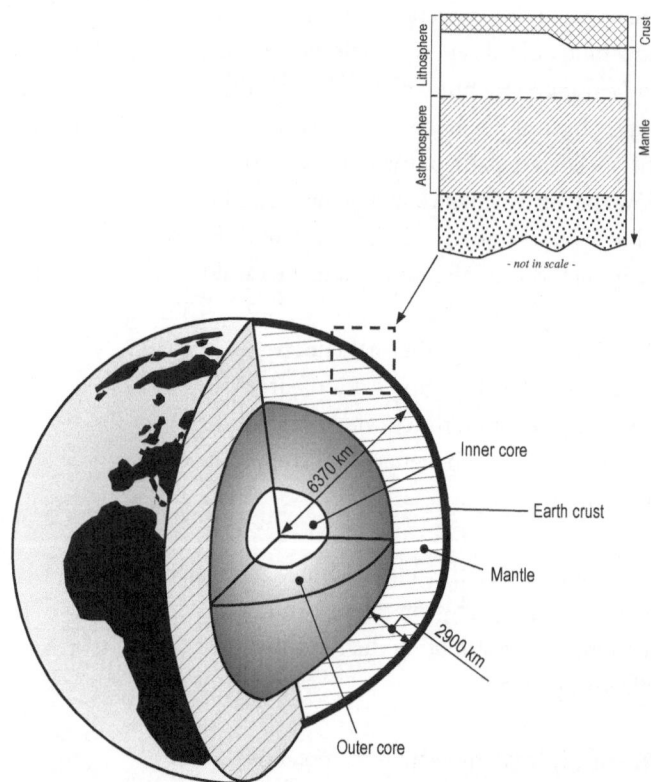

Figure 1.1 The Earth's crust, mantle and core. Top right: a section through the crust and the uppermost mantle

been limited to areas in which geological conditions permit a carrier (water in the liquid phase or steam) to 'transfer' the heat from deep hot zones to or near the surface, thus giving rise to geothermal resources, but innovative techniques in the near future may offer new perspectives in this sector.

There are examples in many areas of life of practical applications preceding scientific research and technological developments, and the geothermal sector is no exception. In the early part of the nineteenth century the geothermal fluids were already being exploited for their energy content. A chemical industry was set up in that period in Italy (in the zone now known as Larderello) to extract boric acid from the hot waters emerging naturally or from specially drilled shallow boreholes. The boric acid was obtained by evaporating the hot fluids in iron boilers, using wood from nearby forests as fuel. In 1827 Francesco Larderel, founder of this industry, developed a system for utilizing the heat of the boric fluids in the evaporation process, rather than burning wood from the rapidly depleting forests. Exploitation of the natural steam for its mechanical energy began at much the same time. The geothermal steam was used to raise liquids in primitive gas lifts and later in reciprocating and centrifugal pumps and winches, all of which were connected with drilling activity or in the local boric acid industry. Between 1850 and 1875 the factory at Larderello held the monopoly in Europe for boric acid production. Between 1910 and 1940 the low-pressure steam in this area of Tuscany was brought into use to heat the industrial and residential buildings and greenhouses. In 1928 Iceland, another pioneer in the utilization of geothermal energy, also began exploiting its geothermal fluids (mainly hot waters) for domestic heating.

The first attempt at generating electricity from geothermal steam was made at Larderello in 1904. The success of this experiment indicated the industrial value of geothermal energy and marked the beginning of a form of exploitation that was to develop significantly from then on. Electricity generation at Larderello was a commercial success. By 1942 the installed geothermoelectric capacity had reached 127,650 kW$_e$. Several countries followed the example set by Italy. The first geothermal wells in Japan were drilled at Beppu in 1919 and in the United States at The Geysers, California, in 1921. In 1958 a small geothermal power plant began operating in New Zealand, in 1959 in Mexico, in 1960 in the United States, and in many other countries in the years to follow.

1.1.2 Present status of geothermal utilization

After the Second World War many countries were attracted by geothermal energy, considering it to be economically competitive with other forms of energy. It did not have to be imported, and in some cases it was the only energy source available locally. The countries that utilize geothermal energy to *generate electricity* are listed in Table 1.1, which also gives the installed geothermal electric capacity in 1995 (6,833 MW$_e$), in 2000 (7,974 MW$_e$), and the increase between 1995 and the year 2000 (Huttrer, 2001). The same table also reports the total installed capacity in 2004 (8,806.45 MW$_e$). The geothermal power installed in the developing countries in 1995 and 2000 represents 38 and 47 per cent of the world total, respectively.

The utilization of geothermal energy in developing countries has exhibited an interesting trend over the years. In the five years between 1975 and 1979 the geothermal electric capacity installed in these countries increased from 75 to 462 MW$_e$; by the end of the next five-year period (1984) this figure had reached 1495 MW$_e$, showing a rate of increase during these two periods of 500 per cent and 223 per cent, respectively (Dickson and Fanelli, 1988). In the next sixteen years, from 1984 to 2000, there was a further increase in the total of almost 150 per cent.

Geothermal power plays a fairly significant role in the energy balance of some areas, and of the developing countries in particular, as can be inferred from the data reported in Table 1.2, which shows the percentage of geothermal power with respect to total electric power installed in some of these countries, relative to 1998.

As regards *non-electric applications* of geothermal energy, Table 1.3 gives the installed capacity (15,145 MW$_t$) and energy use (190,699 TJ/yr) worldwide for the year 2000. There are now fifty-eight countries reporting direct uses, compared to twenty-eight in 1995 and twenty-four in 1985. The data reported in this table are always difficult to collect and interpret, and should therefore be used with caution. The most common non-electric use worldwide (in terms of installed capacity) is heat pumps (34.80 per cent), followed by bathing (26.20 per cent), space heating (21.62 per cent), greenhouses (8.22 per cent), aquaculture (3.93 per cent), and industrial processes (3.13 per cent) (Lund and Freeston, 2001).

Table 1.1 Installed geothermal generating capacities world-wide from 1995 to 2000 (from Huttrer, 2001), and in 2004.

Country	1995 (MW$_e$)	2000 (MW$_e$)	1995–2000 (increase in MW$_e$)	% increase (1995–2000)	2004 (MW$_e$)
Australia	0.17	0.17	0	0	0.17
Austria	—	—	—	—	1.25
China	28.78	29.17	0.39	1.35	32
Costa Rica	55	142.5	87.5	159	162.5
El Salvador	105	161	56	53.3	162
Ethiopia	0	8.52	8.52		7
France (Guadeloupe)	4.2	4.2	0	0	15
Guatemala	0	33.4	33.4		29
Iceland	50	170	120	240	202
Indonesia	309.75	589.5	279.75	90.3	807
Italy	631.7	785	153.3	24.3	790.5
Japan	413.705	546.9	133.195	32.2	537
Kenya	45	45	0	0	127
Mexico	753	755	2	0.3	953
New Zealand	286	437	151	52.8	453
Nicaragua	70	70	0	0	77.5
Papua New Guinea	—	—	—	—	6
Philippines	1 227	1 909	682	55.8	1 931
Portugal (Azores)	5	16	11	220	16
Russia	11	23	12	109	81.6
Thailand	0.3	0.3	0	0	0.3
Turkey	20.4	20.4	0	0	20.4
USA	2 816.7	2 228	−588	n/a	2 395
Total	6 833	7 974	1 141	16.7	8806.45

Table 1.2 Electric capacity from geothermal energy out of total electric capacity for some developing countries in 1998 (MW$_e$)

Country	Total electric installed power	Geothermal electric installed power	% of the total power installed
Philippines	11 601	1 861	16.0
Nicaragua	614	70	11.4
El Salvador	996	105	10.5
Costa Rica	1 474	115	7.8
Kenya	889	45	5.1
Indonesia	22 867	589.5	2.6
Mexico	45 615	755	1.7

1.2 NATURE OF GEOTHERMAL RESOURCES

1.2.1 The Earth's thermal engine

The *geothermal gradient* expresses the increase in temperature with depth in the Earth's crust. Down to the depths accessible by drilling with modern technology (that is, over 10,000 m) the average geothermal gradient is about 2.5 to 3 °C/100 m. For example, if the temperature within the first few metres below ground level, which on average corresponds to the mean annual temperature of the external air,

Table 1.3 Non-electric uses of geothermal energy in the world (2000): installed thermal power (in MW$_t$) and energy use (in TJ/yr)

Country	Power (MW$_t$)	Energy (TJ/yr)	Country	Power (MW$_t$)	Energy (TJ/yr)
Algeria	100	1 586	Japan	1 167	26 933
Argentina	25.7	449	Jordan	153.3	1 540
Armenia	1	15	Kenya	1.3	10
Australia	34.4	351	Korea	35.8	753
Austria	255.3	1 609	Lithuania	21	599
Belgium	3.9	107	Macedonia	81.2	510
Bulgaria	107.2	1 637	Mexico	164.2	3 919
Canada	377.6	1 023	Nepal	1.1	22
Caribbean Islands	0.1	1	Netherlands	10.8	57
Chile	0.4	7	New Zealand	307.9	7 081
China	2 282	37 908	Norway	6	32
Colombia	13.3	266	Peru	2.4	49
Croatia	113.9	555	Philippines	1	25
Czech Republic	12.5	128	Poland	68.5	275
Denmark	7.4	75	Portugal	5.5	35
Egypt	1	15	Romania	152.4	2 871
Finland	80.5	484	Russia	308.2	6 144
France	326	4 895	Serbia	80	2 375
Georgia	250	6 307	Slovak Republic	132.3	2 118
Germany	397	1 568	Slovenia	42	705
Greece	57.1	385	Sweden	377	4 128
Guatemala	4.2	117	Switzerland	547.3	2 386
Honduras	0.7	17	Thailand	0.7	15
Hungary	472.7	4 086	Tunisia	23.1	201
Iceland	1 469	20 170	Turkey	820	15 756
India	80	2 517	United Kingdom	2.9	21
Indonesia	2.3	43	USA*	3 766	20 302
Israel	63.3	1 713	Venezuela	0.7	14
Italy	325.8	3 774	Yemen	1	15
TOTAL				15 145	190 699

*During 2001 these figures increased to 4,200 MW$_t$ and 21,700 TJ/yr (Lund and Boyd, 2001).
Source: Lund and Freeston, 2001.

is 15 °C, then we can reasonably assume that the temperature will be about 65° to 75 °C at 2,000 m depth, 90° to 105 °C at 3,000 m and so on for a further few thousand metres. There are, however, vast areas in which the geothermal gradient is far from the average value. In areas in which the deep rock basement has undergone rapid sinking, and the basin is filled with geologically 'very young' sediments, the geothermal gradient may be lower than 1 °C/100 m. On the other hand, in some 'geothermal areas' the gradient is more than ten times the average value.

The difference in temperature between deep hotter zones and shallow colder zones generates a conductive flow of heat from the former towards the latter, with a tendency to create uniform conditions, although, as often happens with natural phenomena, this situation is never actually attained. The mean *terrestrial heat* flow of continents and oceans is 65 and 101 mWm^{-2}, respectively, which, when weighted by area, yields a global mean of 87 mWm^{-2} (Pollack et al., 1993). These values are based on 24,774 measurements at 20,201 sites covering about 62 per cent of the Earth's surface. Empirical estimators, referenced to geological map units, enabled heat flow to be estimated in areas without measurements.

The temperature increase with depth, as well as volcanoes, geysers, hot springs and so on, are in a sense the visible or tangible expression of the heat in the interior of the Earth, but this heat also engenders other phenomena that are less discernible by humans, but of such magnitude that the Earth has been compared to an immense 'thermal engine'. We will try to describe these phenomena, referred to collectively as the *plate tectonics* theory, in simple terms, and to show their relationship with geothermal resources.

Our planet consists of a *crust*, which reaches a thickness of about 20 to 65 km in continental areas and about 5 to 6 km in oceanic areas; a *mantle*, which is roughly 2,900 km thick; and a *core*, about 3,470 km in radius (Figure 1.1). The physical and chemical characteristics of the crust, mantle and core vary from the surface of the Earth to its centre. The outermost shell of the Earth, known as the *lithosphere*, is made up of the crust and the upper layer of the mantle. Ranging in thickness from less than 80 km in oceanic zones to over 200 km in continental areas, the lithosphere behaves as a rigid body. Below the lithosphere is the zone known as the *asthenosphere*, 200 to 300 km in thickness, and of less rigid or 'more plastic' behaviour. In other words, on a geological scale in which time is measured in millions of years, this part of the Earth behaves in much the same way as a fluid in certain processes.

Because of the difference in temperature between the different parts of the asthenosphere, convective movements and, possibly, convective cells were formed some tens of millions of years ago. Their extremely slow movement (a few centimetres per year) is maintained by the heat produced continually by the decay of the radioactive elements and the heat coming from the deepest parts of the Earth. Immense volumes of deep hotter rocks, less dense and lighter than the surrounding material, rise with these movements towards the surface, while the colder, denser and heavier rocks near the surface tend to sink, reheat and rise to the surface once again, in a process very similar to what happens to water boiling in a pot or kettle.

In zones where the lithosphere is thinner, and especially in oceanic areas, the lithosphere is pushed upwards and broken by the very hot, partly molten material ascending from the asthenosphere, in correspondence to the ascending branch of convective cells. It is this mechanism that created and still creates the *spreading ridges* that extend for more than 60,000 km beneath the oceans, emerging in some places (Azores, Iceland) and even creeping between continents, as in the Red Sea. A relatively tiny fraction of the molten rocks upwelling from the asthenosphere emerges from the crests of these ridges and, in contact with the seawater, solidifies to form a new oceanic crust. Most of the material rising from the asthenosphere, however, divides into two branches that flow in opposite directions beneath the lithosphere. The continual generation of new crust and the pull of these two branches in opposite directions has caused the ocean beds on either side of the ridges to drift apart at a rate of a few centimetres per year. Consequently, the area of the ocean beds (the oceanic lithosphere) tends to increase.

The ridges are cut perpendicularly by enormous fractures (in some cases a few thousand kilometres in length) called *transform faults*. These phenomena lead to a simple observation: since there is apparently no increase in the Earth's surface with time, the formation of new lithosphere along the ridges and the spreading of the ocean beds must be accompanied by a comparable shrinkage of the lithosphere in other parts of the globe. This is indeed what happens in *subduction zones*, the largest of which are indicated by huge ocean trenches, such as those extending along the western margin of the Pacific Ocean and the western coast of South America. In the subduction zones the lithosphere folds downwards, plunges under the adjacent lithosphere and re-descends to the very hot deep zones, where it is 'digested' by the mantle and the cycle begins all over again. Part of the lithospheric material returns to a molten state and may rise to the surface again through fractures in the crust. As a consequence, *magmatic arcs* with numerous volcanoes are formed parallel to the trenches, on the opposite side from the ridges. Where the trenches are located in the ocean, as in the Western Pacific, these magmatic arcs consist of chains of volcanic islands; where the trenches run along the margins of

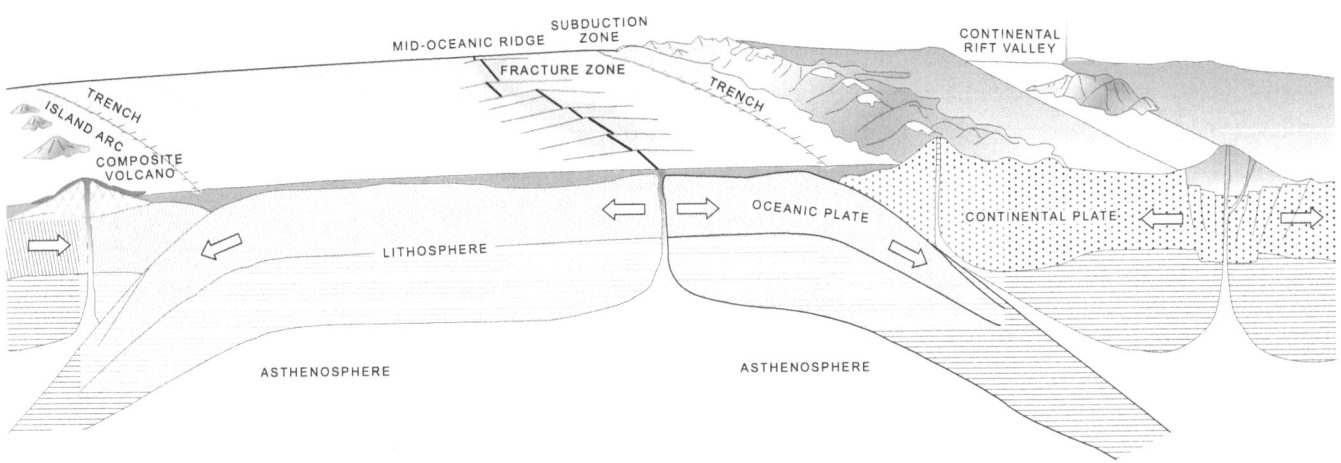

Figure 1.2 Schematic cross-section showing plate tectonic processes

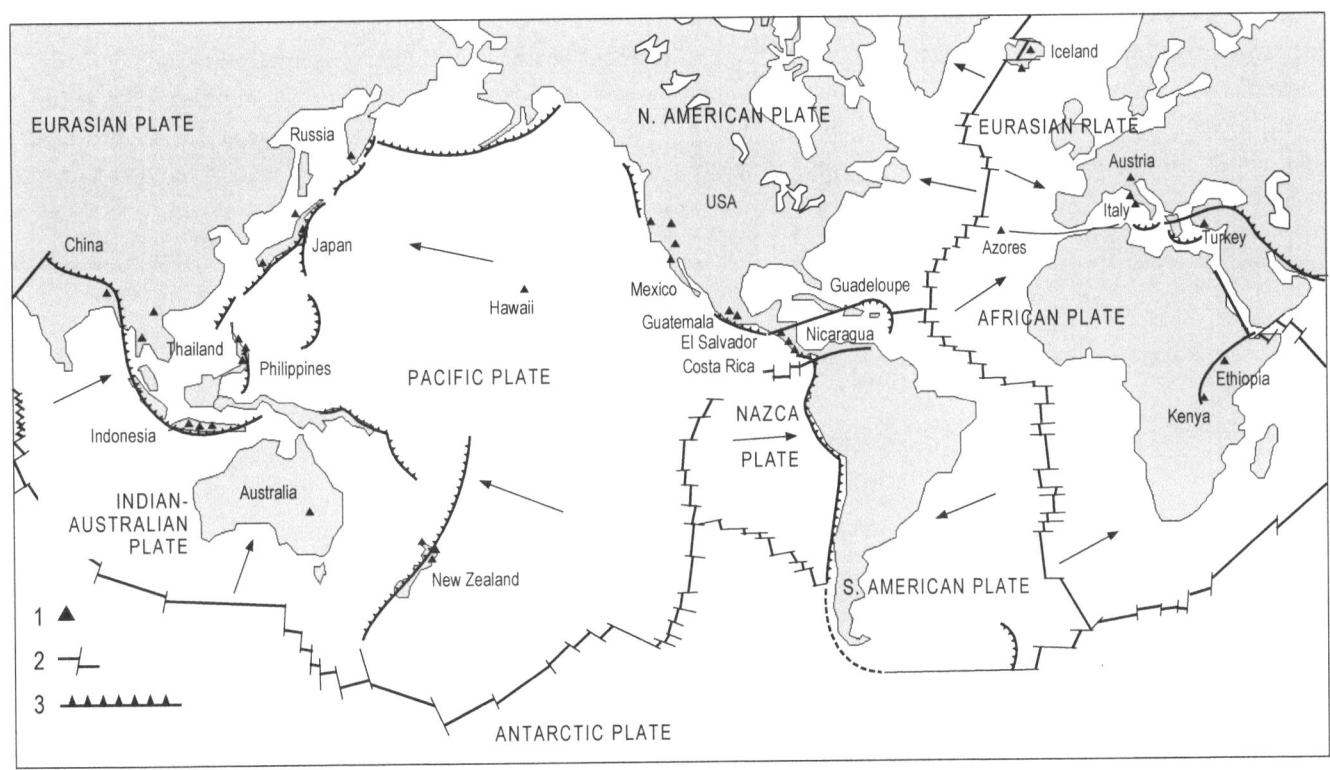

Figure 1.3 World pattern of plates, oceanic ridges, oceanic trenches, subduction zones and geothermal fields. Arrows show the direction of movement of the plates towards the subduction zones. (1) Geothermal fields producing electricity; (2) mid-oceanic ridges crossed by transform faults (long transversal fractures); (3) subduction zones, where the subducting plate bends downwards and melts in the asthenosphere

continents, the arcs consist of chains of mountains with numerous volcanoes, such as the Andes. Figure 1.2 illustrates the phenomena we have just described.

Spreading ridges, transform faults and subduction zones form a vast network that divides our planet into six immense and several other smaller lithospheric areas or *plates* (Figure 1.3). Because of the huge tensions generated by the Earth's thermal engine and the asymmetry of the zones producing and consuming lithospheric material, these plates drift slowly up against one another, shifting position continually.

The margins of the plates correspond to weak, densely fractured zones of the crust, characterized by an intense seismicity, by a large number of volcanoes and, because of the ascent of very hot materials towards the surface, by a high terrestrial heat flow. As shown in Figure 1.3, the most important geothermal areas are located around plate margins.

1.2.2 Geothermal systems

Geothermal systems can therefore be found in regions with a normal or slightly above normal geothermal gradient, and especially in regions around plate margins where the geothermal gradients may be significantly higher than the average value. In the first case the systems will be characterized by low temperatures, usually no higher than 100 °C at economic depths; in the second case the temperatures could cover a wide range from low to very high, and even above 400 °C.

What is a geothermal system and what happens in such a system? It can be described schematically as '*convecting water in the upper crust of the Earth, which, in a confined space,* *transfers heat from a heat source to a heat sink, usually the free surface*' (Hochstein, 1990). A geothermal system is made up of three main elements: a *heat source*, a *reservoir* and a *fluid*, which is the carrier that transfers the heat. The heat source can be either a very high-temperature (>600 °C) magmatic intrusion that has reached relatively shallow depths (5 to 10 km) or, as in certain low-temperature systems, the Earth's normal temperature, which, as we explained earlier, increases with depth. The reservoir is a volume of hot permeable rocks from which the circulating fluids extract heat. The reservoir is generally overlaid with a cover of impermeable rocks and connected to a surficial recharge area through which the meteoric waters can replace or partly replace the fluids that escape from the reservoir through springs or are extracted by boreholes. The geothermal fluid is water, in the majority of cases meteoric water, in the liquid or vapour phase, depending on its temperature and pressure. This water often carries with it chemicals and gases such as carbon dioxide (CO_2) and hydrogen sulfide (H_2S). Figure 1.4 is a greatly simplified representation of an ideal geothermal system.

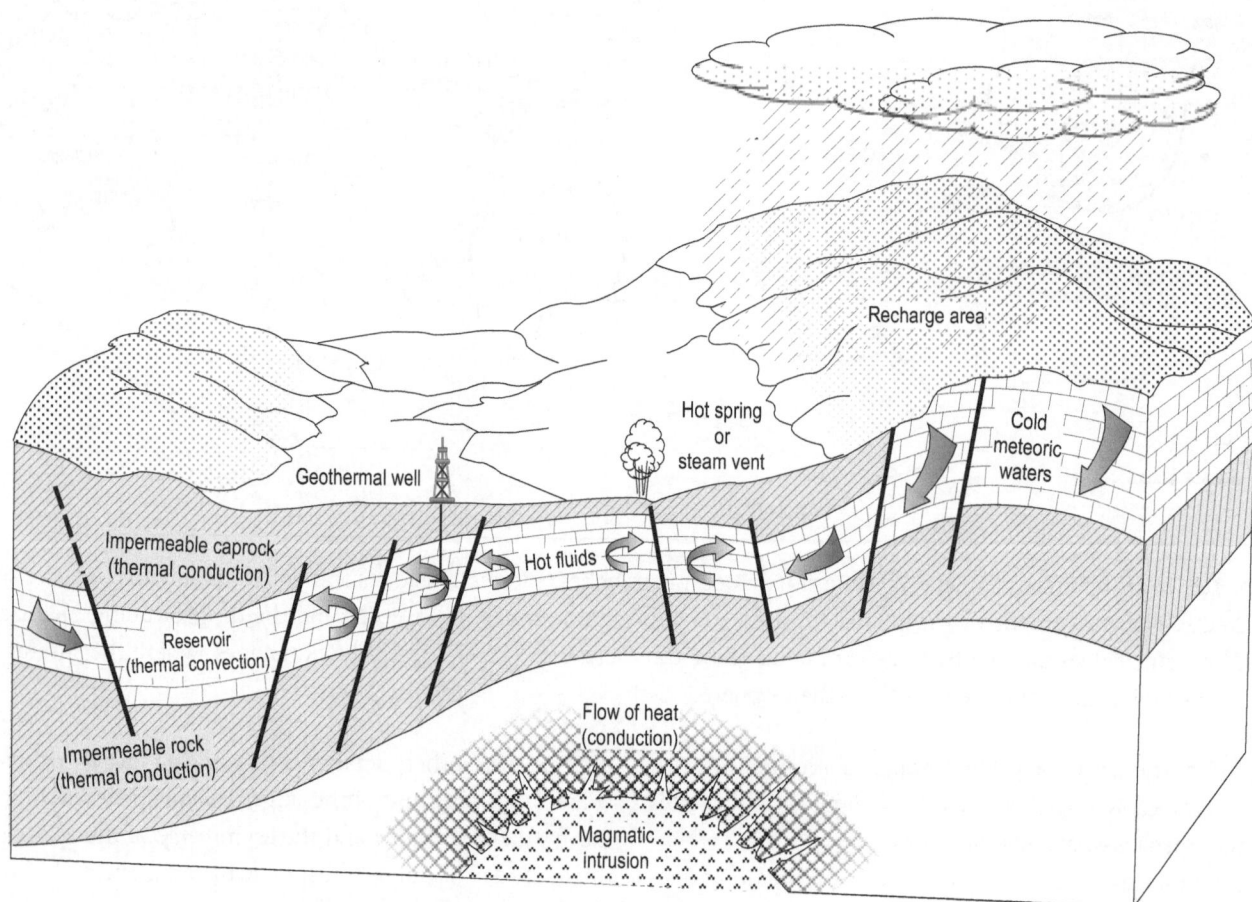

Figure 1.4 Schematic representation of an ideal geothermal system

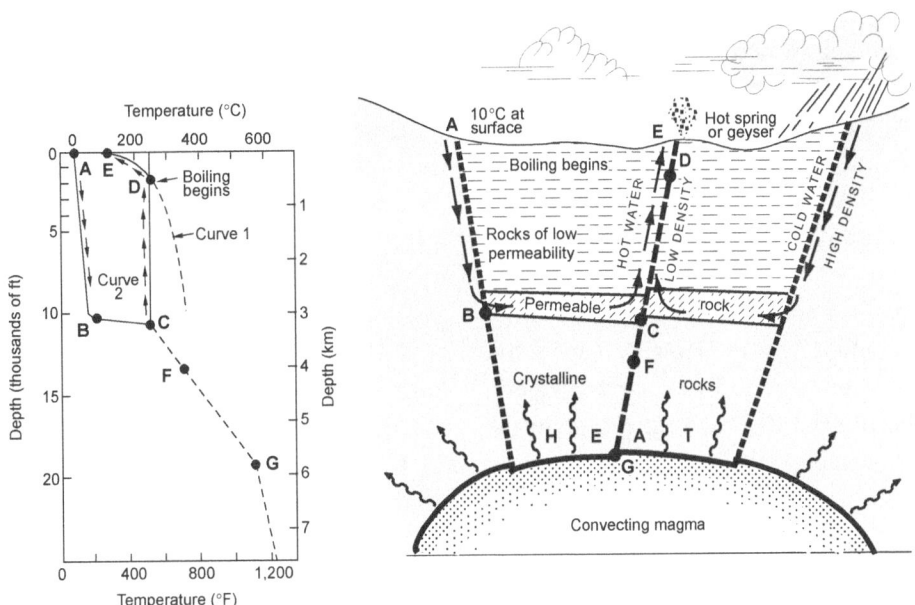

Figure 1.5 Model of a geothermal system. Curve 1 is the reference curve for the boiling point of pure water. Curve 2 shows the temperature profile along a typical circulation route from recharge at point A to discharge at point E
Source: White, 1973.

The mechanism underlying geothermal systems is, by and large, governed by *fluid convection*. Figure 1.5 describes schematically the mechanism in the case of an intermediate-temperature hydrothermal system. Convection occurs because of the heating and consequent thermal expansion of fluids in a gravity field; heat, which is supplied at the base of the circulation system, is the energy that drives the system. Heated fluid of lower density tends to rise and to be replaced by colder fluid of high density, coming from the margins of the system. Convection, by its nature, tends to increase temperatures in the upper part of a system as temperatures in the lower part decrease (White, 1973).

The phenomenon we have just described may seem quite a simple one, but the reconstruction of a good model of a real geothermal system is by no means easy to achieve. It requires skill in many disciplines and a vast experience, especially when dealing with high-temperature systems. Geothermal systems also occur in nature in a variety of combinations of geological, physical and chemical characteristics, thus giving rise to several different types of system.

Of all the elements of a geothermal system, the heat source is the only one that need be natural. Provided conditions are favourable, the other two elements could be 'artificial'. For example, the geothermal fluids extracted from the reservoir to drive the turbine in a geothermal power plant could,

after their utilization, be injected back into the reservoir through specific *injection wells*. In this way the natural recharge of the reservoir is integrated with an artificial recharge. For many years now re-injection has been adopted in various parts of the world as a means of drastically reducing the impact on the environment of power plant operations.

Artificial recharge through injection wells can also help to replenish and maintain 'old' or 'exhausted' geothermal fields. For example, in The Geysers field in California, United States, one of the biggest geothermal fields in the world, production began to decline dramatically at the end of the 1980s because of a lack of fluids. Thanks to the Santa Rosa Geysers Recharge Project, however, 41.5 million litres per day of tertiary treated waste-water will be pumped from the Santa Rosa regional sewage treatment plant and other cities through a 66 km pipeline to The Geysers field, where it will recharge the reservoir through specially drilled boreholes.

In the so-called *hot dry rock* (HDR) projects, which were experimented with for the first time at Los Alamos, New Mexico, United States, in 1970, both the fluid and the reservoir are artificial. High-pressure water is pumped through a specially drilled well into a deep body of hot, compact rock, causing its *hydraulic fracturing*. The water permeates these artificial fractures, extracting heat from the surrounding rock, which acts as a natural reservoir. This 'reservoir' is later penetrated by a second well, which is used to extract the heated water. The system therefore consists of: (a) the borehole used for hydraulic fracturing through which cold water is injected into (b) the artificial reservoir, and (c) the borehole used to extract the hot water. The entire system, complete with surface utilization plant, could form a closed loop (Garnish, 1987).

The Los Alamos project was the forerunner for other similar projects in Australia, France, Germany, Japan and the United Kingdom. After a period of relative neglect, these projects have been given renewed impulse from the discoveries, first, that deep rocks have a certain degree of natural fracturation, and second, that the methodologies and technologies adopted will depend on local geologic conditions. Currently, the most advanced research in the HDR sector is being conducted in Japan and under the European Project at Alsace (France). Several projects launched in Japan in the 1980s (at Hijiori, Ogachi and Yunomori), and heavily financed by the Japanese government and industry, have produced interesting results both from the scientific and industrial standpoints. The European HDR project, on the other hand, has been implemented over a number of phases, including the drilling of two wells, one of which has reached bottomhole at 5,060 m. Very promising results have been obtained from their geophysical surveys and hydraulic tests, and the European Project seems, for the moment, to be the most successful (Tenzer, 2001).

1.3 DEFINITION AND CLASSIFICATION OF GEOTHERMAL RESOURCES

There is still no standard international terminology in use throughout the geothermal community, which is unfortunate, as this would greatly facilitate mutual comprehension. The following are some of the most common definitions and classifications in this discipline.

According to Muffler and Cataldi (1978), when we speak generically about geothermal resources, what we are usually referring to is what should more accurately be called the *accessible resource base* – all of the thermal energy stored between the Earth's surface and a specified depth in the crust, beneath a specified area and measured from local

mean annual temperature. The accessible resource base includes the *useful accessible resource base* (= Resource) – that part of the accessible resource base that could be extracted economically and legally at some specified time in the future (less than 100 years). This category includes the *identified economic resource* (= Reserve) – that part of the resources of a given area that can be extracted legally at a cost competitive with other commercial energy sources, and that is known and characterized by drilling or by geochemical, geophysical and geological evidence. Figure 1.6 illustrates in graphic form these and other terms that may be used by geothermal specialists.

The most common criterion for classifying geothermal resources is, however, that based on the *enthalpy* of the geothermal fluids that act as the carrier transporting heat from the deep hot rocks to the surface. Enthalpy, which can be considered more or less proportional to temperature, is used to express the heat (thermal energy) content of the fluids,

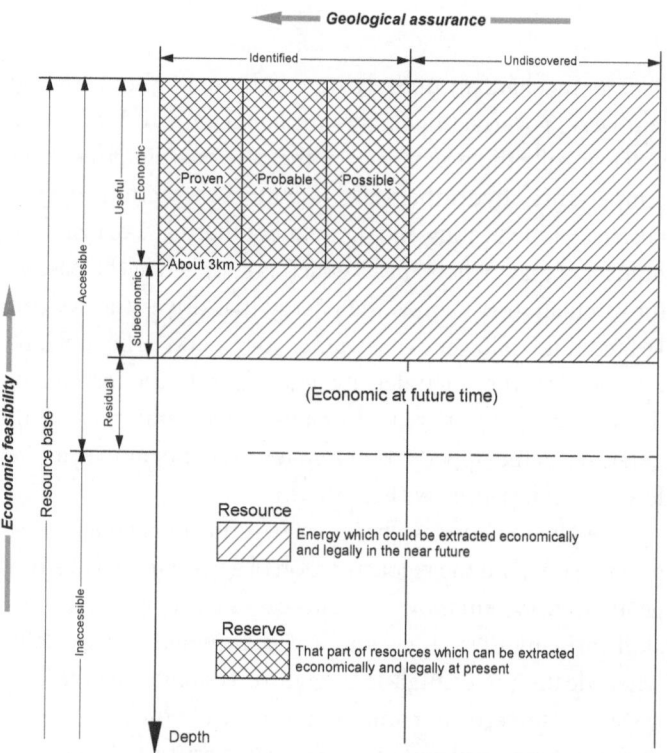

Figure 1.6 Diagram showing the different categories of geothermal resources. The vertical axis is the degree of economic feasibility; the horizontal axis is the degree of geological assurance

Source: Muffler and Cataldi, 1978.

Table 1.4 Classification of geothermal resources based on temperature (°C)

	Muffler and Cataldi (1978)	Hochstein (1990)	Benderitter and Cormy (1990)	Nicholson (1993)	Axelsson and Gunnlaugsson (2000)
Low enthalpy resources	<90	<125	<100	≤150	≤190
Intermediate enthalpy resources	90–150	125–225	100–200	—	—
High enthalpy resources	>150	>225	>200	>150	>190

and gives a rough idea of their 'value'. The resources are divided into low, medium and high enthalpy (or temperature) resources, according to criteria that are generally based on the energy content of the fluids and their potential forms of utilization. Table 1.4 reports the classifications proposed by a number of authors. A standard method of classification, as with terminology, would avoid confusion and ambiguity, but until such a method exists we must indicate the temperature values or ranges involved case by case, since terms such as low, intermediate and high are meaningless at best, and frequently misleading.

Frequently a distinction is made between water- or liquid-dominated geothermal systems and vapour-dominated (or dry steam) geothermal systems (White, 1973). In *water-dominated systems* liquid water is the continuous, pressure-controlling fluid phase. Some vapour may be present, generally as discrete bubbles. These geothermal systems, whose temperatures may range from <125 to >225 °C, are the most widely distributed in the world. Depending on temperature and pressure conditions, they can produce hot water, water and steam mixtures, wet steam, and in some cases dry steam. In *vapour-dominated systems* liquid water and vapour normally coexist in the reservoir, with vapour as the continuous, pressure-controlling phase. Geothermal systems of this type, the best known of which are Larderello in Italy and The Geysers in California, are somewhat rare, and are high-temperature systems. They normally produce dry to superheated steam.

The terms *wet*, *dry* and *superheated* steam, which are used frequently by geothermists, need some explanation for readers who do not have an engineering background. To make it as simple as possible, let us take the example of a pot filled with liquid water in which pressure can be kept constant at 1 atm (101.3 kPa). If we then heat the water, it will begin

boiling once it reaches a temperature of 100 °C (boiling temperature at a pressure of 1 atm) and will pass from the liquid to the gas (vapour) phase. After a certain time the pot will contain both liquid and vapour. The vapour coexisting with the liquid, and in thermodynamic equilibrium with it, is 'wet steam'. If we continue to heat the pot and maintain the pressure at 1 atm, the liquid will evaporate entirely and the pot will contain steam only. This is what we call 'dry steam'. Both wet and dry steam are called 'saturated steam'. Finally, increasing the temperature to, say, 120 °C, and keeping the pressure at 1 atm, we will obtain 'superheated steam' with a superheating of 20 °C, that is, 20 °C above the vaporization temperature at that pressure. At other temperatures and pressures, of course, these phenomena also take place in the underground, in what one author many years ago called 'nature's tea-kettle'.

Another division between geothermal systems is that based on the *reservoir equilibrium state* (Nicholson, 1993), considering the circulation of the reservoir fluid and the mechanism of heat transfer. In the *dynamic systems* the reservoir is continually recharged by water that is heated and then discharged from the reservoir, either to the surface or into underground permeable formations. Heat is transferred through the system by convection and circulation of the fluid. This category includes high-temperature (>150 °C) and low-temperature (<150 °C) systems. In the *static systems* (also known as *stagnant* or *storage* systems) there is only minor or no recharge to the reservoir, and heat is transferred only by conduction. This category includes low-temperature and geopressured systems. The *geopressured systems* are characteristically found in large sedimentary basins (for example, Gulf of Mexico, United States) at depths of 3 to 7 km. The geopressured reservoirs consist of permeable sedimentary rocks, included within impermeable low-conductivity strata, containing pressurized hot water that remained trapped at the moment of deposition of the sediments. The hot water pressure approaches lithostatic pressure, greatly exceeding the hydrostatic pressure. The geopressured reservoirs can also contain significant amounts of methane. The geopressured systems could produce thermal and hydraulic energy (pressurized hot water) and methane gas. These resources have been investigated extensively, but so far there has been no industrial exploitation.

Geothermal field is a geographical definition, usually indicating an area of geothermal activity at the Earth's surface. In cases without surface activity this term may be used to indicate the area at the surface corresponding to the geothermal reservoir below (Axelsson and Gunnlaugsson, 2000).

As geothermal energy is usually described as *renewable* and *sustainable*, it is important to define these terms. Renewable describes a property of the energy source, whereas sustainable describes how the resource is utilized.

The most critical factor for the classification of geothermal energy as a renewable energy source is the rate of energy recharge. In the exploitation of natural geothermal systems, energy recharge takes place by advection of thermal water on the same timescale as production from the resource. This justifies our classification of geothermal energy as a renewable energy resource. In the case of hot dry rocks, and some of the hot water aquifers in sedimentary basins, energy recharge is only by thermal conduction; because of the slow rate of the latter process hot dry rocks and some sedimentary reservoirs should be considered as finite energy resources (Stefansson, 2000).

The *sustainability in consumption* of a resource is dependent on its original quantity, its rate of generation and its rate of consumption. Consumption can obviously be sustained over any period of time in which a resource is being created faster than it is being depleted. The term 'sustainable development' is used by the World Commission on Environment and Development to indicate development that '*meets the needs of the present generation without compromising the needs of future generations*'. In this context, sustainable development does not imply that any given energy resource needs to be used in a totally sustainable fashion, but merely that a replacement for the resource can be found that will allow future generations to provide for themselves, despite the fact that the particular resource has been depleted. Thus, it may not be necessary that a specific geothermal field be exploited in sustainable

fashion. Perhaps we should direct our geothermal sustainability studies towards reaching and then sustaining a certain overall level of geothermal production at a national or regional level, both for electric power generation and direct heat applications, for a certain period (say 300 years), by bringing new geothermal systems on-line as others are depleted (Wright, 1998).

1.4 EXPLORATION

1.4.1 Objectives of exploration

The objectives of *geothermal exploration* are (Lumb, 1981):

- To identify geothermal phenomena.
- To ascertain that a useful geothermal production field exists.
- To estimate the size of the resource.
- To determine the type of geothermal field.
- To locate productive zones.
- To determine the heat content of the fluids that will be discharged by the wells in the geothermal field.
- To compile a body of basic data against which the results of future monitoring can be viewed.
- To determine the pre-exploitation values of environmentally sensitive parameters.
- To acquire knowledge of any characteristics that might cause problems during field development.

The relative importance of each objective depends on a number of factors, most of which are tied to the resource itself. These include anticipated utilization, the technology available and economics, as well as situation, location and time, all of which affect the exploration programme. For example, the preliminary reconnaissance of geothermal manifestations assumes much greater importance in a remote, unexplored area than in a well-known area; estimating the size of the resource may be less important if it is to be used in a small-scale application that obviously requires much less heat than is already discharging naturally; if the energy is to be used for district heating or some other application needing low-grade heat, then a high-temperature fluid is no longer an important objective (Lumb, 1981).

To reach these objectives a geothermal exploration manager can draw on a large number of methods and technologies, many of which are in current use and have already been widely experimented with in other sectors of research. The techniques and methodologies that have proved successful in mineral and oil or gas exploration will not, however, necessarily be the best solution in geothermal exploration. Conversely, techniques of little use in oil exploration could turn out to be ideal tools in the search for natural heat (Combs and Muffler, 1973).

1.4.2 Exploration methods

Geological and hydrogeological studies are the starting point of any exploration programme, and their basic functions are to identify the location and extension of the areas worth investigating in greater detail, and to recommend the most suitable exploration methods for these areas. Geological and hydrogeological studies have an important role in all subsequent phases of geothermal research, right up to the siting of both exploratory and producing boreholes. They also provide the background information for interpreting the data obtained with the other exploration methods, and finally, for constructing a realistic model of the geothermal system and assessing the potential of the resource. The information obtained from the geological and hydrogeological studies may also be used in the production phase, providing valuable information for the reservoir and production engineers. A good exploration programme and an efficient co-ordination of the research can appreciably reduce the duration and cost of exploration.

Geochemical surveys (including isotope geochemistry) are a useful means of determining whether the geothermal system is water- or vapour-dominated, of estimating the minimum temperature expected at depth, of estimating the homogeneity of the water supply, of inferring the chemical characteristics of the deep fluid, and of determining the source of recharge water. Valuable information can also be obtained on the type of problems that are likely to arise during the re-injection phase and plant utilization (for example, changes in fluid composition, corrosion and scaling on pipes and plant installations, environmental impact) and how to avoid or combat them. The geochemical survey consists of sampling and chemical and/or isotope analyses of the water and gas from geothermal manifestations (hot springs, fumaroles and so on) or wells in the study area. As a geochemical survey provides useful data for planning exploration, and its cost is relatively low compared with other more sophisticated methods, such as geophysical surveys, geochemical techniques should

be utilized as much as possible before proceeding with other more expensive methodologies.

Geophysical surveys are directed at obtaining indirectly, from the surface or from depth intervals close to the surface, the physical parameters of deep geological formations. These physical parameters include:

- temperature (thermal survey)
- electrical conductivity (electrical and electromagnetic methods)
- propagation velocity of elastic waves (seismic survey)
- density (gravity survey)
- magnetic susceptibility (magnetic survey).

Some of these techniques, such as seismics, gravity and magnetics, which are traditionally adopted in oil research, can give valuable information on the shape, size, depth and other important characteristics of the deep geological structures that could constitute a geothermal reservoir, but they give little or no indication as to whether these structures actually contain the fluids that are the primary objective of research. These methodologies are, therefore, more suited to defining details during the final stages of exploration, before the exploratory wells are sited. Information on the existence of geothermal fluids in the geological structures can be obtained with the electrical and electromagnetic prospectings, which are more sensitive than the other surveys to the presence of these fluids and to variations in temperature. These two techniques have been applied widely with satisfactory results. The magnetotelluric method (MT), which exploits the electromagnetic waves generated by solar storms, has been greatly improved over the last few years, and now offers a vast spectrum of possible applications, despite the fact that it requires sophisticated instrumentation and is sensitive to background noise in urbanized areas. The main advantage of the magnetotelluric method is that it can be used to define deeper structures than are attainable with the electric and the other electromagnetic techniques. The controlled source audiomagnetotelluric method (CSAMT), developed recently, uses artificially induced waves instead of natural electromagnetic waves. The penetration depth is shallower with this technique, but it is quicker, cheaper, and provides far more detail than the classic MT method.

Thermal techniques (temperature measurements, determination of geothermal gradient and terrestrial heat flow) can often provide a good approximation of the temperature at the top of the reservoir.

All geophysical techniques are expensive, although some more than others. They cannot be used indiscriminately in any situation or condition, as a method that produces excellent results in one geological environment may give very unsatisfactory results in another. In order to reduce costs, it is therefore very important that the geophysical method(s) be selected very carefully beforehand by geophysicists working in close collaboration with geologists (Meidav, 1998).

Drilling of exploratory wells represents the final phase of any geothermal exploration programme, and is the only means of determining the real characteristics of the geothermal reservoir and thus of assessing its potential (Combs and Muffler, 1973). The data provided by exploratory wells should be capable of verifying all the hypotheses and models elaborated from the results of surface exploration, and of confirming that the reservoir is productive and contains enough fluids of adequate characteristics for the use for which it is intended. Siting of the exploratory wells is therefore a very delicate operation.

1.4.3 Exploration programme

Before a geothermal exploration programme is drawn up, all existing geological, geophysical and geochemical data must be collected and integrated with any data available from previous studies on water, minerals and oil resources in the study area and adjacent areas. This information frequently plays an important role in defining the objectives of the geothermal exploration programme, and could lead to a significant reduction in costs.

The exploration programme is usually developed on a step-by-step basis: *reconnaissance*, *pre-feasibility* and *feasibility*. During each of these phases we gradually eliminate the less interesting areas and concentrate on the most promising ones. The methods used also become progressively more sophisticated and more detailed as the programme develops. The size and budget of the entire programme should be proportional to its objectives, to the importance of the resources we expect to find, and to the planned forms of utilization. The programme schedule should be flexible and reassessed as the results come in from the various surveys of each phase; similarly the geological–geothermal model should be progressively updated and improved. These periodic reassessments of the programme should ideally eliminate any operations that are no longer necessary and insert others, according to the results attained at each stage. Clearly any reduction in the

number and size of the prospectings will lead to a decrease in costs, and also a corresponding increase in the risk of error or failure. Conversely, by decreasing the risk of error we increase the overall cost. The economic success of a geothermal exploration programme hinges on finding the proper balance between the two.

1.5 UTILIZATION OF GEOTHERMAL RESOURCES

Electricity generation is the most important form of utilization of high-temperature geothermal resources (>150 °C). The medium-to-low temperature resources (<150 °C) are suited to many different types of application. The classical Lindal diagram (Lindal, 1973) (Figure 1.7), which shows

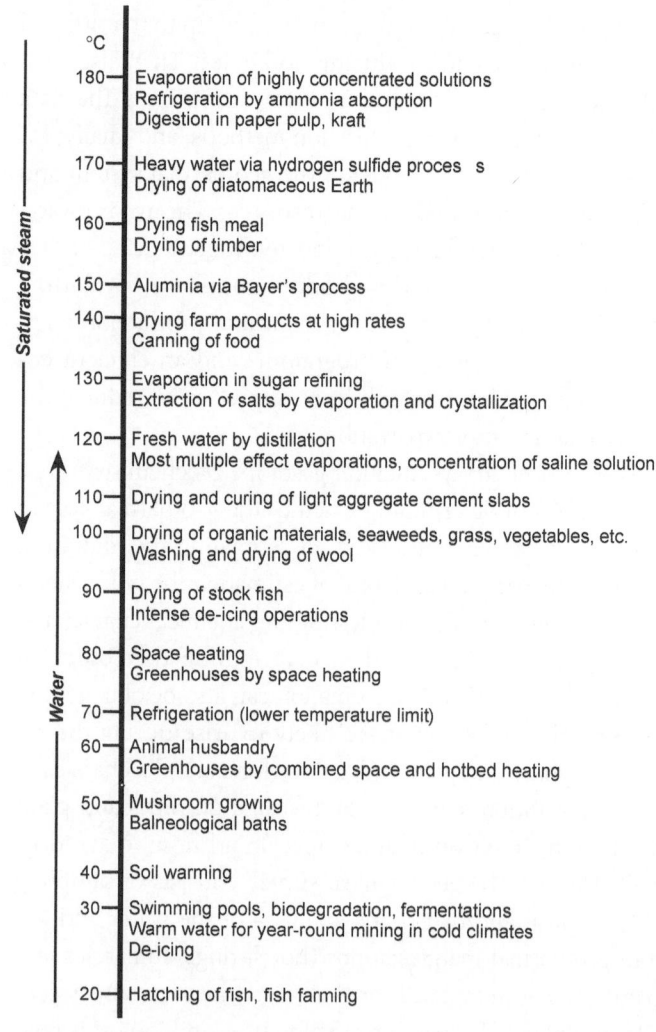

Figure 1.7 The Lindal diagram
Source: Lindal, 1973.

the possible uses of geothermal fluids at different temperatures, still holds valid, but the generation of electric energy in binary cycle plants can now be added above 85 °C. The lower limit of 20 °C is exceeded only in very particular conditions, or by the use of heat pumps. The Lindal diagram emphasizes two important aspects of the utilization of geothermal resources (Gudmundsson, 1988): first, with cascading and combined uses it is possible to enhance the feasibility of geothermal projects, and second, the resource temperature may limit the possible uses. Existing designs for thermal processes can, however, be modified for geothermal fluid utilization in certain cases, thus widening its field of application.

1.5.1 Electricity generation

Electricity generation mainly takes place in conventional steam turbines and binary plants, depending on the characteristics of the geothermal resource.

Conventional steam turbines require fluids at temperatures of at least 150 °C, and are available with either atmospheric (back-pressure) or condensing exhausts. Atmospheric exhaust turbines are simpler and cheaper. The steam, direct from dry steam wells or, after separation, from wet wells, is passed through a turbine and exhausted to the atmosphere. With this type of unit, steam consumption (from the same inlet pressure) per kilowatt-hour produced is almost double that of a condensing unit. However, the atmospheric exhaust turbines are extremely useful as pilot plants, or stand-by plants, in the case of small supplies from isolated wells, and for generating electricity from test wells during field development. They are also used when the steam has a high non-condensable gas content (>12 per cent in weight). The atmospheric exhaust units can be constructed and installed very quickly and put into operation in little more than thirteen to fourteen months from their order date. This type of machine is usually available in small sizes (2.5 to 5 MW$_e$).

The condensing units, having more auxiliary equipment, are more complex than the atmospheric exhaust units, and the bigger sizes can take twice as long to construct and install. The specific steam consumption of the condensing units is, however, about half that of the atmospheric exhaust units. Condensing plants with a capacity of 55 to 60 MW$_e$ are very common, but recently plants of 110 MW$_e$ have also been constructed and installed.

Generating electricity from low-to-medium temperature geothermal fluids and from the waste hot waters coming from the separators in water-dominated geothermal fields has made considerable progress since improvements were made in binary fluid technology. The *binary plants* utilize a secondary working fluid, usually an organic fluid (typically n-pentane) that has a low boiling point and high vapour pressure at low temperatures, compared with steam. The secondary fluid is operated through a conventional Rankine cycle: the geothermal fluid yields heat to the secondary fluid through heat exchangers, in which this fluid is heated and vaporizes; the vapour produced drives a normal axial flow turbine, is then cooled and condensed, and the cycle begins again. When suitable secondary fluids are selected, binary systems can be designed to utilize geothermal fluids in the temperature range 85 to 170 °C. The upper limit depends on the thermal stability of the organic binary fluid, and the lower limit on technical–economic factors: below this temperature the size of the heat exchangers required would render the project uneconomical. Apart from low-to-medium temperature geothermal fluids and waste fluids, binary systems can also be utilized where flashing of the geothermal fluids should preferably be avoided (for example, to prevent well sealing). In this case, down-hole pumps can be used to keep the fluids in a pressurized liquid state, and the energy can be extracted from the circulating fluid by means of binary units.

Binary plants are usually constructed in small modular units of a few hundred kW$_e$ to a few MW$_e$ capacity. These units can then be linked up to create power plants of a few tens of megawatts. Their cost depends on a number of factors, but particularly on the temperature of the geothermal fluid produced, which influences the size of the turbine, heat exchangers and cooling system. The total size of the plant has little effect on the specific cost, as a series of standard modular units is joined together to obtain larger capacities.

Binary plant technology is a very cost-effective and reliable means of converting into electricity the energy available from water-dominated geothermal fields (below 170 °C). A new binary-fluid cycle has recently been developed, called the Kalina, which utilizes a water-ammonia mixture as the working fluid. This fluid is expanded, in superheated conditions, through the high-pressure turbine, and then reheated before entering the low-pressure turbine. After the second expansion the saturated vapour moves through a recuperative boiler before being condensed in a water-cooled condenser. The Kalina cycle is estimated to be up to 40 per cent more efficient than existing geothermal binary power plants.

Small mobile plants, conventional or not, cannot only reduce the risk inherent to drilling new wells but also, what is more important, can help in meeting the energy requirements of isolated areas. The standard of living of many communities could be considerably improved were they able to draw on local sources of energy. Electricity could facilitate many apparently banal, but extremely important, operations such as pumping water for irrigation and freezing fruit and vegetables for longer conservation.

The convenience of the small mobile plants is most evident for areas without ready access to conventional fuels, and for communities that it would be too expensive to connect to the national electric grid, despite the presence of high-voltage transmission lines in the vicinity. The expense involved in serving these small bypassed communities is prohibitive, since the step-down transformers needed to tap electricity from high-voltage lines cost more than US$675,000 each, installed, and the simplest form of local distribution of electricity, at 11 kV using wooden poles, costs a minimum of US$20,000 per kilometre (US$, 1994). By comparison, the capital cost (US$, 1998) of a binary unit is of the order of 1,500 to 2,500 US$/kW installed, excluding drilling costs. The demand for electric capacity per person at off-grid sites will range from 0.2 kW in less-developed areas to 1.0 kW or higher in developed areas. A 100 kW_e plant could serve 100 to 500 people. A 1,000 kW_e plant would serve 1,000 to 5,000 people (Entingh et al., 1994).

1.5.2 Direct heat uses

Direct heat use is one of the oldest, most versatile and also the most common form of utilization of geothermal energy (see Table 1.3). Bathing, space and district heating, agricultural applications, aquaculture and some industrial uses are the best-known forms of utilization, but heat pumps are the most widespread (12.5 per cent of the total energy use in 2000). There are many other types of utilization, on a much smaller scale, some of which are unusual.

Space and district heating has made great progress in Iceland, where the total capacity of the operating geothermal district heating system had risen to about 1,200 MW_t by the end of 1999 (Figure 1.8), but systems are also widely distributed in the East European countries, as well as in the United States, China, Japan, France and so on.

Geothermal district heating systems are capital-intensive. The main costs are initial investment costs, for production and injection wells, down-hole and transmission pumps, pipelines and distribution networks, monitoring and control

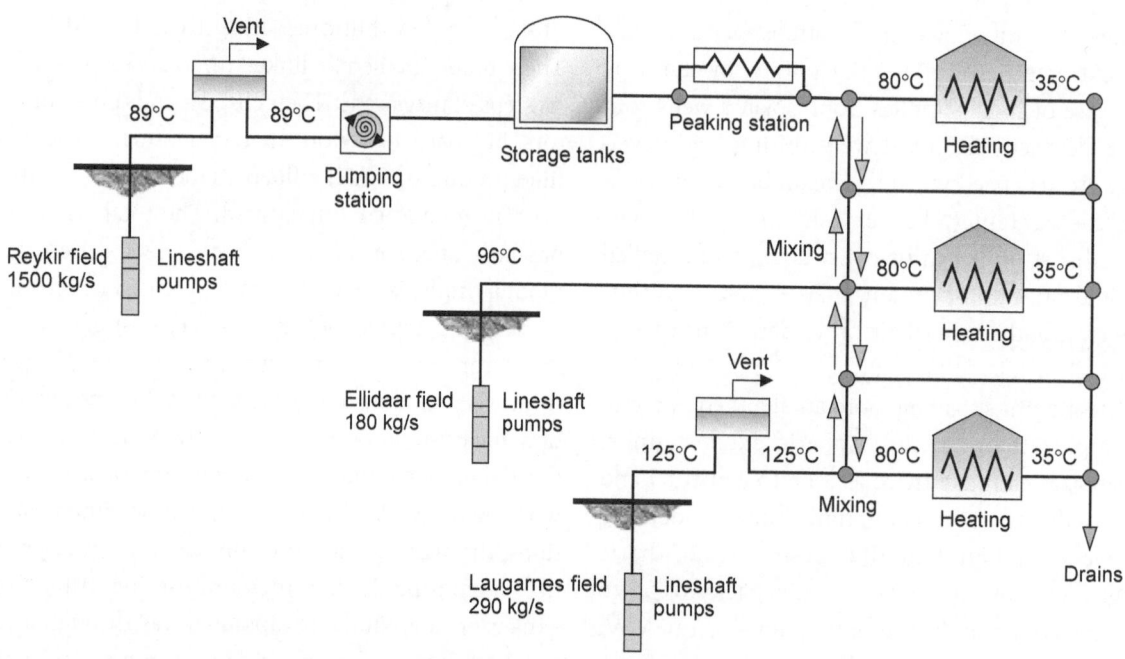

Figure 1.8 Simplified flow diagram of the geothermal district heating system of Reykjavik
Source: Gudmundsson, 1988.

equipment, peaking stations and storage tanks. Operating expenses, however, are comparatively lower than in conventional systems, and consist of pumping power, system maintenance, control and management. A crucial factor in estimating the initial cost of the system is the thermal load density, or the heat demand divided by the ground area of the district. A high heat density determines the economic feasibility of a district heating project, since the distribution network is expensive. Some economic benefit can be achieved by combining heating and cooling in areas where the climate permits. The load factor in a system with combined heating and cooling would be higher than the factor for heating alone, and the unit energy price would consequently improve (Gudmundsson, 1988).

Space cooling is a feasible option where absorption machines can be adapted to geothermal use. The technology of these machines is well known, and they are readily available on the market. The absorption cycle is a process that utilizes heat instead of electricity as the energy source. The refrigeration effect is obtained by utilizing two fluids: a refrigerant, which circulates, evaporates and condenses, and a secondary fluid or absorbent. For applications above 0 °C (primarily in space and process conditioning), the cycle uses lithium bromide as the absorbent and water as the refrigerant. For applications below 0 °C an ammonia/water cycle is adopted, with ammonia as the refrigerant and water as the absorbent. Geothermal fluids provide the thermal energy to drive these machines, although their efficiency decreases with temperatures lower than 105 °C.

Geothermal 'space conditioning' (heating and cooling) has expanded considerably since the 1980s, following on the introduction and widespread use of heat pumps. The various systems of heat pumps available permit us to economically extract and utilize the heat content of low-temperature bodies, such as the ground and shallow aquifers, ponds and so on (Sanner, 2001) (see, for example, Figure 1.9). As our engineering readers will already know, heat pumps are machines that move heat in a direction opposite to that in which it would tend to go naturally, that is, from a cold space or body to a warmer one. A heat pump is effectively nothing more than a refrigeration unit (Rafferty, 1997). Any refrigeration device (window air conditioner, refrigerator, freezer and so on) moves heat from a space (to keep it cool) and discharges that heat at higher temperatures. The only difference between a heat pump and a refrigeration unit is the desired effect, cooling for the refrigeration unit and heating for

Ground Coupled Heat Pumps (GCHP)
(closed loop heat pumps)

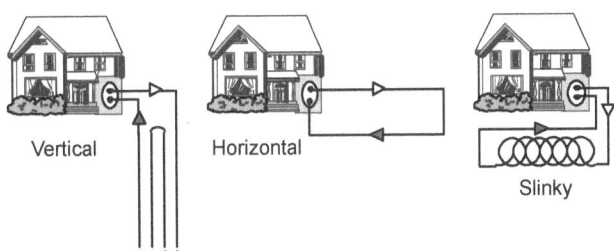

Groundwater Heat Pumps (GWHP)
(open loop heat pumps)

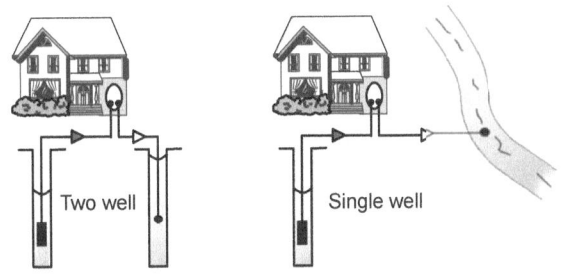

Surface Water Heat Pumps (SWHP)
(lake or pond heat pumps)

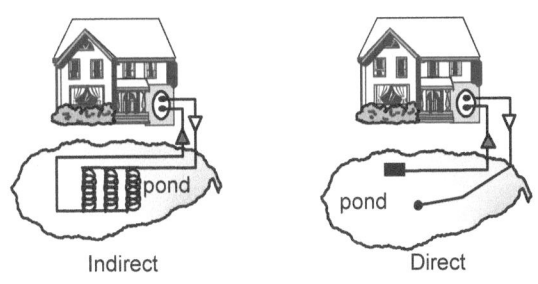

Figure 1.9 Simplified schemes of ground source heat pumps

the heat pump. A second distinguishing factor of many heat pumps is that they are reversible and can provide either heating or cooling in the space. The heat pumps, of course, need energy to operate, but in suitable climatic conditions and with a good design, the energy balance will be a positive one. Ground-coupled and groundwater heat pump systems have now been installed in great numbers in twenty-seven countries, for a total thermal capacity of 6,875 MW$_t$ (in 2000). The majority of these installations are in the United States (4,800 MW$_t$), Switzerland (500 MW$_t$), Sweden (377 MW$_t$), Canada (360 MW$_t$), Germany (344 MW$_t$) and Austria (228 MW$_t$) (Lund, 2001). Aquifers and soils with temperatures in the 5 to 30 °C range are being used in these systems.

The *agricultural applications* of geothermal fluids consist of open-field agriculture and greenhouse heating. Thermal water can be used in open-field agriculture to irrigate and/or heat the soil. The greatest drawback in irrigating with warm waters is that, to obtain any worthwhile variation in soil temperature, such large quantities of water are required at temperatures low enough to prevent damage to the plants that the fields would be flooded. One possible solution to this problem is to adopt a subsurface irrigation system coupled to a buried-pipeline soil-heating device. Heating the soil in buried pipelines without the irrigation system could decrease the heat conductivity of the soil, because of the drop in humidity around the pipes, and consequent thermal insulation. The best solution seems to be to combine soil heating and irrigation. The chemical composition of the geothermal waters used in irrigation must be monitored carefully to avoid adverse effects on the plants. The main advantages of temperature control in open-field agriculture are: (a) it prevents any damage ensuing from low environmental temperatures, (b) it extends the growing season, increases plant growth, and boosts production, and (c) it sterilizes the soil (Barbier and Fanelli, 1977).

The most common application of geothermal energy in agriculture is, however, in *greenhouse heating*, which has been developed on a large scale in many countries. The cultivation of vegetables and flowers out of season, or in an unnatural climate, can now draw on a widely tested technology. Various solutions are available for achieving optimum growth conditions, based on the optimum growth temperature of each plant (Figure 1.10), and on the quantity of light, the CO_2 concentration in the greenhouse environment, the humidity of the soil and air, and air movement.

The walls of the greenhouse can be made of glass, fibreglass, rigid plastic panels or plastic film. Glass panels are more transparent than plastic and will let in far more light, but will provide less thermal insulation, are less resistant to shocks, and are heavier and more expensive than the plastic panels. The simplest greenhouses are made of single plastic films, but recently some greenhouses have been constructed with a double layer of film separated by an air space. This system reduces the heat loss through the walls by 30 to 40 per cent, and thus greatly enhances the overall efficiency of the greenhouse. Greenhouse heating can be accomplished by forced circulation of air in heat exchangers, hot-water circulating pipes or ducts located in or on the floor, finned units located along the walls and under benches, or a combination of these methods. Exploitation of geothermal heat in greenhouse heating can considerably reduce their operating costs, which in some cases account for 35 per cent of the product costs (vegetables, flowers, house plants and tree seedlings).

Farm animals and aquatic species, as well as vegetables and plants, can benefit in quality and quantity from optimum conditioning of their environmental temperature (Figure 1.11). In many cases geothermal waters could be used profitably in a combination of *animal husbandry* and geothermal greenhouses. The energy required to heat a breeding installation is about 50 per cent of that required for a greenhouse of the same surface area, so cascade utilization could be adopted. Breeding in a temperature-controlled

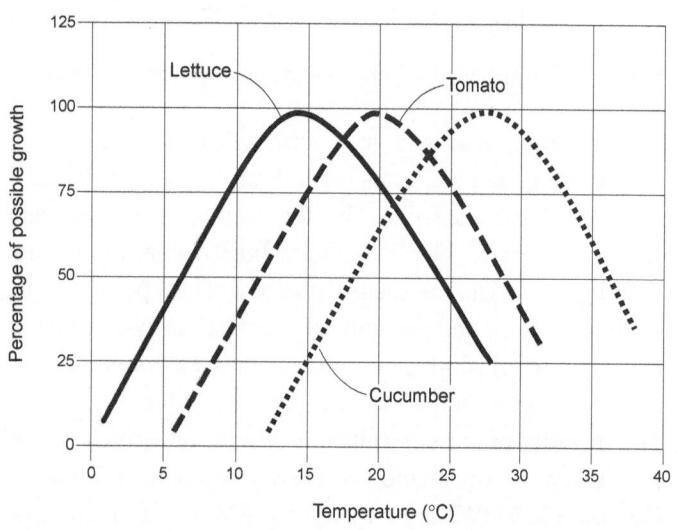

Figure 1.10 Growth curves for some crops
Source: Beall and Samuels, 1971.

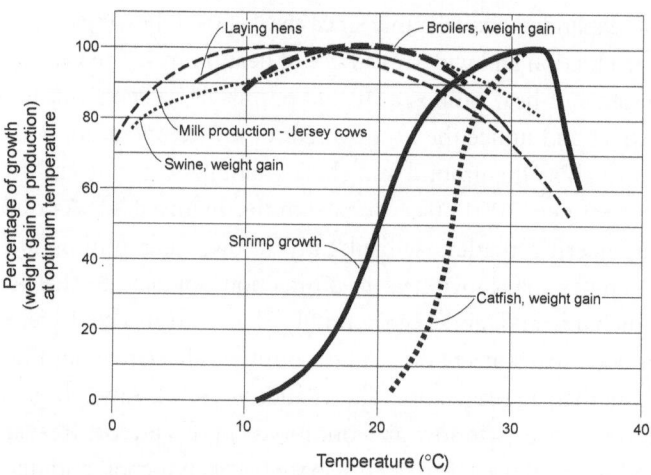

Figure 1.11 Effect of temperature on growth or production of food animals
Source: Beall and Samuels, 1971.

environment improves animal health, and the hot fluids can also be utilized to clean, sanitize and dry the animal shelters and waste products (Barbier and Fanelli, 1977).

Aquaculture, which is the controlled breeding of aquatic forms of life, is gaining worldwide importance nowadays, because of an increasing market demand. Control of the breeding temperatures for aquatic species is of much greater importance than for land species, as can be seen in Figure 1.11, which shows that the growth curve trend of aquatic species is very different from that of land species. By maintaining an optimum temperature artificially we can breed more exotic species, improve production and even, in some cases, double the reproductive cycle (Barbier and Fanelli, 1977). The species that are typically raised are carp, catfish, bass, tilapia, mullet, eels, salmon, sturgeon, shrimp, lobster, crayfish, crabs, oysters, clams, scallops, mussels and abalone.

Aquaculture also includes alligator and crocodile breeding, as tourist attractions and for their skins, which could prove a lucrative activity. Past experience in the United States has shown that, if its growth temperature is maintained at about 30 °C, an alligator can be grown to a length of about 2 m in three years, whereas alligators bred under natural conditions will reach a length of only 1.2 m over the same period. These creatures have been bred on farms in Colorado and Idaho for some years now, and the Icelanders are planning something similar.

The temperatures required for aquatic species are generally in the 20 to 30 °C range. The size of the installation will depend on the temperature of the geothermal source, the temperature required in the fishponds and the heat losses from the latter.

The cultivation of *Spirulina* can also be considered a form of aquaculture. This single-celled, spiral-shaped, blue-green micro-algae is frequently called 'super-food' because of its nutrient density; it has also been proposed to solve the problem of famine in the poorest countries of the world, although at the moment it is being marketed as a nutritional food supplement. Spirulina is now being farmed in a number of tropical and sub-tropical countries, in lakes or artificial basins, where conditions are ideal for its fast and widespread growth (a hot, alkaline environment rich in CO_2). Geothermal energy has already been used successfully to provide the heat needed to grow Spirulina throughout the year in temperate countries.

The entire temperature range of geothermal fluids, whether steam or water, can be exploited in *industrial applications*,

as shown in the Lindal diagram (Figure 1.7). The different possible forms of utilization include process heating, evaporation, drying, distillation, sterilization, washing, de-icing and salt extraction. Industrial process heat has applications in nineteen countries (Lund and Freeston, 2001), where the installations tend to be large and energy consumption high. Examples also include concrete curing, bottling of water and carbonated drinks, paper and vehicle parts production, oil recovery, milk pasteurization, the leather industry, chemical extraction, CO_2 extraction, laundry use, diatomaceous Earth drying, pulp and paper processing, and borate and boric acid production. There are also plans to utilize low-temperature geothermal fluids to de-ice runways and disperse fog in some airports. A cottage industry has developed in Japan that utilizes the bleaching properties of the H_2S in geothermal waters to produce innovative and much admired textiles for ladies' clothing. In Japan they have also tested a technique for manufacturing a lightweight 'geothermal wood' that is particularly suited to certain types of construction. During treatment in the hot spring water the polysaccharides in the original wood hydrolyse, rendering the material more porous and thus lighter.

1.5.3 Economic considerations

The elements that have to be considered in any cost estimate, whether they are assigned to plant or operating costs, and the price of the 'products' of geothermal energy, are all more numerous and more complicated than for other forms of energy. All these elements must nevertheless be evaluated carefully before a geothermal project is launched. We can only offer a few indications of a more general character, which, together with information on local conditions and on the value of the geothermal fluids available, should help the potential investor to reach a decision.

A resource-plant system (geothermal power facility) consists of the geothermal wells, the pipelines carrying the geothermal fluids, the utilization plant, and frequently a re-injection system as well. The interaction of all these elements bears heavily on investment costs, and must therefore be subjected to careful analysis. To give an example, in the generation of electricity, a plant that discharges to the atmosphere is the simplest solution, and is therefore cheaper than a condensing plant of the same capacity. It will, however, require almost twice as much steam as the condensing plant to operate, and consequently twice as many wells to feed it.

Since wells are very expensive, a condensing power plant is effectively a cheaper option than the discharge-to-atmosphere plant. The latter is, in fact, usually chosen for reasons other than economy.

Geothermal fluids can be transported over fairly long distances in thermally insulated pipelines. In ideal conditions, the pipelines can be as long as 60 km. However, the pipelines, the ancillary equipment needed (pumps, valves and so on), and their maintenance are all quite expensive, and could weigh heavily on the capital cost and operating costs of a geothermal plant. The distance between the resource and the utilization site should therefore be kept as small as possible.

The capital cost of a geothermal plant is usually higher, and sometimes much higher, than that of a similar plant run on a conventional fuel. Conversely, the energy driving a geothermal plant costs far less than the energy from a conventional fuel, and corresponds to the cost of maintaining the geothermal elements of the plant (pipelines, valves, pumps, heat exchangers and so on). The savings in energy costs should recover the higher capital outlay. The resource-plant system should therefore be designed to last long enough to amortize the initial investment and, wherever possible, even longer.

Appreciable savings can be achieved by adopting integrated systems that offer a higher utilization factor (for example, combining space heating and cooling) or cascade systems, where the plants are connected in series, each utilizing the wastewater from the preceding plant (for example, electricity generation + greenhouse heating + animal husbandry).

In order to reduce maintenance costs and shutdowns, the technical complexity of the plant should be on a level that is accessible to local technical personnel or to experts who are readily available. Highly specialized technicians or the manufacturers should ideally be needed only for large-scale maintenance operations or major breakdowns.

Finally, if the geothermal plant is to produce consumer products, a careful market survey must be carried out beforehand to guarantee an outlet for these products. The necessary infrastructures for the economic transport of the end product from the production site to the consumer should already exist, or be included in the initial project.

The foregoing observations can be applied to any form of utilization of geothermal energy and any local conditions, and are therefore of a purely qualitative nature. For a quantitative idea of investments and costs we recommend the *World Energy Assessment Report*, prepared by UNDP, UN-DESA and the World Energy Council, and published

Table 1.5 Energy and investment costs for electric energy production from renewables

	Current energy cost	Potential future energy cost	Turnkey investment cost
	US cents/kWh	US cents/kWh	US$/kWh
Biomass	5–15	4–10	900–3 000
Geothermal	2–10	1–8	800–3 000
Wind	5–13	3–10	1 100–1 700
Solar:			
photovoltaic	25–125	5–25	5 000–10 000
thermal electricity	12–18	4–10	3 000–4 000
Tidal	8–15	8–15	1 700–2 500

Source: Fridleifsson, 2001.

Table 1.6 Energy and investment costs for direct heat production from renewables

	Current energy cost	Potential future energy cost	Turnkey investment cost
	US cents/kWh	US cents/kWh	US$/kWh
Biomass (including ethanol)	1–5	1–5	250–750
Geothermal	0.5–5	0.5–5	200–2 000
Wind	5–13	3–10	1 100–1 700
Solar heat low temperature	3–20	2–10	500–1 700

Source: Fridleifsson, 2001.

in 2000. The WEA data are given in Tables 1.5 and 1.6, which also compare geothermal energy with other renewable forms of energy (Fridleifsson, 2001).

1.6 ENVIRONMENTAL IMPACT

During the 1960s, when our environment was healthier than it is today and we were less aware of any threat to the Earth, geothermal energy was still considered a 'clean energy'. There is actually no way of producing or transforming energy into a form that can be utilized by humankind without making some direct or indirect impact on the environment. Even the oldest and simplest form of producing thermal energy, that is, burning wood, has a detrimental effect, and deforestation, one of the major problems in recent years, first began when our ancestors cut down trees to cook their food and heat their houses. Exploitation of geothermal energy also has an impact on the environment, but there is no doubt that it is one of the least polluting forms of energy.

1.6.1 Sources of pollution

In most cases the degree to which geothermal exploitation affects the environment is proportional to the scale of its exploitation (Lunis and Breckenridge, 1991). Table 1.7 summarizes the probability and relative severity of the effects on the environment of developing geothermal direct-use projects. Electricity generation in binary cycle plants will affect the environment in the same way as direct heat uses. The effects are potentially greater in the case of conventional back-pressure or condensing power plants, especially as regards air quality, but can be kept within acceptable limits.

Any modification to our environment must be evaluated carefully, in deference to the relevant laws and regulations (which in some countries are very severe), but also because an apparently insignificant modification could trigger a chain of events whose impact is difficult to fully assess beforehand. For example, a mere 2 to 3 °C increase in the temperature of a body of water as a result of discharging the wastewater from a utilization plant could damage its ecosystem. The plant and animal organisms that are most sensitive to temperature variations could gradually disappear, leaving a fish species without its food source. An increase in water

temperature could impair development of the eggs of other fish species. If these fish are edible and provide the necessary support for a fishing community, then their disappearance could be critical for the community at large.

The first perceptible effect on the environment is that of *drilling*, whether the boreholes are shallow ones for measuring the geothermal gradient in the study phase, or exploratory/producing wells. Installation of a drilling rig and all the accessory equipment entails the construction of access roads and a drilling pad. The latter will cover an area ranging from 300 to 500 m^2 for a small truck-mounted rig (max. depth 300 to 700 m) to 1,200 to 1,500 m^2 for a small-to-medium rig (max. depth of 2,000 m). These operations will modify the surface morphology of the area and could damage local plants and wildlife. Blowouts can pollute surface water; blowout preventers should be installed when drilling geothermal wells where high temperatures and pressures are anticipated (Lunis and Breckenridge, 1991). During drilling or flow tests undesirable gases may be discharged into the atmosphere. The impact on the environment caused by drilling mostly ends once drilling is completed.

The next stage, installation of the pipelines that will transport the geothermal fluids, and construction of the *utilization plants*, will also affect animal and plant life and the surface morphology. The scenic view will be modified, although in some areas such as Larderello (Italy) the network of pipelines criss-crossing the countryside and the power-plant cooling towers have become an integral part of the panorama and are indeed a famous tourist attraction.

Environmental problems also arise during plant operation. Geothermal fluids (steam or hot water) usually contain *gases* such as carbon dioxide (CO_2), hydrogen sulfide (H_2S), ammonia (NH_3), methane (CH_4), and trace amounts of other gases, as well as *dissolved chemicals* whose concentrations usually increase with temperature. For example, sodium chloride (NaCl), boron (B), arsenic (As) and mercury (Hg) are a source of pollution if discharged into the environment. Some geothermal fluids, such as those utilized for district heating in Iceland, are freshwaters, but this is very rare. The wastewaters from geothermal plants also have a higher temperature than the environment and therefore constitute a potential thermal pollutant.

Air pollution may become a problem when generating electricity in conventional power plants. H_2S is one of the main pollutants. The odour threshold for H_2S in air is about 5 parts per billion by volume, and subtle physiological effects

Table 1.7 Probability and severity of potential environmental impact of direct-use projects

Impact	Probability of occurring	Severity of consequences
Air quality pollution	L	M
Surface water pollution	M	M
Underground pollution	L	M
Land subsidence	L	L to M
High noise levels	H	L to M
Well blowouts	L	L to M
Conflicts with cultural and archaeological features	L to M	M to H
Social-economic problems	L	L
Chemical or thermal pollution	L	M to H
Solid waste disposal	M	M to H

L = Low; M = Moderate; H = High.
Source: Lunis and Breckenridge, 1991.

can be detected at slightly higher concentrations (Weres, 1984). Various processes, however, can be adopted to reduce emissions of this gas. CO_2 is also present in the fluids used in the geothermal power plants, although much less CO_2 is discharged from these plants than from fossil-fuelled power stations: 13 to 380 g for every kWh of electricity produced in the geothermal plants, compared with the 1,042 g/kWh of the coal-fired plants, 906 g/kWh of oil-fired plants and 453 g/kWh of natural gas-fired plants (Fridleifsson, 2001). Binary cycle plants for electricity generation and district-heating plants may also cause minor problems, which can virtually be overcome simply by adopting closed-loop systems that prevent gaseous emissions.

Discharge of wastewaters is also a potential source of chemical pollution. Spent geothermal fluids with high concentrations of chemicals such as boron, fluoride or arsenic should be treated, re-injected into the reservoir, or both. However, the low-to-moderate temperature geothermal fluids used in most direct-use applications generally contain low levels of chemicals, and the discharge of spent geothermal fluids is seldom a major problem. Some of these fluids can often be discharged into surface waters after cooling (Lunis and Breckenridge, 1991). The waters can be cooled in special storage ponds or tanks to avoid modifying the ecosystem of natural bodies of water (rivers, lakes and even the sea).

Extraction of large quantities of fluids from geothermal reservoirs may give rise to *subsidence*, that is, a gradual sinking of the land surface. This is an irreversible phenomenon, but by no means catastrophic, as it is a slow process distributed over vast areas. Over a number of years the lowering of the land surface could reach detectable levels, in some cases of the order of a few tens of centimetres and even metres, and should be monitored systematically, as it could damage the stability of the geothermal buildings and any private homes in the neighbourhood. In many cases subsidence can be prevented or reduced by re-injecting the geothermal wastewaters.

The withdrawal and/or re-injection of geothermal fluids may trigger or increase the frequency of *seismic events* in certain areas. However these are microseismic events that can only be detected by means of instrumentation. Exploitation of geothermal resources is unlikely to trigger major seismic events, and so far has never been known to do so.

The noise associated with operating geothermal plants could be a problem where the plant in question generates electricity. During the production phase there is the higher pitched noise of steam travelling through pipelines and the occasional vent discharge. These are normally acceptable. At the power plant the main noise pollution comes from the cooling tower fans, the steam ejector, and the turbine 'hum' (Brown, 2000). The noise generated in direct heat applications is usually negligible.

1.7 FINAL CONSIDERATIONS

The thermal energy present in the underground is enormous. A group of experts has estimated (Table 1.8) the

Table 1.8 Geothermal potential worldwide

	High-temperature resources suitable for electricity generation		Low-temperature resources suitable for direct use in million TJ/yr of heat (lower limit)
	Conventional technology in TWh/yr of electricity	Conventional and binary technology in TWh/yr of electricity	
Europe	1 830	3 700	>370
Asia	2 970	5 900	>320
Africa	1 220	2 400	>240
North America	1 330	2 700	>120
Latin America	2 800	5 600	>240
Oceania	1 050	2 100	>110
World potential	11 200	22 400	>1 400

Source: International Geothermal Association, 2001.

geothermal potential of each continent in terms of high- and low-temperature resources (International Geothermal Association, 2001).

If exploited correctly, geothermal energy could certainly assume an important role in the energy balance of many countries. In certain circumstances even small-scale geothermal resources are capable of solving numerous local problems and of raising the living standards of small isolated communities.

In the preceding paragraphs, we have highlighted the various benefits that can accrue from utilizing this form of energy. It would be a mistake, however, to assume that it can solve all energy problems and that its exploitation will be a success in any location and under any conditions. We certainly have no wish to discourage potential users of geothermal energy; quite the contrary, it is our mission and our wish to promote this energy form as much as possible. But at the same time we have to emphasize the importance of making a careful evaluation, with the backing of reliable data, of the physical, technical, economic and social situation, before taking any action. We should also like to recall a few other important points that must be considered beforehand.

Exploitation nearly always triggers certain physical and chemical processes in the underground that, added to naturally occurring processes, could lead to the depletion of the geothermal resources. In order to avoid unpleasant surprises at a later date, it is therefore of crucial importance that we begin by modelling the geothermal system, to obtain some indication of how long our resources can be expected to last. The smaller the resources and the financial investment involved, the simpler this model will be. This preliminary evaluation and a rational programme for developing the resources should guarantee that the funds spent on research and construction of the utilization plants will prove a worthwhile investment. Even in the simplest, and admittedly somewhat rare, case of the utilization of thermal fluids emerging naturally at the ground surface (a hot spring), where no artificial extraction by wells or pumps is involved, a few preliminary studies should still be carried out on the quality of the fluid and on the surrounding hydrogeological conditions. Fluids that exhibit scaling or corrosion properties will create severe problems in a utilization plant within a short time. The extraction of large quantities of underground water or any other operations that modify the hydrological conditions in the area could also change the characteristics of the resource and even lead eventually to its depletion.

The engineering of geothermal plants and the materials used in their construction are often more expensive than in plants that use traditional fuels. Similarly, maintenance operations can in some cases prove more expensive.

The environmental impact of geothermal energy can be kept within fairly acceptable limits and is generally less harmful than that of other energy sources. Nevertheless, these effects cannot be ignored altogether, and their elimination or mitigation may require expensive operations and the installation of auxiliary plants.

All these factors (the need for accurate preliminary evaluations, the higher cost of the utilization plants and of their maintenance, environmental impact), and others that we have already discussed, could weigh against geothermal energy in a balance of the pros and cons of different energy sources. Their influence on the final choice could, however, be reduced by basing the geothermal programme and operative decisions on the information and recommendations of qualified geothermal experts. It would be difficult to find geologists, engineers or teams of consultants willing to admit that they are incapable of planning and implementing a geothermal project. Yet the research and utilization of geothermal energy actually pose a series of specific problems that can be anticipated and overcome only by individuals with the relevant know-how and experience.

There are many factors that must be considered before opting for the geothermal solution and launching a research and development programme, and not all of them are positive ones. Having warned you, as a potential user, of the pitfalls lying in wait for the unwary, we feel that it is only right that we also present some of the positive aspects that make geothermal energy such a valuable asset for many nations:

- Geothermal is a 'national' energy that could, in favourable circumstances, lead to a reduction in the import of the more expensive conventional fuels or, on the contrary, an increase in the export of the latter.
- In certain areas and in certain situations geothermal energy may be the only energy source available.
- The 'fuel' itself costs nothing.
- The utilization of geothermal energy contributes to the reduction of greenhouse gas, as recommended by the UN Framework Convention on Climate Change (Kyoto, 1997).

REFERENCES

Armstead, H. C. H. 1983. *Geothermal Energy*. London, E. & F. N. Spon.

Axelsson, G.; Gunnlaugsson, E. 2000. Background: Geothermal utilization, management and monitoring. In: *Long-term monitoring of high- and low-enthalpy fields under exploitation*, pp. 3–10. Japan, WGC 2000 Short Courses.

Barbier, E.; Fanelli, M. 1977. Non-electrical uses of geothermal energy. *Prog. Energy Combustion Sci.*, Vol. 3, pp. 73–103.

Beaeall, S. E.; Samuels, G. 1971. The use of warm water for heating and cooling plant and animal enclosures. *Oak Ridge National Laboratory*, ORNL-TM-3381.

Benderitter, Y.; Cormy, G. 1990. Possible approach to geothermal research and relative costs. In: M. H. Dickson and M. Fanelli (eds.), *Small Geothermal Resources: A Guide to Development and Utilization*, pp. 59–69. New York, UNITAR.

Brown, K. L. 2000. Impacts on the physical environment. In: K. L. Brown (ed.), *Environmental Safety and Health Issues in Geothermal Development*, pp. 43–56. Japan, WGC 2000 Short Courses.

Bullard, E. C. 1965. Historical introduction to terrestrial heat flow. In: W. H. K. Lee (ed.), *Terrestrial Heat Flow*, pp. 1–6. American Geophysical Un., Geophysical Monographs, Series 8.

Combs, J.; Muffler, L. P. J. 1973. Exploration for geothermal resources. In: P. Kruger and C. Otte (eds.), *Geothermal Energy*, pp. 95–128. Stanford, Conn., Stanford University Press.

Dickson, M. H.; Fanelli, M. 1988. Geothermal R. & D. in developing countries: Africa, Asia and the Americas. *Geothermics*, No. 17, pp. 815–77.

Entingh, D. J.; Easwaran, E.; McLarty, L. 1994. *Small geothermal electric systems for remote powering*. Washington D.C., US DoE, Geothermal Division.

Fridleifsson, I. B. 2001. Geothermal energy for the benefit of the people. *Renewable and Sustainable Energy Reviews*, Vol. 5, pp. 299–312.

Garnish, J. D. (ed.). 1987. Proceedings of the First EEC/US Workshop on Geothermal Hot-Dry Rock Technology, *Geothermics*, Vol. 16, pp. 323–461.

Gudmundsson, J. S. 1988. The elements of direct uses. *Geothermics*, Vol. 17, pp. 119–36.

Hochstein, M. P. 1990. Classification and assessment of geothermal resources. In: M. H. Dickson and M. Fanelli (eds.), *Small Geothermal Resources: A Guide to Development and Utilization*, pp. 31–57. New York, UNITAR.

Huttrer, G. W. 2001. The status of world geothermal power generation 1995-2000. *Geothermics*, Vol. 30, pp. 7–27.

International Geothermal Association. 2001. Report of the IGA to the UN Commission on Sustainable Development, Session 9 (CSD-9), New York, April 2001.

Lindal, B. 1973. Industrial and other applications of geothermal energy. In: H. C. H. Armstead (ed.), *Geothermal Energy*, pp. 135–48. Paris, UNESCO.

Lubimova, E. A. 1968. Thermal history of the Earth. In: *The Earth's Crust and Upper Mantle*, pp. 63–77, American Geophysics Un., Geophysics Monographs, Series 13.

Lumb, J. T. 1981. Prospecting for geothermal resources. In: L. Rybach and L. J. P. Muffler, (eds.), *Geothermal Systems, Principles and Case Histories*, pp. 77–108. New York, J. Wiley & Sons, New York.

Lund, J. W. 2001. Geothermal heat pumps: an overview. *Bulletin Geo-Heat Center*, Vol. 22, No. 1, pp. 1–2.

Lund, J. W.; Boyd, T. L. 2001. Direct use of geothermal energy in the United States: 2001. *Geothermal Resources Council Transactions*, Vol. 25, pp. 57–60.

Lund, J. W.; Freeston, D. 2001. Worldwide direct uses of geothermal energy 2000. *Geothermics*, Vol. 30, pp. 29–68.

LUNIS, B.; BRECKENRIDGE, R. 1991. Environmental considerations. In: P. J. Lienau and B. C. Lunis (eds.), *Geothermal Direct Use, Engineering and Design Guidebook*, pp. 437–45. Klamath Falls, Ore., Geo-Heat Center.

MEIDAV, T. 1998. Progress in geothermal exploration technology. *Bulletin Geothermal Resources Council*, Vol. 27, No. 6, pp. 178–81.

MUFFLER, P.; CATALDI, R. 1978. Methods for regional assessment of geothermal resources. *Geothermics*, Vol. 7, pp. 53–89.

NICHOLSON, K. 1993. *Geothermal Fluids*. Berlin, Springer Verlag.

POLLACK, H. N.; HURTER, S. J.; JOHNSON, J. R. 1993. Heat flow from the Earth's interior: Analysis of the global data set. *Rev. Geophys.*, Vol. 31, pp. 267–80.

RAFFERTY, K. 1997. An information survival kit for the prospective residential geothermal heat pump owner. *Bull. Geo-Heat Center*, Vol. 18, No. 2, pp. 1–11.

SANNER, B. 2001. Shallow geothermal energy. *Bulletin Geo-Heat Center*, Vol. 22, No. 2, pp. 19–25.

STACEY, F. D.; LOPER, D. E. 1988. Thermal history of the Earth: a corollary concerning non-linear mantle rheology. *Phys. Earth. Planet. Inter*. Vol. 53, pp. 167–74.

STEFANSSON, V. 2000. The renewability of geothermal energy. *Proc. World Geothermal Energy, Japan*. On CD-ROM.

TENZER, H. 2001. Development of hot dry rock technology. *Bulletin Geo-Heat Center*, Vol. 32, No. 4, pp. 14–22.

WERES, O. 1984. *Environmental protection and the chemistry of geothermal fluids*. California, Lawrence Berkeley Laboratory, LBL 14403.

WHITE, D. E. 1973. Characteristics of geothermal resources. In: P. Kruger and C. Otte (eds.), *Geothermal Energy*, pp. 69–94. Stanford, Conn., Stanford University Press.

WRIGHT, P. M. 1998. The sustainability of production from geothermal resources. *Bull. Geo-Heat Center*, Vol. 19, No. 2, pp. 9–12.

SELF-ASSESSMENT QUESTIONS

1. The terrestrial heat flow that can be measured at the surface derives mainly from:
(a) What is left in the Earth of its original heat, that is, of the heat that existed at the time our planet was formed.
(b) The decay of the radioactive isotopes inside the Earth.
(c) The heat from friction caused by the movement of crustal plates against one another.

2. In 2000, the total geothermoelectric capacity installed in the world had reached 7,974 MW_e. What percentage of this figure was installed in the developing countries?
(a) 22 per cent; (b) 35 per cent; (c) 47 per cent

3. What was the total world capacity relative to non-electric uses of geothermal energy in 2000?
(a) About 10,000 MW_t; (b) About 15,000 MW_t;
(c) About 20,000 MW_t

4. A well drilled vertically has the following characteristics:
– elevation of wellhead: 655 m above sea level
– depth: 870 m
– stratigraphy: mainly Pliocenic clay
– bottomhole temperature: 53 °C
– mean annual outdoor temperature: 13 °C (surface temperature).

What is the geothermal gradient? What temperature can be expected at a depth of 1,350 m?

5. The zones of the Earth that are most likely to contain high-temperature geothermal areas, and consequently the most important geothermal fields, are distributed according to a certain pattern. Where are these zones located? If possible, cite some examples.

6. What are the main elements of a 'geothermal system'? Which of these elements can be either partly or totally artificial? What physical phenomenon governs the mechanism of a geothermal system?

7. What is 'superheated steam'?

8. Reservoir temperature is an essential parameter for defining even the simplest of models of a geothermal resource. Name the methods that can give us indications of the temperature in the reservoir.

9. A socio-economic study demonstrated that, in a certain area, it would be feasible to install a fish farm, greenhouses, a district heating system and a plant for drying vegetables. Since the well drilled in the area produces water at 80 °C, it was decided that the cost of the well and its maintenance should be spread over a number of utilization schemes in cascade. Specify which schemes could effectively be adopted and their order in the cascade system, in decreasing order of utilization temperature.

10. Heat pumps are now widely used in a number of countries. Are thermal waters necessary for their utilization? What is the minimum water temperature required?

11. The most effective way of reducing the impact on the environment of the effluent from a geothermal power plant is to inject it back underground in re-injection wells. Wherever possible these fluids are injected back into the reservoir from which we extract the fluids feeding the geothermal power plant. If this operation (re-injection) is not carried out correctly, it can create serious problems. What are these problems?

12. Why does an increase in the efficiency of a geothermal power plant lead to a reduction in the impact on the environment?

13. What is the main characteristic of a geopressured reservoir? What type of energy could these reservoirs provide, assuming they could be put into production?

14. What is the main difference between 'renewable' and 'sustainable' with reference to a resource? Define these two terms with reference to geothermal energy.

ANSWERS

1.
(b) Terrestrial heat flow is caused mainly by the decay of the radioactive isotopes of uranium, thorium and potassium. The other sources play a minor role.

2. (c).

3. (b).

4. The data needed to calculate the geothermal gradient are surface temperature (the mean annual ambient temperature is acceptable), bottomhole temperature, and well depth. The geothermal gradient (g) is calculated as the difference between temperature at bottomhole (t_b) and temperature at ground level (t_g), divided by the depth of the well (d), i.e. ($t_b - t_g$)/$d = g$. Therefore, g in this case is 0.046 °C/m. If the geothermal gradient remains constant below well depth, then we can extrapolate a temperature of 62 °C at 1,350 m.

5. Most geothermal fields are on or near the margins of lithospheric plates, where the molten material and magmatic bodies can ascend more easily from deep zones to near the surface. This phenomenon can clearly be observed on the western margin of the Pacific Ocean, where there are elongated areas rich in geothermal fields, such as in Japan, the Philippines and Indonesia. Other examples of this phenomenon are Iceland, at the contact between the Eurasian and North American plates, and Central America, at the margin between the Cocos and Caribbean plates.

6. A geothermal system is made up of a heat source, a reservoir and a fluid (water and/or steam). Of these three elements, the heat source (energy) clearly must be natural, while the other two could, in some cases, be artificial. A geothermal system is governed by convection.

7. In simple terms, 'superheated steam' is a fluid (steam) that has received more heat than is actually required to complete the phase change from water to vapour.

8. Geophysical and geochemical methods are mainly used. Geophysics can provide us with the temperature at the 'top' of the reservoir, that is, on its upper surface, by means of measurements of the geothermal gradient (or heat flow) in specially drilled boreholes or in existing wells. This method is called thermal prospecting. Geochemistry, by means of chemical or isotope thermometers, can give us a reasonable estimate of the temperature of the fluids in the reservoir.

9. Vegetable drying plants usually require temperatures of at least 100 °C, so they are not recommended in this case. The other schemes could utilize the geothermal water as follows: district heating, from 80 to 45 °C; greenhouse heating, from 45 to 30 °C; finally, heating of ponds for fish farming with the residual water at 30 °C.

10. Hot water is not necessary in order to use heat pumps. The 'heat source' could be a non-thermal aquifer, a surface body of water (such as a pond, lake, or so on), or even the ground itself. There are heat pump installations currently exploiting 'heat sources' at temperatures between 5 and 30 °C.

11. Re-injection must be preceded by a careful study of the reservoir and the fluids circulating within it. Errors in siting the re-injection well could in fact produce a cooling of the reservoir and seriously jeopardize production.

12. Simply because the higher the efficiency of the power plant, the less geothermal fluid is needed for its operation and, consequently, less effluent to be disposed of in the environment.

13. The main feature of a geopressured reservoir is the pressure of the fluid it contains, which is always near the lithostatic value. These reservoirs are potential sources of both thermal and mechanical energy, as they produce high-temperature and high-pressure fluids.

14. The term 'renewable' is used to describe the resource itself, whereas 'sustainable' is used to describe how the resource is utilized. In other words, a resource is considered renewable when it is recharged during exploitation. A resource is considered sustainable if, using adequate precautions, it can be exploited without creating problems for future generations.

CHAPTER 2

Electricity Generation

Roger B. Hudson
PB Power, Auckland, New Zealand

AIMS

1. To show that geothermal energy can in many cases be used to generate electricity; that there are various types of conventional power plants that can operate with geothermal fluid, and that the plants designed and constructed specifically for geothermal fluids have now been thoroughly tried and tested and that their efficiency is satisfactory.

2. To show that electricity of geothermal origin can be competitive compared to electricity produced from conventional fuels, provided that we are well aware of the particular characteristics of this form of energy and its constraints.

3. To provide a broad overview of the main technical and economic characteristics of geothermal generating plants.

OBJECTIVES

When you have completed this chapter you should be able to:

1. Discuss the general concepts of geothermal power plants.

2. Describe the basic technical features of the plant type considered:
– atmospheric exhaust conventional steam turbine
– condensing exhaust conventional steam turbine
– binary plant
– biphase rotary separator turbo-alternator.

3. Give indicative costs and scheduling information for geothermal power plants.

4. Discuss the main economic elements relevant to geothermal plants and the associated power planning aspects.

The approach taken in preparing this chapter is to provide only the basic technical background of the plant types. Emphasis has been placed on detailing the economic and technical characteristics that influence the choice of an appropriate plant.

2.1 TECHNICAL FEATURES OF PLANT OPTIONS

2.1.1 Atmospheric exhaust conventional steam turbine

Atmospheric exhaust (back-pressure) turbines are the simplest, and have the lowest capital cost, of all geothermal cycles. With this type of plant, steam is separated from the geothermal discharge and fed through a conventional axial flow steam turbine that exhausts directly to the atmosphere. A simplified schematic of an atmospheric exhaust plant is shown in Figure 2.1.

Such machines consume about twice as much steam per kilowatt of output (for the same inlet pressure) as condensing plants, and are therefore wasteful of energy and costly in wells. Nevertheless, they have their uses as pilot plants, as standby plants, for small local supplies from isolated wells, and for generating electricity from the discharge of test wells during field development.

A further advantage with atmospheric exhaust units is that they can generally be started without the need for an external power supply, as the only essential auxiliary, the lubricating oil pump, can be driven by a small steam turbine.

A sketch of two types of small atmospheric exhaust turbo-alternators offered by Mitsubishi Heavy Industries is shown in Figure 2.2. These units can also be operated with condensing exhaust with appropriate inlet pressures and blading. The choice of Mitsubishi for this example was purely for reasons of convenience and availability of information. The type and performance of equipment offered by alternative manufacturers are comparable with those offered by Mitsubishi.

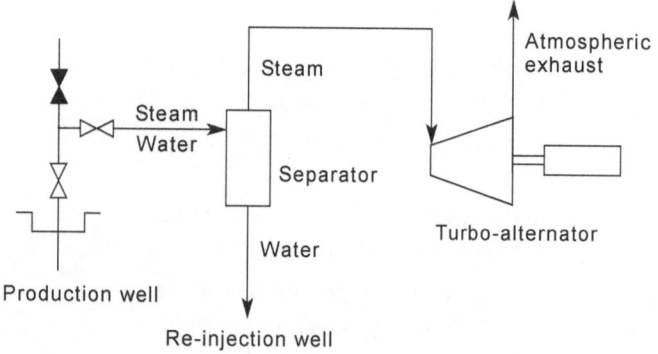

Figure 2.1 Atmospheric exhaust cycle simplified schematic

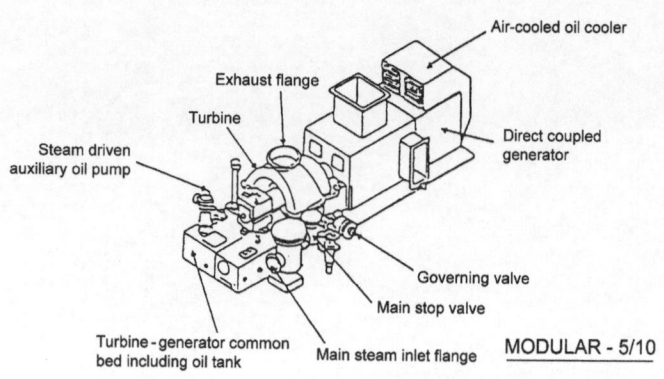

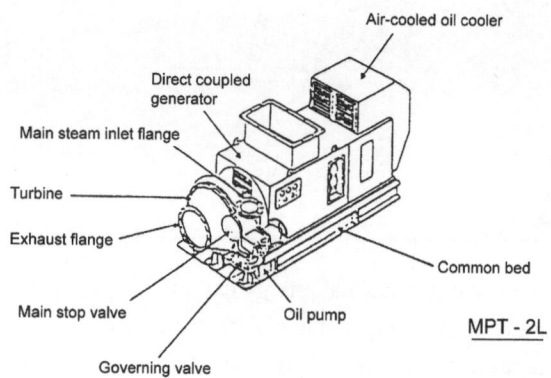

Figure 2.2 Sketch of Mitsubishi Modular-5/10 and MPT-2L mini-turbines

Steam consumption

The steam consumption of the standard atmospheric exhaust units offered by Mitsubishi is shown in Figure 2.3. This graph is based on the ideal case of 0 per cent non-condensable gas content in the steam flow. The steam consumption curves show the rated steam consumption and load that can be achieved for a given rated inlet pressure at the rated design operating point only. They do not indicate the variation of steam flow and load for a given inlet pressure with throttle-controlled load governing. The curves therefore do not represent the performance of any one turbine, but are rather design operating points that can be achieved from the standard turbines with appropriate blading configurations. However the standard turbines are designed to be versatile, and maintain good performance over varying inlet pressures. Generally for a specific standard turbine the isentropic efficiency at the design condition is approximately maintained down to about 50 per cent of rated load with progressively deteriorating efficiency below this point.

The altitude at which the unit is operated has an important effect on the power that is produced from a given inlet

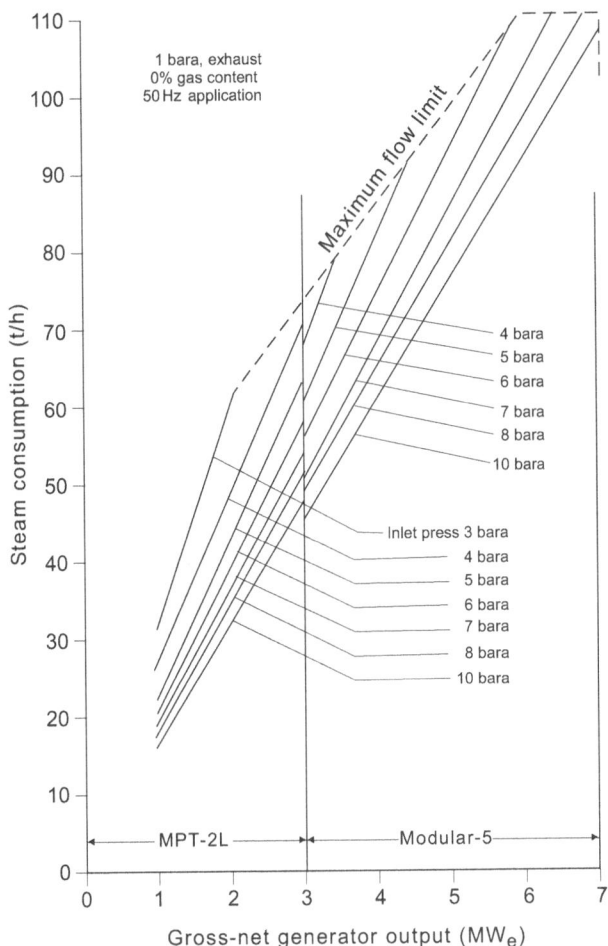

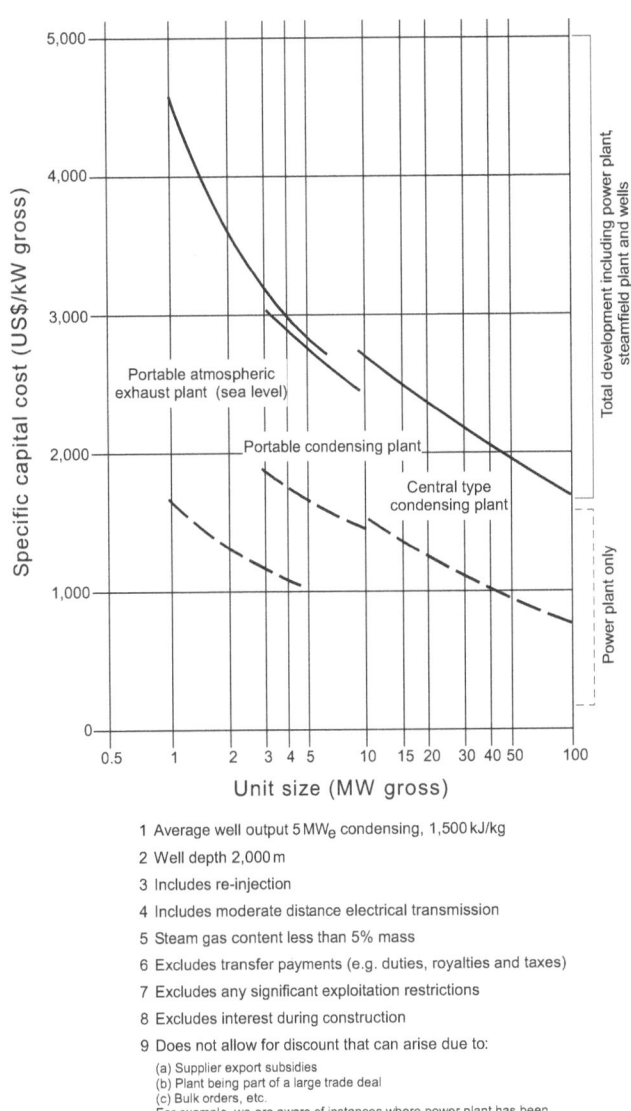

Figure 2.3 Atmospheric-exhaust (sea level) turbine steam consumption curves

1 Average well output 5 MW$_e$ condensing, 1,500 kJ/kg
2 Well depth 2,000 m
3 Includes re-injection
4 Includes moderate distance electrical transmission
5 Steam gas content less than 5% mass
6 Excludes transfer payments (e.g. duties, royalties and taxes)
7 Excludes any significant exploitation restrictions
8 Excludes interest during construction
9 Does not allow for discount that can arise due to:
   (a) Supplier export subsidies
   (b) Plant being part of a large trade deal
   (c) Bulk orders, etc.
   For example, we are aware of instances where power plant has been purchased for approximately half the cost shown
10 Includes design and erection

Figure 2.4 Generalized capital cost for conventional steam turbine geothermal power development (1998)

pressure and steam mass flow. At higher altitudes the lower exhaust pressure, corresponding to the lower atmospheric pressure, results in greater power generation. For example, for the Olkaria field in Kenya (1,950 m above sea level) where the atmospheric pressure is 0.8 bara (1 bara = 1 bar absolute. 1 bar = 1×10^5 Pa), the available power for a given steam mass flow is increased by 10.5 per cent for a typical inlet pressure of 8 bara. This percentage increase is greater if lower inlet pressures are used.

Generalized costing, manufacturing and erection

Figure 2.4 shows a generalized development cost graph for well-head atmospheric exhaust, well-head condensing and central station condensing plants versus unit size. These data must be used with appropriate caution, as the costs are only indicative averages and should never be used to estimate the cost of a specific development.

The manufacturing and erection times given in Table 2.1 would be expected for a typical well-head development.

2.1.2 Condensing exhaust conventional steam turbine

This type of plant is a thermodynamic improvement on the atmospheric exhaust design, and is by far the most common plant concept employed for geothermal power generation. Instead of the steam from the turbine being discharged to the atmosphere, it is discharged to a condensing chamber

Table 2.1 Expected manufacture and erection times for a typical well-head development

	Order date to shipment (months)	Shipment to commissioning (months)	Total (months)
Atmospheric exhaust (2.5 MW)	9	4	13
Atmospheric exhaust (5 MW)	10	4	14

that is maintained at a very low absolute pressure, typically about 0.10 bara. Because of the greater pressure drop across a condensing turbine, approximately twice as much power as with an atmospheric exhaust turbine is generated from a given steam flow, at typical inlet conditions. However, the addition of a condenser and the associated cooling towers and pumping equipment significantly increases the cost of the total plant. In addition, power consumption is required for the main cooling water pumps and cooling tower fans, with total station auxiliary power consumption being typically approximately 4.6 per cent of the gross generation. The reason the exhaust steam must be condensed is that an impractical amount of work would be required to pump the fluid from the low pressure conditions in the condenser if it were not first converted to the liquid state. A simplified schematic of a condensing steam turbine plant is shown in Figure 2.5.

The maximum size of a turbine is limited by the last-stage blade length, which traditionally used to be limited to approximately 660 mm, but in recent years has been extended up to 765 mm for a 50 Hz machine. The steam production pressures appropriate for geothermal generation are typically in the range of 3 bara to 15 bara (see Hudson, 1988,

for technical discussion of separator and turbine inlet pressure optimization), which is considerably lower than for a fossil fuel-fired thermal power plant. Because of the relatively low density of steam at these pressures, and the limiting length of the last-stage blades, the maximum capacity of geothermal condensing turbines was typically limited to 55 to 60 MW for a double-flow turbine with 660 mm last-stage blades. However, machines of 110 MW capacity have recently been manufactured (for example, Wayang Windu Unit 1), using a turbine inlet pressure of 10 bara and 765 mm last-stage blades. This maximum capacity is considerably smaller than available for fossil fuel-fired thermal units, which are commonly 600 to 1,000 MW, and limits the economy of scale that can be achieved for increased unit size with geothermal units.

Turbine sizes less than approximately 35 MW, at typical turbine inlet pressures of approximately 6.5 bara, would generally be single-flow machines (the steam flows in only one direction) and mounted at ground level with an overhead exhaust duct to an adjacent ground-level condenser. Double-flow machines would generally be mounted on an elevated pedestal with an underhung condenser. The isentropic efficiency for a geothermal turbine would typically range between 81 and 85 per cent, with a turbine-generator mechanical efficiency of 96.3 per cent.

The 55 MW double-flow turbine became somewhat of an industry standard during the 1980s and early 1990s. A diagram of the Ansaldo 55 MW double-flow machine is shown in Figure 2.6, and this is typical of the standard approach of the bottom-entry steam pipework and underhung condenser that is commonly used.

A more comprehensive process schematic for a geothermal condensing unit is shown in Figure 2.7 and will be used to discuss some of the pertinent features of the cycle.

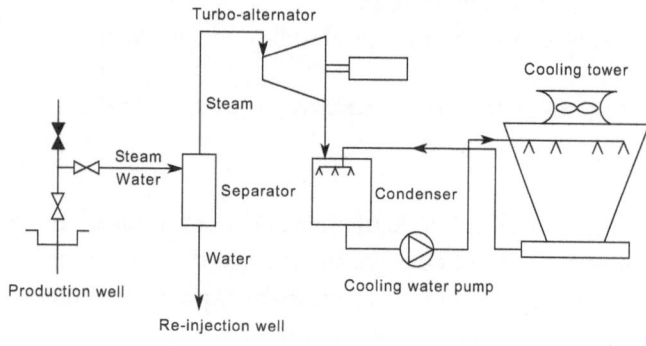

Figure 2.5 Condensing cycle simplified schematic

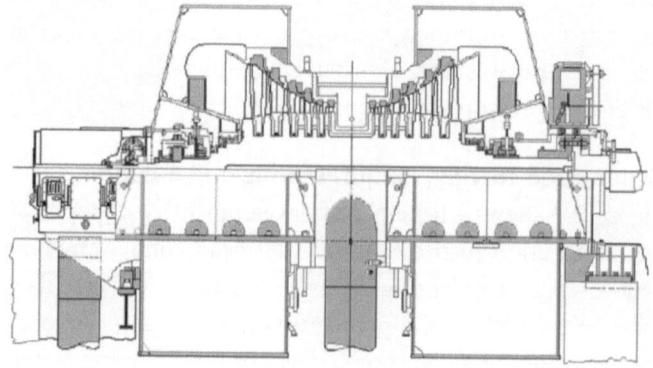

Figure 2.6 Ansaldo 55 MW double-flow geothermal turbine

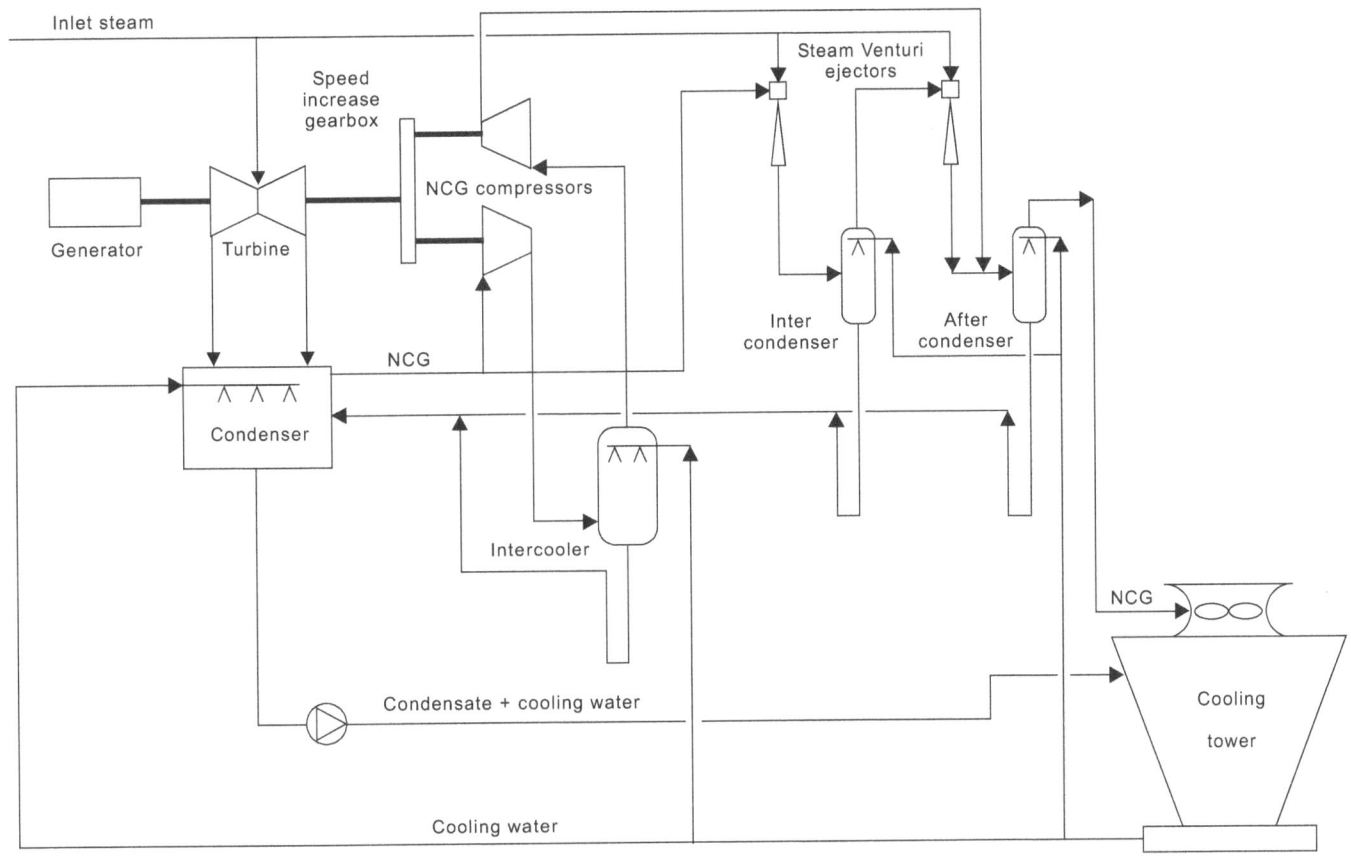

Figure 2.7 Condensing cycle process schematic

Condenser

As there is no need to recover the condensate for reuse in the process cycle, direct contact condensers are generally used for condensing the turbine exhaust steam. These incorporate banks of spray nozzles through which cooling water is passed in order to condense the steam. The condensed steam and cooling water are mixed together in the condenser and pumped directly to the cooling tower for recirculation. The amount of fluid being added to the cooling water by steam condensation typically exceeds that being lost by cooling tower evaporation and drift loss, and generally there will be a requirement to dispose of, typically, approximately 20 per cent of the turbine steam flow as condensate blowdown. The net addition of fluid to the cooling circuit, however, avoids the need for continuous cooling water make-up, as is required for conventional thermal plants (to compensate for evaporation loss and the blowdown flow required to maintain water quality).

In order to reduce the amount of non-condensable gas that dissolves in the cooling water, and also to employ more effective counterflow heat exchange, the last part of the condensation process is undertaken in a separate gas-cooling zone where the steam and gas flow are directed vertically upwards. Typically, approximately 10 per cent of the steam will be condensed in the gas cooling zone, requiring approximately 11 per cent of the total cooling water flow, and therefore only this relatively small proportion of the cooling water flow will be exposed to the very high non-condensable gas partial pressures (which are the driving force for dissolving the gas in the water) that will apply at the end of the condensation path.

Where non-condensable gas content is high, and environmental regulations require a hydrogen sulfide (H_2S) abatement system to be employed, a surface shell-and-tube condenser will generally be used. The condensed steam is still generally mixed in with the recirculating cooling water, after being separately pumped from the condenser; however, this arrangement minimizes the amount of non-condensable gas that dissolves in the cooling water (some of which would be released to the atmosphere during passage of the water through the cooling tower) and also minimizes water carry-over to the gas extraction plant.

Hydrogen sulfide abatement

Where H_2S extraction is required, this is generally undertaken using either the Stretford process or the ARI LO-CAT II® H_2S oxidation process. The latter process has been the most common for recent developments and provides an isothermal, relatively low operating cost method for carrying out the modified Claus reaction:

$$H_2S + 1/2O_2 \rightarrow H_2O + S^o$$

This reaction is accomplished in an aqueous scrubbing system by using a water soluble metal ion, typically an iron (Fe) chelate solution, which is capable of removing electrons from a sulfide ion to form sulfur, and in turn can transfer the electrons to oxygen, from ambient air, in the regeneration process.

Non-condensable gas extraction

The non-condensable gases that are also present in geothermal steam, and which accumulate in the condenser, must be pumped from the condenser separately using gas extraction equipment. Unless H_2S abatement is required, the non-condensable gas is generally disposed of by mixing it with the cooling tower discharge air plume. The appropriate equipment to be used for gas extraction is dependent on the proportion of non-condensable gas present with the steam. At low gas contents (less than 1.5 per cent by weight), steam Venturi gas ejectors are generally the most economic choice. These are simple and reliable but are relatively inefficient. The type of ejector used for condenser gas extraction, requiring relatively high compression ratios to be achieved, is also not able to be throttled; it is therefore necessary to change the ejector nozzles if it is desired to reduce the steam consumption to suit a reduced non-condensable gas flow. Historically, two stages of steam Venturi ejectors in series have generally been used; however, since operation at the compression ratio associated with this duty suffers significant performance fall-off, a more efficient (but higher capital cost) arrangement of three series stages is now often used.

At higher gas contents, the high steam consumption of the comparatively inefficient steam ejectors leads to the selection of higher capital cost, lower auxiliary consumption alternatives. Generally, for non-condensable gas contents between approximately 1 and 3.5 per cent by weight, the most economic option will be a hybrid system involving first-stage steam Venturi ejectors with liquid-ring vacuum pumps for the second-stage compression. The liquid-ring vacuum pump is essentially a constant-volume flow device, so physically large and expensive units would be required if these were also to be used for the first-stage compression.

At non-condensable gas contents above approximately 3.5 per cent by weight it is generally more economic to use multi-stage centrifugal compressors. These are generally coupled directly to the turbine through a gearbox in order to obtain shaft speeds of typically 10,700 rpm for the high-pressure (HP) stages and 5,300 rpm for the low-pressure (LP) stages. At least one stage of intercooling is applied during the compression, and this is generally provided by separate direct contact cooling vessels. The head–flow curve for a centrifugal compressor generally has an initially positive slope at low flow rates, which results in two flow rates applying to the same discharge pressure. This can lead to damaging compressor surging at low flow rates, so a minimum flow recirculation system must be adopted in order to permit stable operation during start-up and low load conditions.

The schematic of a condensing cycle presented in Figure 2.7 has been based on the use of turbine-driven centrifugal gas compressors in order to illustrate a greater range of plant. Steam Venturi ejectors are also required with this arrangement in order to be able to pull vacuum in the condenser during start-up; a typical arrangement for these is therefore also included with this schematic.

For non-condensable gas contents exceeding about 12 per cent mass of the steam, it is generally more economic to use a back-pressure plant because of the large amount of power required to extract the gases from the condenser. This will generally involve a conventional steam turbine exhausting to atmospheric pressure (or slightly greater), with additional heat recovery being achieved by an atmospheric condenser (to generate process heat water at approximately 100 °C), or else by a bottoming binary plant.

Cooling tower and main cooling water pumps

When direct contact condensers are used, care must be taken in selecting the type of packing for the cooling tower, as fouling will typically result from the deposition of contaminants from the cooling water. For this reason, splash packing has historically been adopted, as opposed to high-performance film-type packing, which suffers significant reduction in heat transfer performance with fouling. High-performance film-type packings can, however, offer significant cost savings for the cooling system, so their use can be considered in applications where the chemistry of the cooling water is relatively benign and tight control of the water quality during operation can be assured.

Because of the relatively low inlet temperature for a geothermal power plant, the thermal cycle efficiency is similarly relatively low, being typically 18 per cent for a 6.5 bara inlet, 0.1 bara exhaust turbine. This compares with typically 34 per cent for a 600 MW, reheat coal-fired unit, or 58 per cent for a modern combined cycle plant. Consequently, the amount of heat that has to be rejected in the condenser and cooling system is proportionally larger than for a conventional thermal power plant, and a proportionally larger

amount of the station cost applies to the condensing and cooling water system. This has an impact on the economics for cooling system equipment selection, the most notable point being that the most economic arrangement for the main cooling water pumps is generally found to be 2 × 50 per cent pumps, as opposed to 2 × 100 per cent hot-well pumps for a conventional thermal power plant.

Steam consumption

The steam consumption of a geothermal condensing unit is shown in Figure 2.8. This graph has been based on the ideal case of 0 per cent non-condensable gas content in the steam flow, and assuming a typical condensing pressure of 0.1 bara. As for the atmospheric exhaust case, these curves do not represent the performance of any one turbine, but are

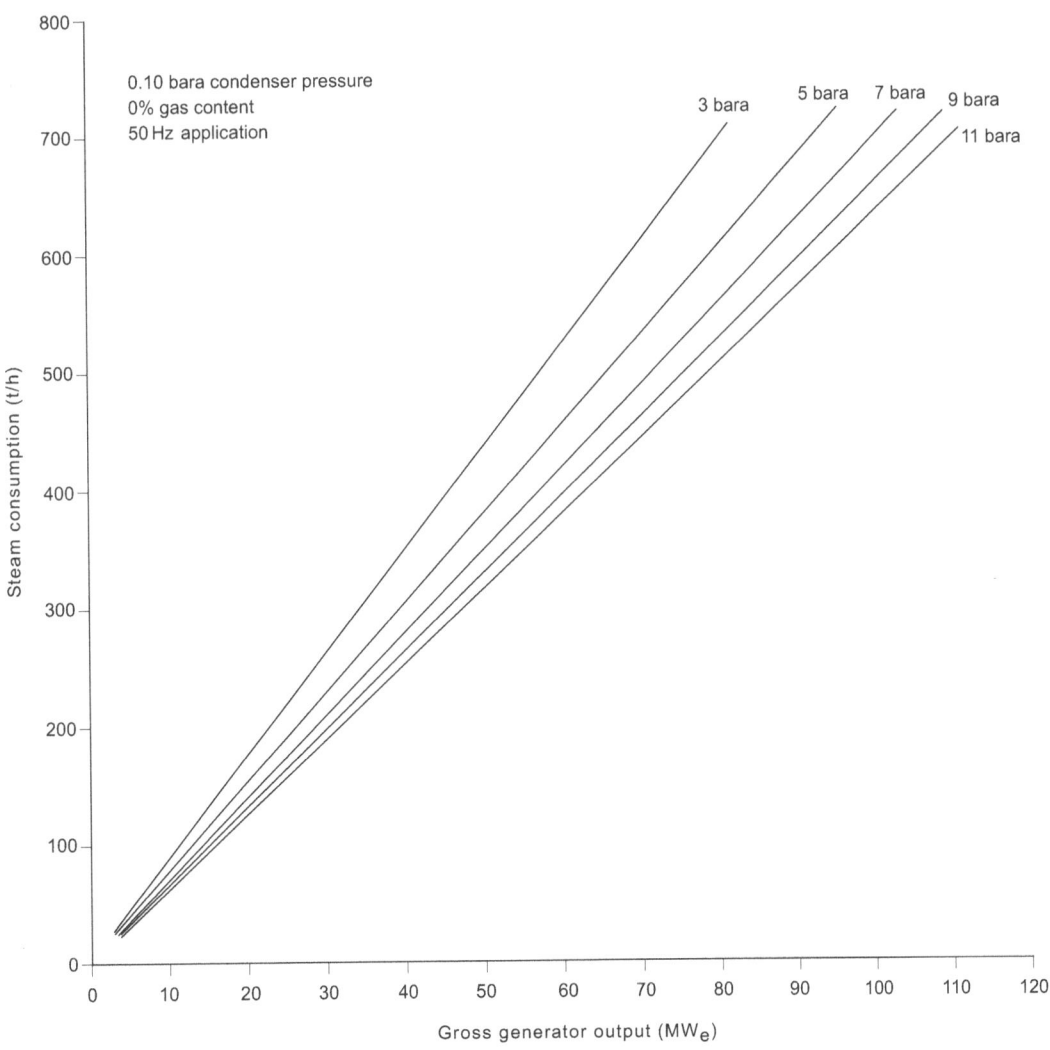

Figure 2.8 Condensing turbine steam consumption curves

rather design operating points that can be achieved from a turbine with appropriate blading configuration.

An important point to note is that it is not appropriate to compare the steam consumption of an atmospheric exhaust turbine with that of a condensing turbine at the same inlet pressure, because the optimum separation pressure (and therefore turbine inlet pressure) for an atmospheric exhaust plant is higher than that for a condensing plant. For those readers wishing to pursue this point further, a technical discussion of this matter is included in Hudson (1988).

Generalized costing, manufacturing and erection

A generalized development cost graph for wellhead atmospheric exhaust, wellhead condensing and central station condensing plants versus unit size has been shown previously in Figure 2.4.

The manufacture and erection times given in Table 2.2 would be expected for a typical development.

Table 2.2 Manufacture and erection times for a typical development

	Order date to shipment (months)	Shipment to commissioning (months)	Total (months)
Condensing (5 MW)	11	5	16
Condensing (10 MW)	13	6	19
Condensing (55 MW)	15	9	24
Condensing (110 MW)	16	10	26

2.1.3 Binary plant

Geothermal binary fluid technology was developed primarily to generate electricity from low-to-medium temperature resources, and to increase the utilization of thermal resources by recovering waste heat. Unless there are specific exploitation restrictions that must be considered, the most economic generation of electricity from high-temperature geothermal resources is generally achieved from conventional steam turbine plants.

A simplified schematic of a binary system is shown in Figure 2.9. The binary system utilizes a secondary working fluid, typically n-pentane, which, compared with steam, has a low boiling point and high vapour pressure at low temperatures. This secondary fluid is operated through a conventional

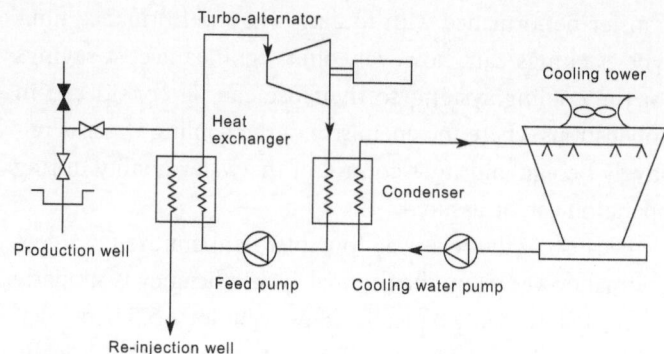

Figure 2.9 Binary cycle simplified schematic

Rankine cycle. By selecting the appropriate working fluid, binary systems can be designed to operate with inlet temperatures in the range 85 to 170 °C. The upper temperature limit is restricted by the thermal stability of the organic binary fluids. The lower temperature limit is primarily restricted by practical and economic considerations, as the required heat exchanger size for a given capacity becomes impractical. Heat is transferred from the geothermal fluid to the binary cycle via heat exchangers, where the binary fluid (or working fluid) is heated and vaporized before being expanded through a turbine to some lower pressure/temperature.

A modification of the binary cycle, known as the Kalina cycle, utilizes a water/ammonia mixture for the process fluid. This process has the benefit that the boiling point of the mixture increases as the evaporation progresses, which reduces the impact of the pinch-point limitation detailed in Appendix A.2, and therefore permits a higher heat exchange effectiveness to be achieved.

Traditionally, binary plants have been small modular units varying in size from a few hundred kilowatts to several megawatts. The cost-effectiveness of the small developments is supported by their modular construction, which facilitates short manufacturing and installation times. Larger developments of 10 to 50 MW can be achieved by bringing a number of modular units together in a common development.

With wells that do not flow spontaneously, or where it is advantageous to prevent flashing of the geothermal fluid (to prevent well calciting for example), down-hole pumps can be used to keep the fluid in a pressurized liquid state. Binary units can then be used to extract energy from the circulating fluid.

The supply of geothermal binary plants worldwide has been dominated by ORMAT® Industries, as a consequence of aggressive pricing and marketing and a sound product.

Although binary plants are principally appropriate for low-temperature applications, these have also on occasion been applied directly to higher-temperature applications, where the commercial opportunity presented itself. However, in recent years ORMAT have progressed to a very thermo-dynamically and economically appropriate combined-cycle concept for high-temperature applications.

With the combined-cycle concept a conventional back-pressure steam turbine is utilized as the topping cycle; this exhausts into a condenser, which is the inlet heat exchanger of a binary plant. The binary unit can thus be used in its appropriate temperature range to extract additional heat from the cycle, and the condensing pressure for the back-pressure turbine can be maintained above atmospheric pressure so that no pumping equipment is required for gas extraction. For a wet-steam field, installing further binary modules on the separated brine flow can also generate additional power. This combined-cycle concept achieves very good overall thermal efficiency, and has now achieved plant costs that are competitive with a conventional flashed steam condensing turbine plant for moderate enthalpy developments (circa 1,500 kJ/kg well discharge).

Binary plant heat exchangers

Heat exchangers are required to heat and evaporate the binary fluid, as well as to de-superheat and condense it during the heat rejection phase of the cycle. The binary fluid would normally be heated and evaporated in two separate units.

Conventional heat exchangers are of the shell-and-tube or plate type. These are physically large, and form a great portion of the cost of a binary plant. Much of the performance/cost trade-off of binary plants lies with the selection of heat exchangers. One of the major disadvantages of hydro-carbons and refrigerants, used as binary fluids, is that they have poor heat transfer characteristics. Scaling often compromises the heat transfer performance of heat exchangers, which in some cases can make certain development strategies impractical. Scale reduces the heat transfer and hydraulic performance of conventional surface heat exchangers, and also gives rise to higher maintenance costs and reduced plant utilization.

Research and state-of-the-art techniques have been directed towards direct contact (mixing) heat exchangers, which are very efficient and much smaller than conventional shell-and-tube types. The different boiling points of the two fluids are used to facilitate their separation after heat exchange.

The main difficulties associated with direct contact heat exchangers include:

- The need to have the primary and secondary fluids at the same pressure.
- The slight solubility of geothermal fluid in the binary fluid and vice versa. This has a polluting effect on both fluids, and can also compromise the performance of the turbine.
- Several test facilities have been built that demonstrate that direct contact technology can be applied to binary systems. As yet, however, the costs and benefits have not been clearly defined.

Net generated electric power

The choice of organic fluid and heat exchanger size are optimized for each application. This involves consideration of the operating temperature and value of the power produced. However, for feasibility study purposes the following formula enables the net generated electric power to be calculated with adequate accuracy:

$$NEP = [(0.18T - 10)ATP]/278$$

where NEP is net electric power (kW), T is inlet temperature of the primary fluid (°C), and ATP is available thermal power (kW).

The available thermal power is the heat available from the geothermal flow, and is conventionally calculated relative to a temperature 10 °C higher than the bottom-cycle temperature. A bottom-cycle temperature of 40 °C is normally assumed. For example, using the above equation it can be readily calculated that a net conversion efficiency of approximately 5.5 per cent can be achieved with a geothermal fluid inlet temperature of 140 °C but only 2.8 per cent for an inlet temperature of 100 °C.

For a combined-cycle binary plant, a combination of the previous steam consumption information for conventional steam turbines, and that presented above for the binary plant module, can be used to determine the combined performance.

The auxiliary power consumption for a binary plant, or combined-cycle binary plant, is significantly greater than for a conventional condensing plant. Typically the auxiliary power consumption for a combined cycle binary plant would be 9.5 to 15 per cent of the total gross generation, depending on the process conditions.

Generalized costing, manufacturing and erection

These figures must be used with appropriate caution, as costs can vary considerably between specific developments. In particular, the inlet temperature of the geothermal fluid has a significant effect on the cost of the binary plant. The inlet temperature influences the size of the turbine, heat exchangers and cooling towers required for a given power output, and these have a dominant effect on the capital cost of the unit. However, plant capacity does not have a significant effect on the specific cost, as larger capacities are generally achieved by using a series of standard 1.2 MW to 5 MW nominal units. The following typical installed cost (excluding wells) would, however, be expected for typical binary installations (1998 US$).

- binary unit, 150 °C geothermal fluid, dry cooling: US$2,500/kW net
- binary unit, 150 °C geothermal fluid, wet cooling: US$2,100/kW net
- binary unit, 175 °C geothermal fluid, wet cooling: US$1,500/kW net
- combined-cycle binary, 7 MW net, dry cooling: US$1,300/kW net
- combined-cycle binary, 50 MW net, dry cooling: US$1,000/kW net

Note that this excludes the cost of wells or geothermal fluid collection system, and the cost of the re-injection system. It assumes short power transmission lines, and excludes duties and taxes.

The manufacturing and erection times given in Table 2.3 would be expected for modular binary and combined-cycle binary installations.

Table 2.3 Expected manufacture and erection times for modular binary and combined-cycle installations

	Order date to taking over (months)
Binary 1.2 MW modules	12–14
Combined-cycle binary, 7 MW net	16
Combined-cycle binary, 50 MW net	24

2.1.4 Biphase rotary separator turbo-alternator

The biphase rotary separator was developed to extract power from a two-phase, steam/water mixture. The unit is made up of three main components, as shown in Figure 2.10: a series of two-phase nozzles, a rotary separator and a liquid turbine. The two-phase nozzle converts part of the enthalpy of the two-phase mixture into fluid kinetic energy. Passing through the nozzle, the mixture is expanded from high inlet pressure to low exit pressure with steam and water droplets intimately mixed. The expanding gas accelerates, entraining liquid droplets with it, thus increasing the kinetic energy of the water and steam. The result is a two-phase jet with high kinetic energy.

The two-phase jet is directed tangentially onto the inner surface of a drum-shaped rotary separator. The separator rotates at a speed close to that of the jet, so there is very little friction loss and the liquid kinetic energy stays high. The high centrifugal acceleration forces the heavier liquid to the wall, resulting in a clean separation from the steam. The separated liquid rotates with the drum, while the steam flows inwardly to an exit port.

The liquid on the nozzle side of the separator rotor passes through liquid transfer holes in the disc to form a liquid layer on the liquid turbine side of the separator rotor. The liquid turbine, which converts the kinetic energy of the liquid to shaft power using the impulse principle, consists of a rotor arm with one turbine element at each end. The liquid turbine

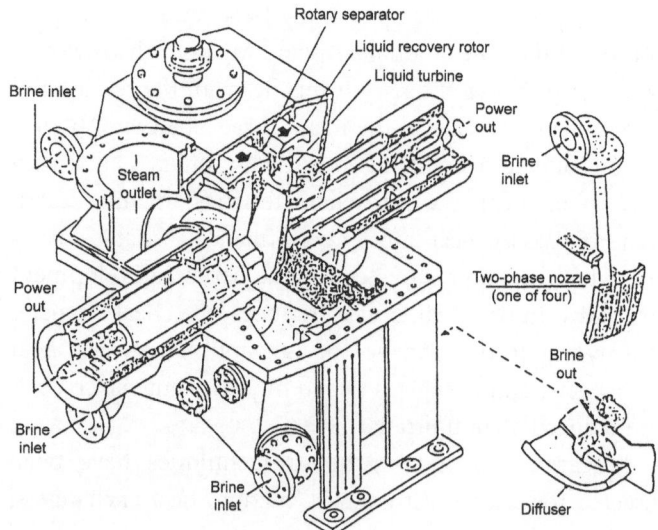

Figure 2.10 Isometric of biphase rotary separator

elements pick up the liquid from the separator rotor, reverse the flow direction 180 degrees, and discharge the liquid onto the liquid transfer rotor. The liquid flow reversal within the turbine element results in a force on the element and a torque to the turbine rotor.

The liquid leaving the liquid turbine elements falls onto the inside of the liquid transfer rotor. This is a rotor, similar in construction to the primary rotor, that is driven by the kinetic energy remaining in the liquid.

Typically, the rotational speed of the liquid turbine (Nu) is about 60 per cent of the rotary separator speed (Ns). Ignoring losses in the turbine element, 96 per cent of the liquid kinetic energy would be extracted. The remaining energy is used to drive the liquid transfer rotor. For typical Nu = 0.6 Ns the liquid transfer rotor speed is Nlt = 0.2 Ns in the same direction as the rotary separator.

The spent liquid is picked up from the liquid transfer rotor by the stationary diffuser. The diffuser has a cross-section that diverges, providing a 3 to 1 expansion ratio. The divergence of the diffuser converts the remaining kinetic energy into pressure to help meet re-injection pressure requirements. If the speed ratio Nu/Ns is raised, more energy can be left in the liquid for pressure recovery by the diffuser. If the speed ratio is varied to 0.75, injection pressure requirements up to 17 bar can be met without the aid of re-injection pumps.

The biphase unit is generally used in conjunction with a conventional turbine, which is driven by the steam that is discharged from the biphase rotary separator.

Topping arrangement

The usual arrangement for the biphase plant is as a topping system for a conventional condensing turbine, as shown in Figure 2.11. However, the biphase rotary separator turbine can also be used to supply an atmospheric exhaust turbine, or discharge directly to atmosphere. The performance of the biphase turbine when arranged as a topping unit for a conventional condensing turbine is illustrated in Figure 2.12. This shows a graph of the ratio of net power from the standard condensing Mitsui-Biphase unit to that of a conventional optimized single-flash condensing unit for various biphase inlet pressures and fluid enthalpies. The geothermal fluid flow rate is constant. The separation pressure for the flash plant is the optimum pressure corresponding to the various fluid enthalpies. One point clearly illustrated by this graph is that, for moderate enthalpy fluids, it is necessary to adopt high biphase inlet pressures to achieve an advantage over the efficiency of a conventional condensing unit. However, this graph does not identify the fact that the available well flow will generally be less at the higher inlet pressures required for the biphase plant. The performance of the biphase plant must therefore be penalized for this effect

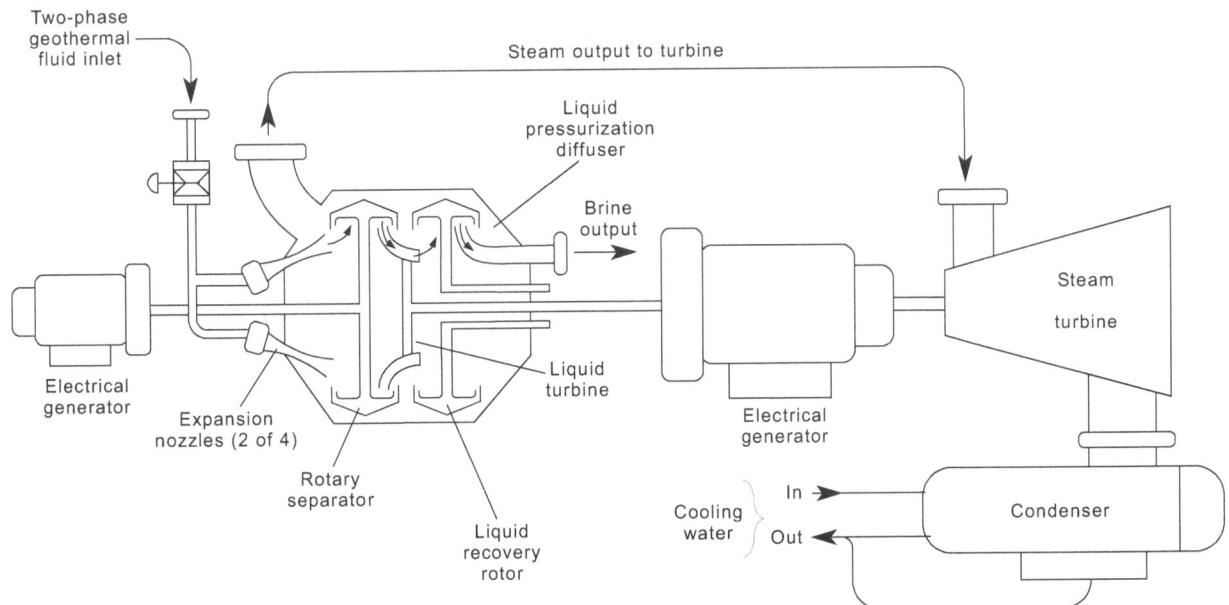

Figure 2.11 Topping system arrangement for biphase turbine with conventional condensing turbine

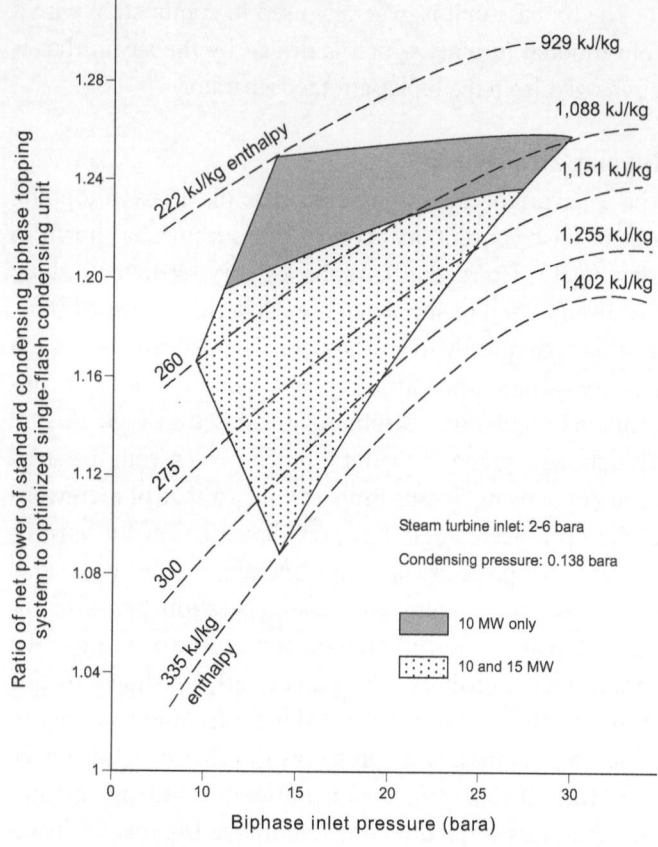

Figure 2.12 Ratio of net power of standard condensing biphase topping system to optimized single-flash condensing unit versus biphase inlet pressure (for constant mass flow)

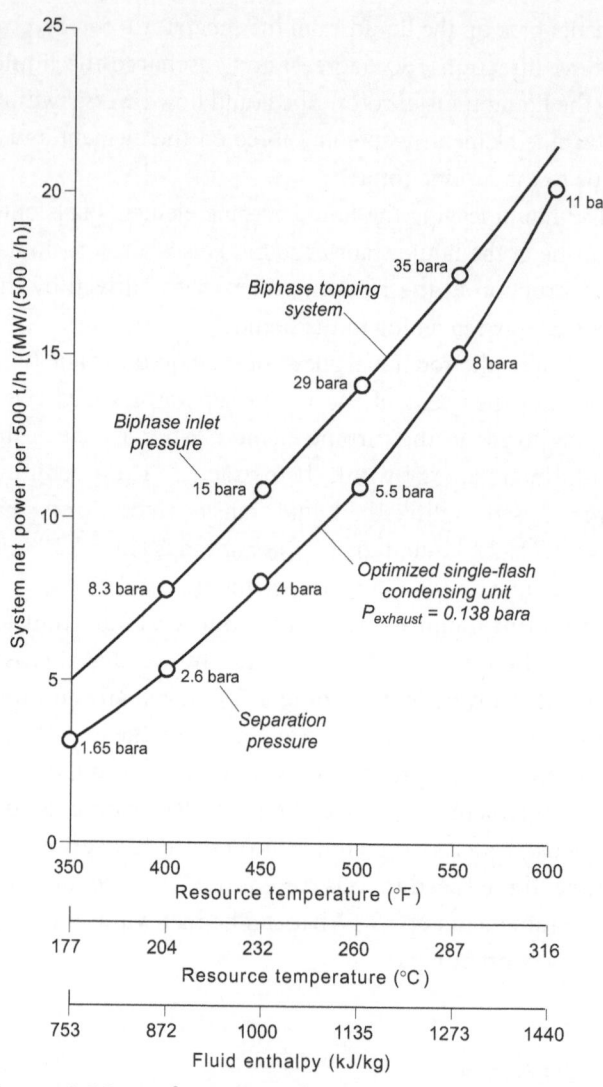

Figure 2.13 Performance of condensing biphase topping system and optimized single-flash condensing unit versus fluid enthalpy

before direct comparison with the economics of other systems can be made.

Figure 2.13 compares the performance of a biphase topping system with an optimized single-flash condensing system for various fluid inlet enthalpies. This graph has been based on a similar graph supplied by the manufacturer. However, we have also calculated the biphase inlet pressures and optimized single-flash pressures that have been used to construct the curves. Again this graph does not identify the reduction in well-flow rate because of the high biphase inlet pressure. The resource temperature presented by the manufacturer on the abscissa of the graph assumes that the source well fluid is totally water. We have calculated the corresponding fluid enthalpies for these water temperatures, which is a more general representation of the fluid condition. It can be seen from this graph that the use of a biphase turbine is only of significant benefit for relatively low enthalpy fluids, that is, where there is a large proportion of water in the geothermal flow.

Bottoming arrangement

An arrangement of the biphase turbine where the previous performance graphs can be used for direct comparison is a 'bottoming system'. With this arrangement the discharged water from a conventional flash separation is passed through a biphase plant, instead of being directly re-injected. With this case the fluid flow rate is constant regardless of the inlet pressure of the bottoming plant. One further advantage of the biphase unit for this application is that the exhaust water from the biphase unit can be discharged at the same or greater pressure than the input fluid, which can be very useful for re-injection purposes. The fluid pressure must be dropped to flash steam for the conventional steam turbine system.

The major disadvantage with biphase bottoming plants is that very large machines are required to handle the high flow rates of low enthalpy fluid. The use of biphase plants for this type of application has clear thermodynamic advantages, but the large size presents a significant cost penalty.

The economics of biphase plants are not well defined, as the only plant of this type that has been operating commercially is the 12.0 MW Desert Peak, Nevada, unit. The biphase turbine contributed 3 MW of generation, but significant maintenance and control stability difficulties were encountered with its operation, and the biphase turbine was subsequently disconnected and the station operated using just the conventional turbine supplied with single flash steam. It would appear, however, that the specific installed cost (cost per installed kW) of a biphase condensing topping plant is similar to a conventional condensing plant. A biphase plant is generally able to generate more power from a geothermal well than a conventional plant, but as discussed previously, this advantage is often restricted to wells with relatively low enthalpy discharges where the appropriate biphase inlet pressure is not too high.

Generalized costing, manufacturing and erection

These figures must be used with appropriate caution, as costs can vary considerably between specific developments. In particular, the biphase topping arrangement is only appropriate for low to moderate enthalpies. An important cost advantage of the biphase plant results from not requiring an additional steam/water separator. The following typical installed cost would, however, be expected for a well-head biphase plant with inlet fluid enthalpy of the order of 1,100 kJ/kg (1998 US$):

- biphase unit exhausting directly to atmosphere: US$1,750/kW
- biphase topping unit with atmospheric exhaust turbine: US$1,620/kW
- biphase topping unit with condensing turbine: US$1,490/kW

Note:

- The fluid enthalpy is approximately 1,100 kJ/kg.
- This excludes the cost of wells.
- It assumes a short steam transmission system, simple re-injection system and short power transmission lines.
- It excludes duties and taxes.

Table 2.4 Expected manufacture and erection periods for a typical biphase installation

	Order date to shipment (months)	Shipment to commissioning (months)	Total (months)
Biphase unit exhausting directly to atmosphere (3 MW)	9	4	13
Biphase topping unit with atmospheric exhaust turbine (7.5 MW)	10	4	14
Biphase topping unit with condensing turbine (12 MW)	13	6	19

The manufacturing and erection periods detailed in Table 2.4 would be expected for a typical biphase installation.

2.2 WELL-HEAD GENERATING UNITS

As their name suggests, well-head units are sited next to a production well pad and are generally supplied with hot water or steam from one or several production wells. Several features distinguish geothermal well-head generating units from a central geothermal power plant. One is that the gross output of well-head units is generally less than 10 MW. Another is that well-head generating units have very short steam lines, whereas central geothermal power plants characteristically have long, interconnecting pipelines to transport the steam from the various well pads to the central unit.

The most characteristic feature of a modern well-head generation facility is its modular construction. Usually the turbogenerator modules are completely assembled at the factory on a single skid. An inherent advantage of pre-assembled modular units is that the completed assembly is tested in the factory after manufacture, thus ensuring that all components have been confirmed to function correctly and are properly connected. This type of factory testing and running drastically shortens the time needed for field pre-operational testing, start-up, and trouble-shooting. The very small, standard design, portable, atmospheric-exhaust units can be erected and producing power in less than twelve months from the date of order. Further, these units can be relocated to a new well in less than one to two months if required. Six to seven years from the drilling of the first well to generation is usually required for a central plant installed on a new, unproven geothermal resource.

A major associated benefit obtained from operating small generating plants on exploration wells is the very valuable resource testing information that can conveniently be obtained. The separation and re-injection plant generally required for long-term production discharge tests in moderate enthalpy wet fields is part of the standard equipment required for the generating plant. In conjunction with the operation of the unit, data on pressure, temperature, non-condensable gas content and chemical content trends can be obtained. Monitoring the response of the reservoir to the production discharge, from surrounding observation wells, enables quantitative estimates of reservoir parameters such as transmissivity and storativity to be made over a broad region of the reservoir. This additional information enables more accurate predictions of long-term reservoir performance to be made, and therefore reduces the development risk associated with large-scale exploitation of the resource.

2.2.1 Economic considerations regarding small geothermal plants

In this section the economic characteristics of small geothermal generating plants are discussed. This discussion is based primarily on conventional atmospheric exhaust or condensing steam turbines. However, the principles apply similarly to binary or biphase plants.

Economy of scale

Figure 2.4 showed a generalized development cost graph for well-head atmospheric exhaust, well-head condensing and central station condensing plants versus unit size. Again it must be emphasized that these costs are only indicative averages and should never be used to estimate the cost of a specific development. Because of economies of scale, the specific development cost (cost per installed kW) for a small unit is invariably more expensive than for a larger unit. However, the following consequential costs resulting from the use of units that are large compared with the size of the system demonstrate why smaller units are often selected as part of a least-cost development plan.

Matching demand growth
When large units are added to a system there is an expense associated with having greater redundant capacity for a

longer period until the demand growth catches up with the installed capacity.

Reserve capacity
Adequate reserve capacity (above dry-year capability) should be provided in the system to allow for the plant being down for scheduled maintenance and for forced outages. A useful arbitrary criterion that is often used is the sum of the first and third largest units on the system. The use of large units in a small system will mean that the reserve capacity required can become a large proportion of the installed capacity. The cost of providing this additional capacity is an expense that is attributable to the use of the large units.

As an extension to this argument, if we consider a system with equally sized units that each require to be down for maintenance for one month each year, then to be able to supply a constant load with the smallest level of reserve capacity, it is necessary to have thirteen units each sized at 1/12 the system load. Unit sizes less than or greater than this will require greater (or at least equal) reserve capacities.

Spinning reserve
Usually enough units are kept on-line, but operating sufficiently below full load that they are able to immediately pick up the load of the largest unit if it should trip. The use of a smaller number of large units will mean that the units will have to run at a low proportion of full load and consequently at lower efficiency.

Risk
Particularly for the case of geothermal steam field development, which is undertaken in the early stages of exploration when the extent or exploitability of a resource is not well defined, the cost of failure should be considered. This will be less if only the investment in the first small unit of a staged development is lost.

Capacity factor

This factor is expressed as:
average load generated in period (kW)/capacity (kW)
or equivalently
total energy generated in period (kWh)/capacity (kW)
× hours in period (h)

This is a measure of the amount of energy generated from a station relative to the maximum possible energy generation.

Generation options such as hydro, coal-fired steam, gas turbine, diesel engine and geothermal each involve a different balance of fixed costs and variable costs. Fixed costs are those costs that are not influenced by the amount of power actually generated, for example, the initial capital cost of the power station. Variable costs are those costs that are dependent on the amount of power generated, for instance, coal or diesel fuel costs.

The costs of a generation option must be recovered by the energy (kWh) generated by the plant over its economic life. For any generation option, as the capacity factor is reduced and therefore the amount of energy generated is reduced, the proportion of the fixed cost that must be recovered by each unit of energy generated increases. This increases the unit cost of that generation. The unit cost of generation is not affected by the variable cost component, however, as the total variable cost (for example, fuel cost) reduces in proportion to the reduced generation.

Geothermal generation is characterized by having predominantly fixed costs and low variable costs. Therefore, as the capacity factor is reduced, the unit cost of geothermal electricity increases at a far greater rate than for fuel-based options such as coal-fired steam, gas turbine or diesel engine plants. At very low-capacity factors (less than 10 to 15 per cent), a gas turbine plant is invariably the cheapest means of providing the required capacity on a large system, with diesel internal combustion generating sets the cheapest for a small system.

It is for this economic reason alone that geothermal plants are stated as generally only being suitable for high-capacity-factor (base load) operation. From a technical point of view there is absolutely no problem in designing a geothermal station to load, follow and operate at low capacity or load factors. This economic characteristic has obvious implications for the use of geothermal plants to supply small isolated loads. Careful economic comparison with fuel-based alternatives is required for these applications.

One point that has not been specifically mentioned in the above, in order to simplify the discussion, is that the fixed cost that is recovered by the energy generated must also include the opportunity cost of the capital. This opportunity cost is represented by the discount rate used with the discounted cash flow evaluation of relevant options.

Avoided cost

Avoided cost is an important concept that arises when considering the benefit or savings to be achieved if additional generation is imposed on an existing system.

A good example of this is the case of evaluating the cost/benefit of exercising the opportunity of installing a mini-turbine on early exploration wells until they are needed for a subsequent major development. A common mistake is to assume that the benefit of the electricity generated from such a unit is the current price of power (even if this price has been calculated appropriately). This is a serious error, as the savings that are made are only the costs that are avoided by using this generation. These costs are only fuel and other variable operating costs, unless future planned capacity can be deferred (which is not generally relevant). The capital cost components of all the existing plants are not avoided by this additional generation, and must still be recovered.

To take an extreme example, if a power system was completely hydro, the worth of the additional electricity generated would be zero as no savings are achieved by its use, since the water ('fuel') used to generate it has no cost. The avoided cost of such generation can be calculated by determining the cost of the fuel (or other variable operating costs) that is saved by not using a fossil-fuelled plant during that part of the day and year in which the additional generation is being achieved with the hydro plant. If the hydro plant is included in the system, the average year (not dry year) generation estimates for each plant type should be used.

The benefit of deferring future capital expenditure on a new plant is generally not relevant with this type of project, because due to the long construction lead times of major power stations, the expansion programme over the period that the mini-turbine would be used is generally committed at the time of the evaluation.

Economic considerations regarding atmospheric exhaust and condensing plants

The economic treatment of well-head turbines is very dependent on the philosophy for their installation. Where wells have been drilled as part of an exploration and production-drilling programme for a conventional major development, their cost is considered sunk and not avoidable, and will be recovered from the output from the final development. When considering the economic justification of installing well-head units on the exploration or early production wells, it is therefore only appropriate to compare the additional benefits with the additional costs.

The additional benefits are the electricity generated and the reservoir test information, and the additional cost is the cost of installing the well-head units and any additional wells, but not the cost of existing wells. Clearly, when considering the least-cost alternative for adding capacity to a power system, costs already incurred are unavoidable and the least-cost development option is that which incurs the least additional cost. However, if it is planned to develop an undrilled resource using well-head turbines to supply, say, a low demand from a rural electrification project, then any wells drilled for this purpose must be included in the economic evaluation.

With reference to the generalized costing for atmospheric exhaust and condensing plants (which were illustrated in Figure 2.4), it is a notable distinction that the installed capital cost (excluding wells) is generally about 50 per cent greater for a condensing plant than for an atmospheric exhaust plant of similar capacity. If the cost of the wells is also included, however, the installed capital cost is generally slightly less for the condensing plant than for the atmospheric exhaust plant. This is because approximately twice as many wells are required for the latter. Consequently, whenever the additional capacity required to be added means a significant proportion of additional wells have to be drilled, the condensing option generally becomes more attractive. This is, of course, the situation with a premeditated development, and that option is generally used for long-term exploitation of existing wells if there is the demand for the larger capacity possible from a condensing unit. There is also always the preference for the more efficient use of the resource. However, if it is intended to exercise the opportunity to generate electricity by taking advantage of existing wells, then generally the atmospheric exhaust plant is the more economic.

An important distinction that was implied in the previous paragraph is, therefore, whether the development of the additional generation is demand-driven or opportunity-driven. If it is demand-driven and the capacity required is more than can be achieved by, say, installing atmospheric exhaust units on all the existing wells, then the development must be penalized by the cost of obtaining the additional capacity, either through drilling more wells or using an alternative plant. With this situation the more expensive condensing plant may therefore be more economic, as approximately twice the capacity could be achieved from the same existing wells. If the development is only opportunity-driven, then the consideration is whether the cost of exercising the opportunity to generate power from the existing wells is less than the avoided cost of generation that would otherwise have been produced from the balance of the power system.

REFERENCE

HUDSON, R. B. 1988. Technical and economic overview of geothermal atmospheric exhaust and condensing turbines, binary cycle and biphase plant. *Geothermics*, Vol. 17, pp. 51–74.

RECOMMENDED LITERATURE

APPS, J. A. 1977. *The definition of engineering development and research problems relating to the use of geothermal fluids for electric power generation and non-electric heating*. Lawrence Berkeley Laboratory, LBL-7025, UC-66a, TID-4500-R66.

CAMPBELL, R. G. 1995. The power plant. In: R. G. Bloomquist (ed.), *Drafting a Geothermal Project for Funding*, pp. 73–94. Pisa, WGC '95 Pre-Congress Courses.

HIBARA, Y. 1986. How to maintain geothermal steam turbines. *ASME Joint Power Generation Conference*, 86-JPGC-Pwr-17.

KESTIN, J.; DIPIPPO, R.; KHALIFA, H. E. AND RYLEY, D. J. (eds.). 1980. *Sourcebook on the Production of Electricity from Geothermal Energy*, Washington, US Department of Energy.

ORMAT. 1989. *Production of Electrical Energy from Low Enthalpy Geothermal Resources by Binary Power Plants*. Rome, Italy, UNITAR/UNDP Centre on Small Energy Resources.

Proceedings of a Topical Meeting on Small Scale Geothermal Power Plants and Geothermal Power Plant Projects, Reno. 1986. Energy Technology Engineering Center, P.O. Box 1449, Canoga Park, California 91304, United States.

Utilization of Geothermal Energy for Electric Power Production and Space Heating. 1985. Selected papers from UNECE Seminar, Florence, Italy, 1984, *Geothermics*, Vol. 14, pp. 119–497.

APPENDIX A: THERMODYNAMICS OF CYCLES

The objective of including this appendix is simply to provide some examples of important thermodynamic processes in the various plant cycles, in case the reader wishes to go more deeply into some of the general concepts given in the chapter.

A.1 Thermodynamics of flash process

With the advent of sophisticated geothermal power cycles such as the binary and Kalina cycles, the inherent efficiency of the flashed-steam process is often overlooked. The flashed-steam process is in fact both extremely simple and efficient. A temperature entropy diagram for a typical flash process is shown in Figure 2.14, where the reservoir fluid has been taken to be totally liquid at 281 °C with a corresponding enthalpy of 1,241 kJ/kg.

The process involves the reservoir fluid undergoing isenthalpic expansion as it flashes during its passage up the well

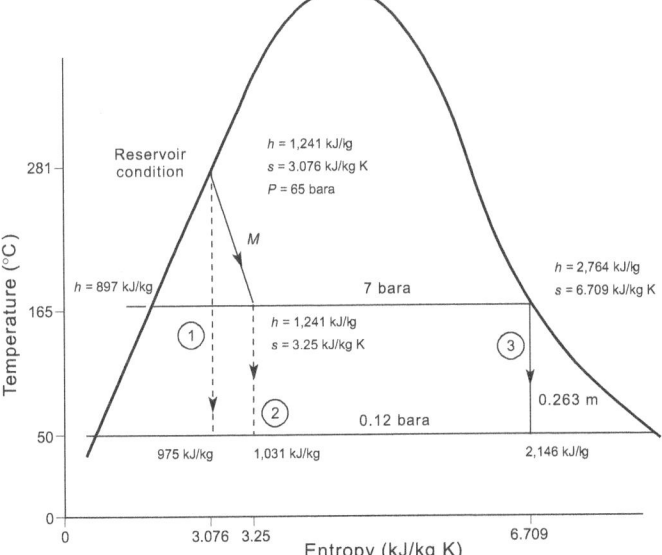

Figure 2.14 Temperature–entropy diagram of typical geothermal flashed-steam process and available isentropic heat drop

to the optimized separation pressure of 7 bara. The separated steam is then passed through a condensing turbine with a condenser pressure of 0.12 bara. Also shown on this diagram is the available isentropic heat drop from various points along this process relative to the condensing temperature of 50 °C. The isentropic process represents the process where the theoretical maximum power is extracted by a heat engine between an initial and final temperature. Three such isentropic processes are shown on Figure 2.14.

The first is the maximum power that could be produced using the reservoir fluid at reservoir conditions. This process is unachievable without down-hole pumps, as the well will not flow unless the surface pressure is significantly less than the reservoir pressure. However, assuming this ideal case we obtain the maximum extractable power output, which has been taken as 100 per cent. The second case involves taking the total water plus steam flow from the separator condition. This results in a maximum extractable power output that is 79 per cent of case one. The third case involves using only the separated steam. This results in a maximum extractable power output of 61 per cent of case one.

This example illustrates the following generalizations regarding the flash process. The exact proportions of the relationships are dependent on the enthalpy of the fluid:

1. It is necessary to waste typically 20 per cent of the available heat drop of the reservoir fluid due to the need to drop the fluid pressure to achieve natural well flow at reasonable flow rates.
2. Of the total fluid in the separator, most of the available heat drop is contained in the steam flow and with this example only 22 per cent is lost with the separated water.

A.2 Thermodynamics of organic Rankine cycle

The use of an organic fluid with the low-temperature Rankine cycle has many advantages over using water. However, it should be noted that the Rankine cycle efficiency for the organic fluids used is little different from that for water/steam between the same two top- and bottom-cycle temperatures. In fact, the cycle efficiency is often slightly less for the organic fluid. The thermodynamic attraction of an organic fluid is that it is able to extract more heat from the geothermal hot source than water. This results from 'pinch-point' heat exchanger limitations, which are primarily a consequence of the organic fluid having a far lower ratio of latent heat of vaporization (at these lower boiling temperatures) versus specific heat capacity than water. As a consequence, even though the cycle efficiency is about the same value in the two cases, the overall efficiency, which is proportional to the generated electrical energy, is considerably higher for the organic fluid. The overall efficiency is, by definition, the cycle efficiency multiplied by the ratio of thermal power extracted to thermal power available from the hot source. The available thermal power is computed using an arbitrary minimum temperature that is 10 °C higher than the bottom temperature of the cycle.

The above points are illustrated in Figure 2.15, which shows temperature entropy diagrams for two Rankine cycles operating between 105 °C and 40 °C using R114 (dichlorotetrafluorethane, CClF2–CClF2) in one case and water in the other. The following processes are involved in the Rankine cycle, as labelled in Figure 2.15.

- a–b: Pressure increase by feed-pump
- b–c: Heating at constant pressure to dry vapour state
- c–d: Isentropic expansion through turbine
- d–a: Condensing.

As shown, the Rankine cycle efficiency (energy extracted in turbine divided by energy supplied to the cycle) is 10.1 per cent for the organic fluid and 10.9 per cent for water/steam. An interesting point is that the turbine exhaust fluid is superheated for the organic case but wet for the water/steam case. This means that the condenser for the organic fluid must de-superheat as well as condense the exhaust. Clearly, the efficiency of the cycle could be improved (at additional cost) by heat exchanging across d–d′: b–b′, that is, using the available de-superheat to preheat the feed-pump discharge.

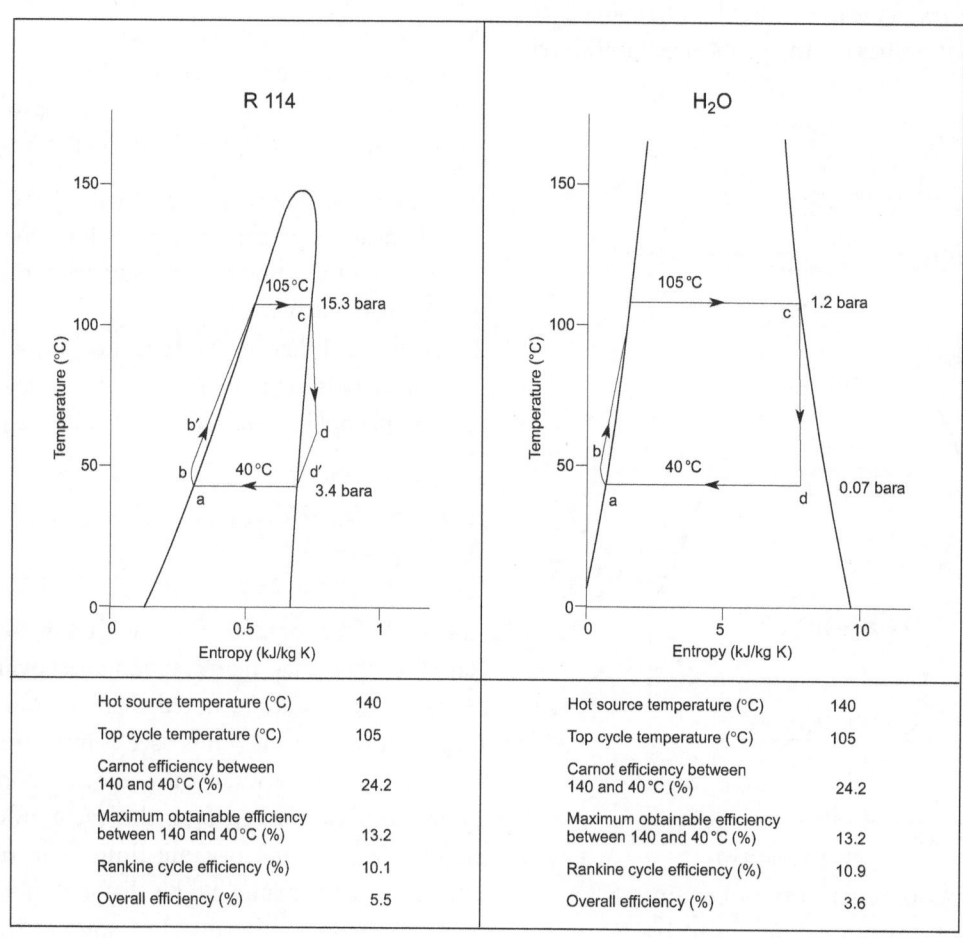

Figure 2.15 Characteristics of organic Rankine cycle using R114N compared with equivalent Rankine cycle using water

The latent heat of vaporization and specific heat capacity for the two fluids at the appropriate temperatures are as shown in Table 2.5.

The effect of the 'pinch-point' heat exchanger limitation for the two cases is shown in Figure 2.16. This example shows a temperature versus thermal power graph for the geothermal fluid heat source and R114 or water/steam working fluid.

For a given mass flow of geothermal fluid, the thermal power transfer with temperature change line is fixed and unalterable for constant pressure conditions. For the geothermal fluid flow shown in Figure 2.16, where 5,900 kW (kJ/s) is transferred with a temperature drop of 49.5 °C, the flow rate can be calculated as 27.8 kg/s using an enthalpy change of 212 kJ/kg from steam tables.

Table 2.5 Latent heat of vaporization and specific heat capacity

	R114	Water/steam
(a) Latent heat of vaporization at 105 °C (kJ/kg)	86.6	2,244
(b) Average specific heat capacity over range 40–105 °C (kJ/kg K)	1.12	4.17
(c) Ratio of (a) to (b)	77	538

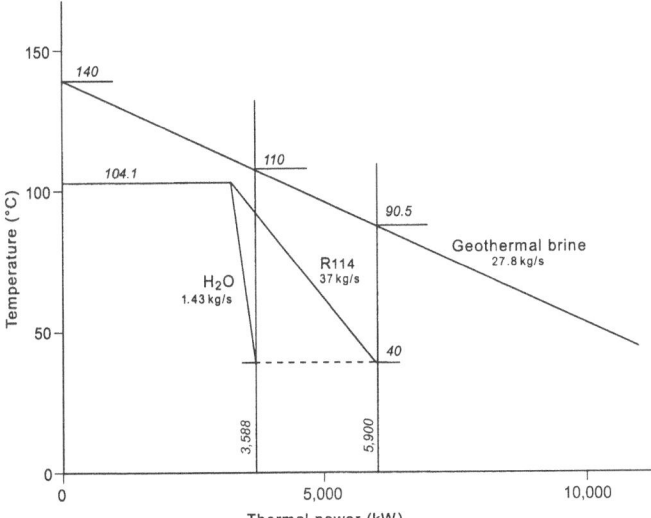

Figure 2.16 Temperature–thermal power diagram for heat exchange using water or R114 as secondary fluid

If we now consider the working fluid boiling at constant pressure and at 105 °C, the thermal power available for boiling (which must occur at a constant saturation pressure for a constant temperature) is found by drawing a constant 105 °C temperature line until it is within about 10 °C (for economical heat transfer) of the geothermal fluid line. This determines that 3,200 kW is available for boiling the working fluid, and therefore determines a maximum flow rate of 37.0 kg/s for the R114 and 1.43 kg/s for the water/steam, using latent heats of vaporization of 86.6 kJ/kg and 2,244 kJ/kg, respectively. With these maximum constrained flow rates it is now possible to determine the additional heat that is required to be transferred to raise the working fluid temperature from 40 °C to 105 °C. Using specific heat capacities of 1.12 kJ/kg K for R114 and 4.17 kJ/kg K for water, the heat transferred is found to be 2,700 kW and 388 kW, respectively. In Figure 2.16 it can be seen that the geothermal fluid can be cooled to 90.5 °C with the R114 but only 110 °C with water/steam, and consequently 64 per cent more heat can be extracted using R114 as the working fluid.

In addition the organic fluids that are used with binary plants have the following further advantages:

- For the organic fluid, the whole expansion takes place outside the saturation curve, while for water, wet steam will flow through the entire turbine. This has the advantage of reducing blade erosion.
- The enthalpy drop is small, and it is possible to design a single-stage turbine with high efficiency, that is at the same time subject to low stresses. For water the enthalpy drop is too high for expansion in a single stage with optimum efficiency. This requires a more complicated, expensive turbine. For a given power, the mass flow rate of organic fluid is proportionally higher, but the size of the equipment is not as large because of the high density of the vapour.
- The density of the organic fluid at the exhaust is low. The volumetric flow rate of steam is about sixteen times higher, and therefore the steam turbine size is considerably larger. This is a severe economic penalty.
- The pressure of the organic fluid is always above atmospheric, so it is not possible to have air leakages into the cycle.

On the other hand, the organic fluid, even if non-toxic and non-flammable, requires a leak-tight plant, which consequently complicates both construction and maintenance.

It should be noted that it is actually thermodynamically more efficient to extract power from the geothermal fluid at the conditions used in the previous example by using a flashed-steam process. For example, the overall efficiency of a single-flash cycle operating at the appropriate optimum separation pressure of 0.70 bara and turbine exhaust pressure of 0.075 bara (40 °C) is 8 per cent. This process would be uneconomic and impractical, however, because of the large size of plant required for the low-density steam, and the problems of extracting the non-condensable gases from such a low-condenser vacuum.

A.3 Thermodynamics of biphase process

The biphase cycle is illustrated on a temperature–entropy diagram in Figure 2.17. Assuming saturated water at point 0, the conventional flashing isenthalpic expansion follows path 0–1. The steam produced is separated, point 2g, and expanded in a steam turbine 2g–3, producing power as shown by a change in enthalpy. However, the expansion of the liquid through a biphase turbine follows path 0–2i. The biphase expansion changes enthalpy from h1 to h2, thus making power available. Separated steam (less than the isenthalpic case for conventional flashing) is ducted to the steam turbine, 2g–3, to produce further power.

The effectiveness of the biphase turbine is primarily dependent on the nozzle isentropic efficiencies, which are generally in the range 60 to 75 per cent. The nozzle efficiencies are quite sensitive to off-design operation.

When comparing the potential power available from the biphase cycle, however, it must be appreciated that the cycle diagram shown in Figure 2.16 is generally not applicable. With a hydrothermal resource, point 0 would

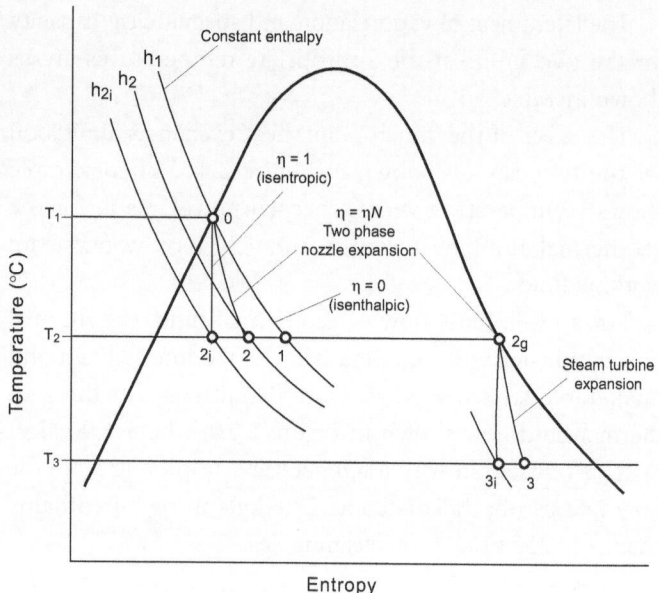

Figure 2.17 Temperature–entropy diagram for biphase rotary separator turbine system

generally represent reservoir conditions, and therefore this pressure and thermodynamic state cannot be achieved at the biphase turbine inlet, as there would be no well flow. The inlet condition for direct admission to a biphase turbine would therefore start somewhere down the isenthalpic path 0–1.

Data provided on biphase turbines often make comparisons with the 'resource utilization' of conventional flash plants, showing the thermodynamic utilization of the well fluid to be improved, but neglecting to account for the reduced well flow due to the high biphase inlet pressures required. The economic penalty of using a well to produce less power, albeit more efficiently, is therefore not identified.

APPENDIX B: PRINCIPAL MANUFACTURERS

B.1 Manufacturers of conventional atmospheric exhaust or condensing steam turbine plants

The principal suppliers of conventional geothermal generation plants are:

- Ansaldo Energia, Italy
- Alstom, France
- Fuji, Japan
- General Electric, United States
- Mitsubishi Heavy Industries, Japan
- Toshiba, Japan.

Elliot (United States), has also made geothermal turbines, but is currently not particularly active in this market.

The Geothermal Power Co. Ltd (United States) provides unused surplus low-pressure marine turbines that are purchased from the US Navy and modified to suit the inlet pressure and flow requirements of the geothermal application.

B.2 Manufacturers of geothermal binary plants

The predominant manufacturer of this plant is ORMAT Industries Inc. (United States). The following list covers other known manufacturers of geothermal binary plants, none of which are currently particularly active in the market.

- Barber-Nichols Engineering, United States
- Borg Warner, York Division, United States
- Electric Power Research Institute, United States
- Mitsubishi Heavy Industries, Japan
- Rotoflow, United States
- Toshiba, Japan
- Turboden, Italy
- Turbonetics Energy Inc., United States.

B.3 Manufacturers of geothermal biphase rotary separator plants

Only one biphase power plant, the 12.0 MW Desert Peak, Nevada, unit has been operated commercially. However the biphase turbine part of the station (3 MW) is no longer being used, and the station is currently operating using only the conventional steam turbine supplied with single-flash steam. This biphase plant was manufactured by Transamerica Delaval Biphase Energy Systems, Santa Monica (United States), which formed an association with Mitsui Engineering and Shipbuilding Co. Ltd for the supply of the steam turbine plant associated with the biphase unit. However, Biphase Energy Systems is no longer in business, and the biphase technology has been taken over by Douglas Energy Co. Ltd, Santa Monica (United States). Currently this is the only manufacturer of the rotary separator biphase plant, although the concept is being considered by other turbine manufacturers.

SELF-ASSESSMENT QUESTIONS

1. (a) What is the typical auxiliary power consumption for a conventional geothermal condensing unit as a percentage of the gross generation?
(b) What is the typical range of auxiliary power consumption for a combined-cycle binary unit as a percentage of the gross generation?

2. What type of gas extraction system would be expected to be the most economic for a condensing unit with 2.5 per cent by weight non-condensable gas content in the steam?

3. If a turbine shaft-driven mechanical compressor is provided for extracting the non-condensable gas from the condenser, is it possible to start up the unit without separate gas extraction equipment?

4. What part of a condensing power plant has to be significantly larger than applicable for a conventional fossil fuel-fired thermal power plant of the same capacity?

5. What are the most obvious differences between well-head generating units and central power plants?

6. There are a few specific advantages with well-head units. Can you describe some of them?

7. A well produces about 170 tonnes per hour of steam, at a temperature of 205 °C and pressure of 5 bara, with a content of 30 per cent by weight of non-condensable gas. This is a real example. What is the most appropriate type of power plant for this well? Can you suggest a system that would permit a more efficient exploitation of the energy supplied by the well?

8. What is the order of magnitude of steam consumption (expressed in kg of steam per kWh generated) of geothermal condensing turbines?

(a) 7–10 kg/kWh; (b) 50–60 kg/kWh; (c) 100–110 kg/kWh

9. The section on *Net generated electric power* gives the equation:

$$NEP = [(0.18T - 10)ATP]/278$$

for calculating net generated electric power. The example that follows shows that we can obtain a net conversion efficiency of 5.5 per cent and 2.8 per cent for primary fluid temperatures of 140 and 100 °C, respectively. What conversion efficiency will we obtain for a fluid with a temperature of 115 °C?

10. Does the performance of a biphase rotary separator turbine arranged as a topping system increase or decrease with respect to the performance of an optimized single-flash condensing unit, when there is an increase in the enthalpy of the geothermal fluid feeding the turbines? What are the effects of varying the inlet pressures of the fluids?

11. Describe some of the main advantages of binary plants.

12. (a) What is the capacity factor?
(b) Why are geothermoelectric power plants more sensitive, from the economic viewpoint, to variations in capacity factor than power plants fed by conventional fuels such as coal or gas?

ANSWERS

1. (a) 4.6 per cent; (b) 9.5 per cent to 15 per cent.

2. A hybrid system with first-stage steam Venturi ejectors and second-stage liquid-ring vacuum pumps.

3. No. The turbine cannot be brought up to speed until a vacuum has been established in the condenser, and the mechanical compressor cannot achieve this if it is driven directly by the turbine shaft.

4. The cooling system, because of the greater proportion of heat rejected in the condenser for a goethermal cycle.

5. There are two main differences. The first is capacity. The capacity of well-head units is generally less than 10 MW. Second, as indicated by their name, well-head units are installed in the immediate vicinity of the production well(s) and have short transmission lines for carrying the geothermal fluid from the well to the plant. The central units, on the other hand, collect the fluids from more than one well, and the fluids are transported by pipelines that are generally several kilometers in length.

6. The well-head units are generally constructed as modular units on a single skid, and are pre-assembled and tested in the factory. This approach drastically reduces the time needed for pre-operational testing, start-up and trouble-shooting in the field before beginning real operations, compared with the time needed for large central units. Their modular design also makes it much easier and faster to transport them from one well site to another, according to production requirements.

The well-head units are also particularly appropriate in the initial development of a geothermal field, as they can provide valuable resource testing information for planning the future development of the field.

7. With these fluid characteristics, the best solution is to install a back-pressure steam turbine exhausting either to atmosphere or to a binary plant (that is, a combined-cycle binary plant). This avoids the high parasitic power consumption that would be required to extract the non-condensable gas from a vacuum condenser. A gas separator would be needed in a condensing plant, precisely because of the high gas content, but this would not be technically advisable or economically feasible at this very high gas content.

For a long-term exploitation of the example well, heat recovery from the exhaust of the atmospheric-exhaust turbine should be considered. An atmospheric condensing heat exchanger could be installed to provide the heat source for a direct use application. Because the condensing heat exchanger, or a binary inlet heat exchanger condenser, is operating at or above atmospheric pressure, the non-condensable gas can simply be bled off from the top of the heat exchanger with no pumping requirement.

8. Steam consumption varies with the characteristics of the fluid and the power plant. Generally it is around 7 to 10 kg/kWh, as can be seen in Figure 2.8.

9. We will obtain a conversion efficiency of 3.8 per cent, that is: $[(115 \times 0.18) - 10]/278 = 3.8$.

10. The performance of a topping system is, on the whole, better than that of an optimized single-flash condensing unit in the case of fluids with a low enthalpy (see Figures 2.12 and 2.13). The performance of a topping system improves with an increase in inlet pressure (see Figure 2.12), but the increase in inlet pressure has a negative effect on well production (flow rate).

11. (a) They enable more heat to be extracted from geothermal fluids by rejecting them at a lower temperature. (b) They can make use of geothermal fluids that occur at much lower temperatures than would be economic for flash utilizations. (c) They confine chemical problems to the heat exchanger alone. (d) They enable use to be made of geothermal fluids that are chemically hostile or that contain high proportions of non-condensable gases. (e) They can accept water/steam mixtures without separation.

12. (a) The capacity factor is, in simple terms, the ratio between the amount of power actually generated in a given period to the maximum power that a plant is capable of generating in that same period.

(b) The cost of electricity produced by a plant is made up of fixed costs and variable costs. Fixed costs, such as the initial capital cost of the plant, are not influenced by the amount of power actually generated, whereas the total variable costs reduce with reduced generation. Since a geothermal project comprises principally fixed costs, a reduction in total generation, implied by a reduced capacity factor, results in the same fixed costs having to be paid from a reduced total quantity of generation. This increases the unit cost of the generation.

CHAPTER 3

Space and District Heating

Einar Tjörvi Elíasson
Reykjavik, Iceland
Halldór Ármannsson and Sverrir Thórhallsson
ÍSOR – Iceland GeoSurvey, Reykjavik, Iceland
María J. Gunnarsdóttir
Icelandic Association of District Heating Companies, Reykjavik, Iceland
Oddur B. Björnsson
Fjarhitun Consulting Engineering Company, Reykjavik, Iceland
Thorbjörn Karlsson
University of Iceland, Reykjavik, Iceland

AIMS

1. To demonstrate that the exploitation of geothermal energy for space and district heating is one of the most convenient and widespread forms of direct utilization of the Earth's heat. Furthermore, that several decades of experience, in a wide variety of situations, have now provided us with the know-how to develop very reliable projects for this type of application.

2. To show that a variety of solutions are possible when designing a geothermal heating system, depending on the local climate, the quality of the geothermal resource and the market demand.

3. To prove that the efficiency of an industrial geothermal system can be quite high, especially if different forms of utilization are combined in either an integrated or cascade scheme.

4. To present the main physical, technical, economic or environmental factors that could hinder or restrict the development of a space or district heating system.

OBJECTIVES

When you have completed this lesson you should be able to:

1. Define the main characteristics of the geothermal fluids that can be used in space or district heating.

2. Discuss the main local factors (climate, population, market and so on) that affect a geothermal heating project.

3. Describe the different geothermal fluid transmission systems, and discuss the operative and economic benefits and disadvantages of all of them.

4. Discuss the different designs for geothermal heating systems, including any auxiliary equipment, and analyse their suitability for different situations.

5. Estimate the order of magnitude of the initial capital investment required for setting out a geothermal heating system, and the subsequent operative costs.

6. Determine the main effects on the environment of adopting a geothermal heating system.

7. Discuss the breakdown of heating costs to the user.

3.1 INTRODUCTION

3.1.1 Preamble

Geothermal hot water was for centuries predominantly used for bathing and washing. The traditional baths of China and Japan in the Far East exemplify this historical fact. So do also the numerous baths established during Roman times and scattered around the Mediterranean Sea and western Europe. The overriding attention paid to geothermal energy as a source for electricity in the 1950s to 1970s shifted after the oil crisis in the 1970s towards a broader application of the resource. This shift in emphasis has widened the prospecting area for geothermal energy from the volcanic regions to other parts of the Earth, where low-temperature resources are more likely.

Geothermal resources are found in all types of rocks: sedimentary, metamorphic and volcanic. On average the formation temperature increases by about 30 °C/km, which means that temperatures of about 100 °C can generally be found at depths of 3 to 4 km. It is not enough to find temperatures suitable for utilization. Locations must be found where water is present at a suitable depth, and the local permeability is sufficient to absorb heat stored in the rocks and carry it up to the surface.

The policy maker is chiefly concerned about the economics of utilizing geothermal hot water and associated financial risks. The economics of the project are directly related to the cost of producing the geothermal hot water, the cost of transferring it to the end users and adapting it to their needs. The cost of disposal of the water, once the heat has been extracted from it, also plays a significant role. Economic viability is, furthermore, dependent upon the size of the market, the population density (heated space per capita) and the intensity of utilization (degree-days per year). Financial risks are principally associated with the drilling for the hot water, that is, the potential success rate and high well costs.

Before a decision is taken to develop geothermal energy at a given location, the economic viability of the project must be shown to compare favourably with, or outdo, the viability attainable using an alternative energy source, for instance conventional sources such as coal, fuel oils and gas, or alternatively, less conventional energy sources such as waste, biomass and solar energy. In comparing the alternative solutions, account must be taken of the relevant environmental ramifications and the indigenous nature of the energy sources. The benefits accruable from the combined use of these diverse resources should also be perused.

3.1.2 Utilization of low-temperature geothermal resources

Space heating by means of geothermal energy is a highly successful present-day application. To date, well over twenty countries have successfully put geothermal energy to this use.

The economic usefulness of warm (hot) water depends primarily upon the temperature and quantity of available water. Hot water at temperatures ranging from 30 to 125 °C has been used for space heating. The range of the water temperature at inlet to the radiator is generally limited to 50 to 80 °C. In some areas (for instance, in Iceland), the geothermal water is of such purity that it may be piped to the users and passed directly through the radiators. More commonly, however, the chemical composition of the water will not allow such direct use. These instances call for heat exchangers to be installed to provide thermal linking while effectively separating the radiator and the geothermal fluid. Geothermal water passed directly through radiators commonly leaves at a temperature of 25 to 40 °C.

It is possible to utilize water at temperatures as low as 35 to 50 °C by employing alternative methods of space heating, for instance radiant panel heating (pipes embedded in the floor or ceiling). Utilization down to temperatures of 20 to 30 °C requires more sophisticated methods. Air-heating systems using geothermal heat exchangers of the water-to-air type are, for example, very suitable. Such installations have, for instance, been employed using heat pumps to boost the temperature of the geothermal water. An interesting feature of heat pump systems is that they can be made reversible so that during hot weather the air heating system changes to an air cooling system. It is also possible to combine some or all of the above heating methods by using them in cascade (series).

We will discuss and describe the various possibilities and aspects of geothermal district heating in more detail in the sections that follow.

3.2 RESOURCE CONSIDERATIONS

3.2.1 Resource development

The time it takes to develop a virgin geothermal resource is typically seven to ten years, from the early exploration work until the resource is ready to produce hot water. Great advances have, however, been made in exploration and interpretation methods during the last decade or so. The resource development time has thus been cut to four to six years depending upon site specifics. Geothermal projects need good planning from the start, with milestones and decision points clearly defined. Bypassing a stage or stages of such work to gain time does increase the associated risk. Good plans are therefore one of the most valuable tools a decision maker has at his or her disposal to monitor the work and to check that the required information for decision making is available. The early phases of geothermal projects are based upon exploration strategy, but the later phases upon conventional construction planning.

3.2.2 Temperature of fluid

The value of the geothermal resource increases, the higher the temperature of the water. Low-enthalpy geothermal fluid temperatures in the range 100 to 150 °C are worthwhile for generation of electricity in a binary plant, particularly when combined with heating applications. The useful temperature drop achievable in the case of heating is typically about 65 to 115 °C. A system temperature of 55 °C is generally accepted as the minimum practicable for direct geothermal heating uses. Below that temperature special thermal boosting facilities are required, for instance fossil-fuelled, natural gas-fuelled or electrically powered boiler plants, or heat from waste burners or heat pumps.

3.2.3 Available flow rates

The available flow rate, together with the water temperature, determines the energy potential of a resource for a district heating application. Hot water is typically delivered to the customer of a geothermal district heating system at temperatures anywhere between 60 to 90 °C and returned at 35 to 50 °C. This return temperature is about 30 °C lower than is typical for conventional fossil-fuelled district heating systems. The efficiency of thermal energy extraction depends upon the water temperature drop achieved at the user end. To give an idea of the power that can be extracted from geothermal water, 360 l/min of hot water give an equivalent of 1 thermal megawatt (1 MW$_t$), when cooled by 40 °C (say from 80 to 40 °C).

Each geothermal well is different in terms of output characteristics, even wells located within the same geothermal

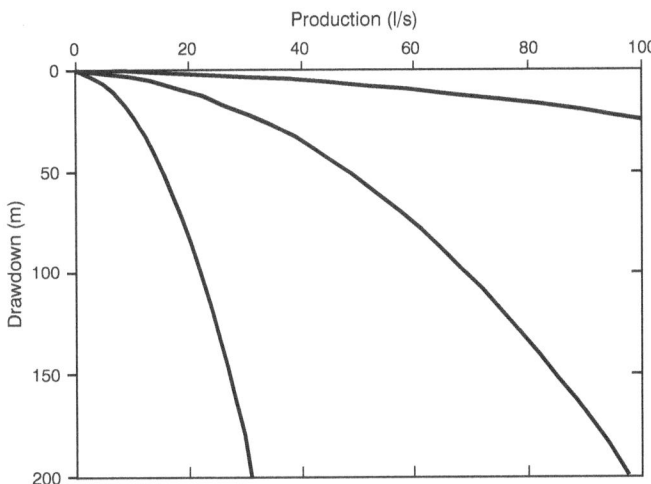

Figure 3.1 Typical well drawdown curves

field (see Figure 3.1). By placing a deep well pump in the wells, the output can be increased by lowering the water level in the well and its immediate formations. In addition to the short-term drop in water level, forced water extraction causes water level drawdown throughout the reservoir. The reservoir drawdown is commonly monitored in separate observation wells. The results generally show a steady pressure decline (reservoir drawdown), with some local variation within the span of a year depending on the rate of extraction (see Figure 3.2).

3.2.4 Chemistry of fluids

Geothermal hot water is richer in minerals than cold groundwater – the higher the temperature, the greater the mineral content. This dependence is used by geoscientists to map geothermal areas and to model groundwater flow. The chemical content of the water must also be known in order to enable engineers to select suitable materials (mild steel, stainless steel, fibreglass, plastics, titanium) for the energy conversion system, and to decide the most viable type of heating system to suit the site specifics. The main chemical constituents that affect this selection are silica, oxygen, chlorides, calcium, magnesium, hydrogen sulfide and the pH of the fluid.

Geothermal hot water that is relatively pure may be used directly both for radiators and as tap water. It must in that case contain less than 50 mg/kg chlorides, less than 150 mg/kg silica, less than 5 µg/kg of oxygen, and otherwise meet the criteria shown in Table 3.1. The use of heat exchangers and

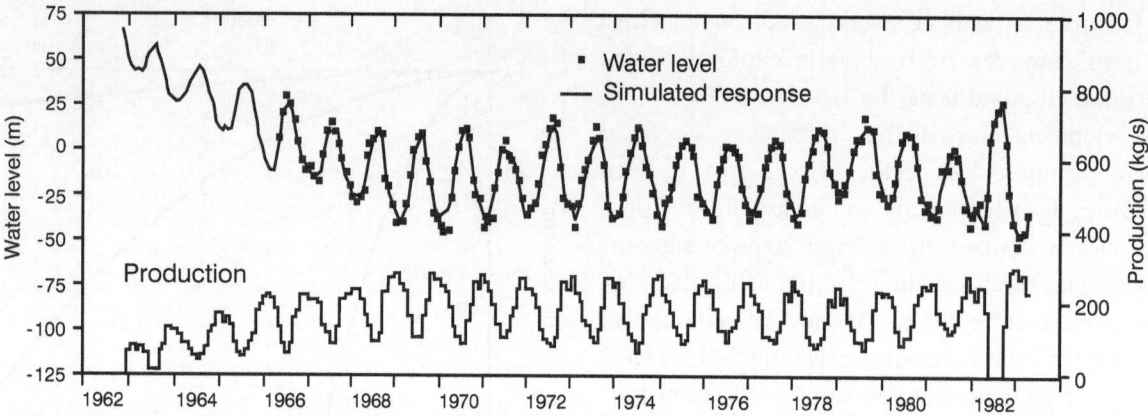

Figure 3.2 Typical reservoir drawdown curves

double-pipe distribution systems may, on the other hand, be necessary if the geothermal water is rich in minerals or contains constituents harmful to the user and/or the system. A closed-loop (two-pipe) system employing heat exchangers is the most common type of geothermal district heating system anywhere, except in Iceland where the geothermal water is more often than not of potable quality.

Changes may take place in the chemistry of the geothermal water after the start of exploitation. In the event of a large drawdown or pressure drop there is a potential danger of inflow of cold groundwater or seawater into the geothermal system, followed by cooling and changes in production characteristics. A change in the chemical composition of the geothermal water often precedes the cooling of the field, and data obtained by chemical monitoring of geothermal fluids may therefore give a warning early enough for preventive action to be taken. Chemical monitoring is often cheaper and more convenient than are direct measurements of temperature and pressure.

Deposition problems are usually not acute in low-temperature utilization, although calcite and sulfide problems are known, but the utilization of high-temperature geothermal systems is fraught with potential deposition problems. Calcite and sulfides may be deposited in wells,

Table 3.1 Maximum permissible levels of chemical concentration in freshwater in comparison with those typical in geothermal effluent

Chemical constituents by weight	Type of usage			Typical for geothermal effluents
	Supply of water for domestic use	Freshwater for fish farming	Water supply for irrigation	
Total dissolved solids (ppm)	1 500	7 000	200–1 000	500–500 000
Chloride Cl (ppm)	600	250	100–200	10–200 000
Sulfate SO_4 (ppm)	400	200	200	1–200
Fluoride F (ppm)	0.8–1.7	1	10	1–200
Boron B (ppm)	30	500	0.5–1	1–1 000
Hydrogen sulfide H_2S (ppm)	0.05	0.3	1	1–300
Arsenic As (ppm)	50	1 000	1 000	1–50
Mercury Hg (ppb)	1	—	—	0.1–10
Lead Pb (ppb)	50	—	—	1–100 000
Cadmium Cd (ppb)	10	—	—	0.1–2 000

and silica deposition may be a severe problem in surface equipment when the high-temperature fluid has cooled down and become supersaturated with silica. Special care has to be taken that geothermal water and cold groundwater do not mix because of the potential danger of magnesium silicate scaling. This can occur in heating systems that rely partly on a source of low-temperature geothermal water and partly on heated groundwater, or there may be natural mixing of cold groundwater and geothermal water near the surface. Such mixing has also been known to take place across heat exchangers, as only a small amount of mixing is needed to cause such deposition. Production characteristics of high-temperature geothermal installations may change because of changes in the reservoir or even within a single well. Drawdown will create a growing steam zone in the upper part of the system, and concurrently the gas content of the steam may increase greatly. Early indications of this tend to be found in chemical changes in the discharged fluids. As high-temperature geothermal fields are often connected to volcanic activity, gas changes need to be monitored carefully as magmatic gas may enter the system and cause undesirable changes. In such systems observed cooling is more often a sign of overexploitation, rather than inflow of cooler water of a different origin into the system.

3.2.5 Distance from potential market

The distance of the geothermal resource from the potential heating market is a very important parameter as regards the technical and economic viability of the heating system. Transmission pipelines of 60 km length have been built with acceptable heat loss values, though shorter transmission distances are much more common and clearly more desirable.

The economic transmission radius, that is, the maximum economical distance between the geothermal field and the heating market, is mainly governed by two groups of parameters. The first group relates to parameters affecting the capital investment in the transmission mains and associated equipment. The most important of these are size and density of heating market, pipe, diameter, thickness, material type, thermal insulation, thermal expansion compensation, pipe anchoring, whether above ground or buried, booster pumps, temperature-boosting equipment, peak load plant, storage tanks and so on. The second group of parameters relates to operating costs. Important in this respect are heat losses in the pipeline (see Figure 3.3), pumping cost, use of oxygen scavengers, use of scaling inhibitors, amount of thermal boosting and type of boosting, amount and type of peak power generation and so on.

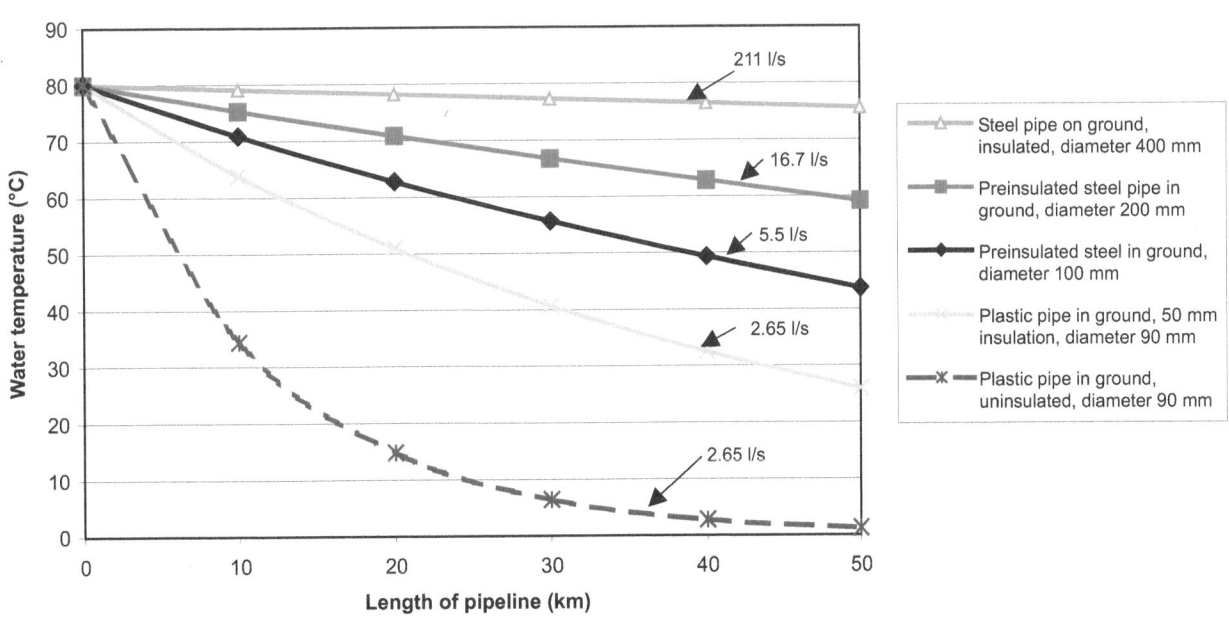

Figure 3.3 Heat loss in pipelines on the basis of 0 °C outdoor temperature

From the above listing of relevant parameters, it is evident that these are interrelated, which complicates the picture further. For large direct-use geothermal heating systems (installed capacity in excess of 200 MW$_t$) based on pumped wells, an economic transmission radius may be of the order of 30 km, whereas for a similar small system (about 10 MW$_t$ installed) the radius may be 7 to 8 km.

3.3 SPACE HEATING (OR COOLING) NEEDS

3.3.1 Climate

For each individual case it is necessary to analyse weather records in order to determine the space heating (or cooling) load required. The outdoor air temperature can vary considerably within the span of a day, from day to day and from season to season. The demand for space heating energy and power is proportional to the difference between the design room temperature, T_i, and the outdoor air temperature. Consequently, the maximum load demand is proportional to the difference between the design value of the room temperature and the minimum outdoor temperature, T_{min}, that the heating system is designed to handle.

A useful tool in determining the energy requirements (load demand) for a system is the so-called 'duration curve' for the mean daily temperature (see Figure 3.4). It depicts the number of days in an average year that have a mean daily temperature lower than any given temperature value.

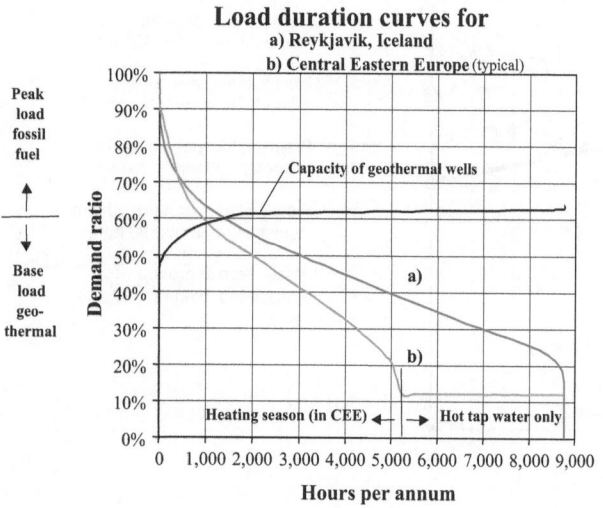

Figure 3.4 Typical duration curve

The area under this curve is proportional to the number of degree-days of required heating, and hence a measure of the energy (or load) requirements. The number of degree-days in a year below a given temperature is defined as the accumulated sum of degrees (mean temperature) below the given temperature value for the total number of the days in the year.

The ratio of the average heat load over the year to the maximum demand or peak heat load of the year is called the 'load factor'. The load factor influences the unit cost of energy delivered to the user. The capital investment for a district heating system is directly related to the capacity (maximum power demand) of the system, and the annual operating costs are to a large extent fixed. This means that the higher the annual load factor, the lower the unit cost for geothermal energy.

Records of the most severe cold waves at the location are needed in order to determine the necessary power (capacity) of the relevant district heating system. A geothermal heating system with the capacity to handle the coldest day of the year would clearly run at a partial load for a greater part of the year, and therefore operate at a very low load factor. A system designed for handling a maximum load represented by an outdoor temperature considerably above the annual minimum would, on the other hand, operate at a much higher load factor. None the less, it would be able to supply the bulk of the heating energy required all year round, and, by virtue of its higher load factor, be much more economical.

Commensurate with the maxim of economics, the design value of the outdoor air temperature should be selected by establishing a criterion that specifies how much the room temperature may drop below the design value during a severe cold spell. A design outdoor temperature value considerably (say 5 to 10 °C) higher than the lowest annual value may safely be selected.

3.3.2 Population, population density

The size and density of the population must be established in each individual area where a district heating system is planned. Since such a system is expected to be in operation for a good number of years, the future population trend (say the twenty- to thirty-year trend) should also be estimated. The design of the district heating system can then be based upon current population values, taking into account projected population trends and planned density of population.

3.3.3 Building types

In order to arrive at a reliable estimate of the specific space heating energy requirements (heat demand), it is necessary to consider the heat loss characteristics of typical buildings in the area, as well as the average space occupied by each person/family.

For evaluation of the inside temperature drop during severe cold waves (or increase during heat waves) sufficient knowledge of the heat storage characteristics of typical buildings specific to the locality is required. In this connection the construction material is an important factor; heavy material such as concrete, cinder blocks and brick have a higher heat capacity than light materials such as lumber and other fibrous materials. The placement of thermal insulation on the walls of buildings is also of importance, since an insulation layer on the outside surface will result in a building with a considerably higher heat storage capacity than one with the insulation layer placed on the inside surface. Infiltration, which depends on building tightness, and thickness of insulation are, however, most important for maintaining a stable inside temperature during all conditions.

It may be possible to reduce the quantity of hot water required for a specific location. This possibility should be explored, for example, by looking into the following:

- building materials and insulation properties as required by existing building codes
- workmanship in the construction and finish of buildings that will affect such building properties as leakage of air through walls and windows.

3.3.4 Techno-economic aspects

When planning a geothermal district heating system the following techno-economic aspects should be considered:

- *Physical aspects*: for instance determination of the area to be supplied, the building density in the area, assessment of the heat loss and heat consumption.
- *Technical aspects*: such as the requirements at the consumer end, for instance, temperature, the transmission and distribution system and the characteristics of the heat source, that is, chemistry of the geothermal fluid, temperature and pressure.
- *Economic aspects*: such as the investments in geothermal wells, collection, transmission and distribution system

and the annual costs related to the operation of the system. This relates to the network configuration and pipe sizing.

3.4 HOT WATER COLLECTION AND TRANSMISSION SYSTEM

3.4.1 Types of district heating systems

Geothermal district heating systems may be divided into two main groups depending on whether the geothermal water is used directly in the house heating systems (secondary system) or indirectly by transferring the geothermal heat to the secondary system through heat exchangers. In the latter case the geothermal water is confined to the primary system.

In the following, the main emphasis is laid on district heating systems with a large number of individual users. In general, however, the solutions described also apply to single users, whether constituting a single home or a large building complex.

Figure 3.5 shows the main types of district heating systems in use. Temperature values indicate typical values under design conditions (for instance, −15 °C outdoor temperature in Iceland). Some of these, such as the heating water temperature leaving the radiators, will change in accordance with the heating load. If the heating load is decreased (outdoor temperature $T_0 > -15\,°C$), the volume of water flowing through the radiators will be reduced, thus lowering the flow velocity, which results in increased cooling of the water. The water temperature leaving the radiators is, therefore, lowered below the design value of 40 °C, the difference increasing as the outdoor temperature increases.

Geothermal water in the primary system only

The geothermal water is, in this case, separated from the district heating water by heat exchangers. This may be necessary because of the chemical characteristics, or the high temperature and pressure of the geothermal fluid used. Such an arrangement also offers more flexibility if, for instance, the geothermal sources become temporarily inoperable.

The purpose of the heat exchangers is to transfer the heat from the geothermal to the heating system medium while keeping the two separated.

The district heating system and the hot tap-water system are each separate closed-loop networks in the residential unit

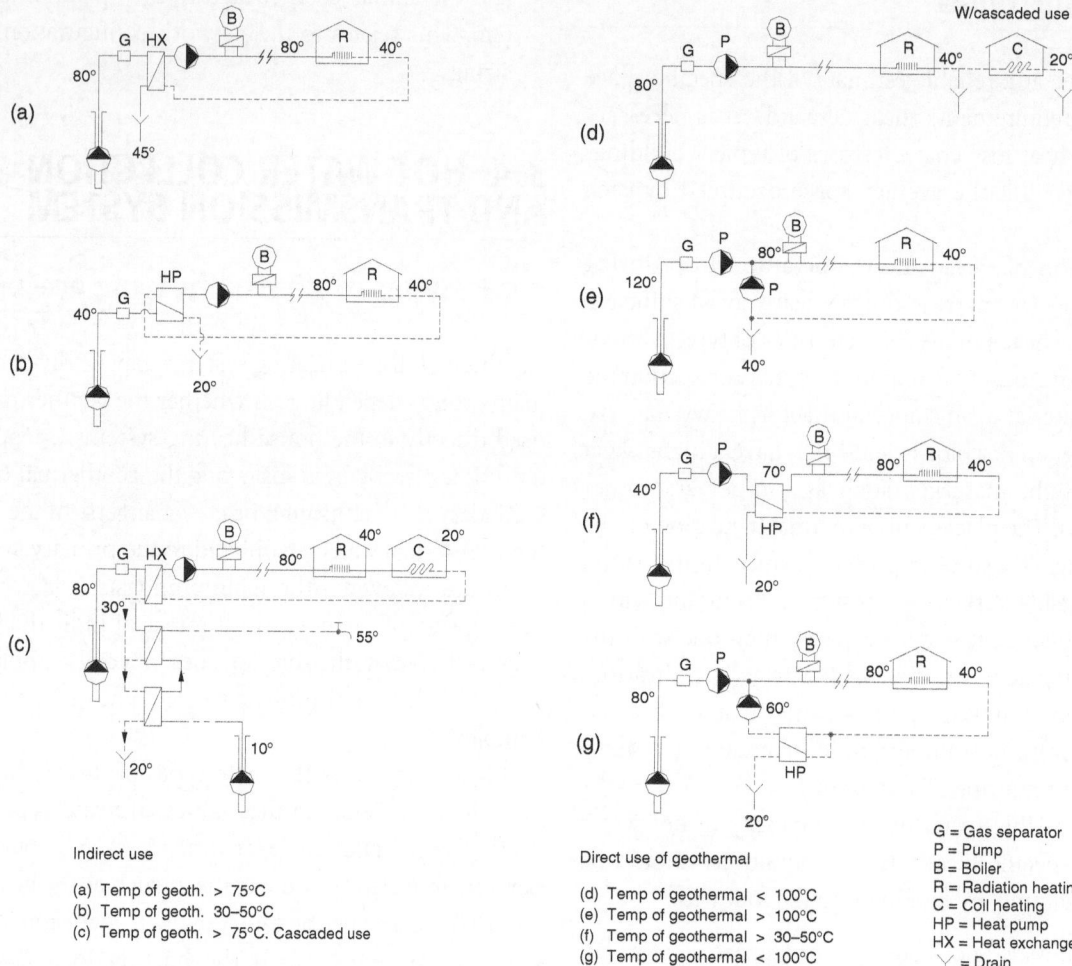

Figure 3.5 Main types of district heating system

or apartment in question. After passing through the house heating systems, the secondary system heating water is collected and transmitted in pipes to the heat exchangers, where it is reheated to the required supply temperature. The secondary side of the hot tap-water heat exchanger is connected to the municipal cold water supply system, which is then heated to the required temperature for the hot water taps (55 to 60 °C, see Figure 3.5c).

Direct use in the house (secondary system)

In some locations (for example Reykjavik, the capital of Iceland), the quality of the geothermal fluid is such as to allow the water to be used directly in the house heating systems (without any fluid separation by heat exchangers).

In such cases the following technical solutions have been used:

- *Single-pipe system*: this single-pass or once-through system uses the geothermal fluid (temperature below 100 °C) directly, for both the hot tap water and radiators. The spent fluid from the radiators is discharged to waste (Figure 3.5d). It should be pointed out, however, that strict statutes are expected in the near future on the maximum allowable hot tap-water temperature. These restrictions will likely call for drastic changes in this direct use practice, such as to make the installation of heat exchangers in the hot tap-water system mandatory.

- *Dual-pipe system*: a part of the spent return water from the house system (temperature 30 to 40 °C) is collected and mixed with the geothermal supply water (temperature 100 to 130 °C) to obtain a constant supply temperature (for example, 80 °C), irrespective of weather conditions. The excess return water is discharged to waste or injected back into the reservoir (re-injected).

Peak load plant

It is generally found to be more economical to install fossil-fuelled boilers to handle the usually very brief periods of peak heat demand, than to provide geothermal capacity sufficient for all load situations. A boiler plant requires a comparatively low capital investment but is expensive to operate. This makes boiler plants for heat generation typically more economical for intermittent peak load applications, than would be the provision of additional geothermal well(s).

A geothermal well is generally found to be more expensive than a boiler plant of similar thermal capacity. Heat pumps may alternatively be employed to boost the thermal base load capacity.

3.4.2 Pipe systems

The source of geothermal fluid to be used for district heating systems is, in most cases, located some distance from the heating market, although geothermal water may also be found within the market area. A transmission pipeline is therefore needed to transport the geothermal fluid from the geothermal field to the end users. Collection mains are required to interconnect the geothermal wells, collect the fluid from each one and transport it to a heat or distribution centre in or close to the geothermal field. In the heat centre the fluid is transferred to the main transmission pipeline, which links the geothermal field and the distribution network at the consumer end.

Collection and main transmission pipeline

The transmission pipeline diameter is dictated by local conditions such as the grade of the land over which the pipeline is laid and the available power for pumping. An approximate rule of thumb is to design the diameter of the pipe so that the pressure drop in a straight section of pipe at maximum rate of flow is the order of 0.05 to 0.10 MPa/km.

Transmission pipelines may be of assorted makes and types (see Figure 3.6). The various alternatives usually differ greatly in both cost and durability. The most common pipeline material is carbon steel, but various plastic materials (polypropylene, polybutylene and polyethylene) may be used for smaller size pipelines. The use of plastic pipes can lead to corrosion damage in downstream metallic components as a consequence of atmospheric oxygen diffusing into the water through the pipe walls. Also, special care should be taken to select the correct pressure class for the expected temperature and pressure conditions. Asbestos-cement pipes have been used in long pipelines when there is an abundance of hot water and the cost of pumping the hot water is low.

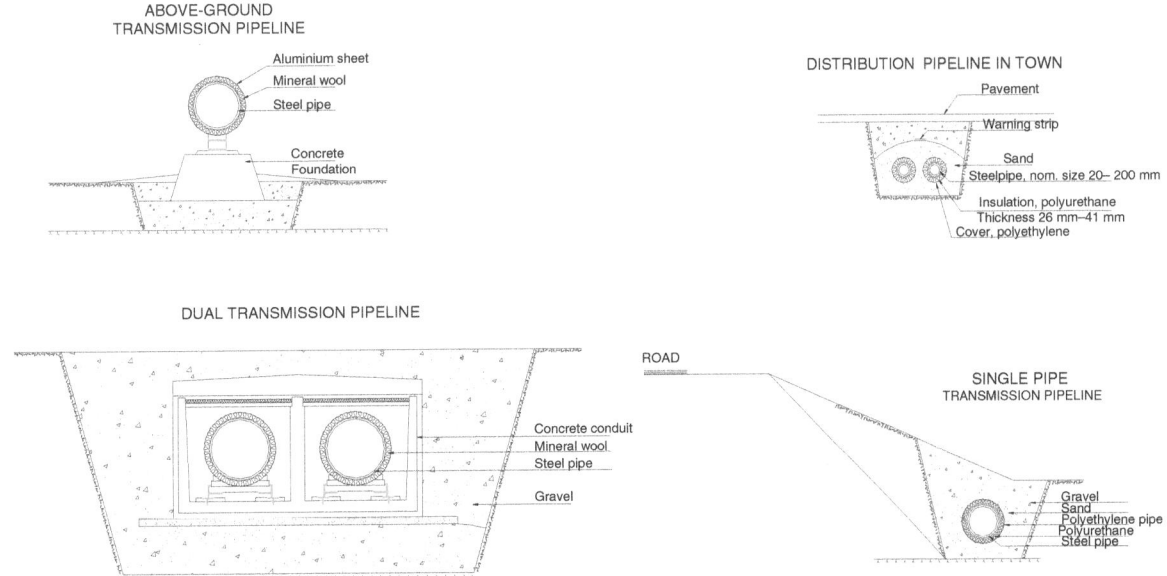

Figure 3.6 Types of pipeline used for geothermal district heating systems

These pipes are attractive because of the low initial cost compared with carbon steel pipe, but have lost popularity in recent years because of the hazard to health associated with cutting, assembling and disassembling the asbestos cement pipes. A brief overview of the various pipelines commonly used in transmission, collection and distribution mains is given in Table 3.2.

Distribution network

The distribution network is a very important part of the district heating system. Care in the planning and design of the distribution network is of great importance for the successful operation and proper functioning of the heating system. Distribution systems are commonly of two types, either a single-pipe system or a system with two pipes, supply and return, that is, a closed-loop system. The former may be employed where the chemistry of the geothermal fluid with respect to corrosion, scaling and water quality permits direct use. The fluid is discharged to waste or injected back into the reservoir (re-injected) after use. The latter type of system must be used where direct use of geothermal fluid is not possible because of the chemical composition, quality and/or temperature of the water.

The distribution network may in general be divided into mains or trunk lines, branch lines and house lines. The trunk lines or mains constitute, as implied by the name, the main pipelines going out from a pumping station. One or more trunk lines may be needed, depending on local conditions and size of network.

Table 3.2 Common types of pipeline

Type of pipe	Application	Advantages	Disadvantages	Cost ratio
1. Buried non-insulated polypropylene or polybutylene pipes	• Transmission pipeline • Water temp. <90 °C • Size range 20–200 mm • Abundance of hot water • Low water production cost	• Low investment cost • Installation ease	• Permits diffusion of oxygen through pipe walls • High heat loss	Single pipe 30–40%
2. Carbon steel pipe mounted above ground and insulated using mineral/glass wool sheathed in sheet metal jacket	• Collection and transmission pipelines • In locations allowing surface pipelines • Size range >200 mm	• High durability • Maintenance cost low • No temp. limitation	• Investment cost high in sizes below 200 mm relative to pipe alternative 4	Single pipe 80–130%
3. Carbon steel pipe mounted above ground and insulated using mineral/glass wool sheathed in spirally wound sheet metal jacket	• Collection and transmission pipelines • In urban locations allowing surface pipelines • Size range >200 mm	• High durability • Maintenance cost low • Upper limit of water temperature 120 °C	• Investment cost high in sizes below 200 mm relative to pipe alternative 4 • Max. water temp. 120 °C	Single pipe 80–130%
4. Buried carbon steel pipe pre-insulated using polyurethane foam encased in a protective sheath of polyethylene	• Distribution pipelines for size range 20 to 150 mm • Transmission pipelines in range of sizes below 900 mm	• High durability • Maintenance cost low • Protective sheath is watertight • Installation ease	• Investment cost high in sizes above 200 mm • Max. water temp. 120 °C • More vulnerable to external damage than alternative 5	Single pipe 100% Double pipe 100%
5. Insulated carbon steel pipe carried in a concrete conduit	• Transmission pipeline • Distribution network • Short, large-dia. pipes that have to be buried in built-up areas • Large consumer market	• High durability • Maintenance cost low • No temp. limitation	• Investment cost high for all pipe sizes	Single pipe 120–200% Double pipe 80–150%

The usual arrangement is to lay the trunk line in such a fashion that the branch lines jut out from the trunk lines as branches on a tree, and the line keeps decreasing in diameter towards the end farthest from the reservoir. Lines forming closed loops are also employed because they are considered to have a certain advantage over the branched type. The advantages of the closed-loop type are, for example:

- Less danger of complete sections of a town having to be closed off during repair work.
- A reduced likelihood of low water-pressure for consumers who are located farthest from the pumping station.

Closed-loop systems are clearly more expensive to construct than simple branch systems, which is probably the reason they are not more frequently used. Also, a higher temperature drop may be experienced.

Service wells, typically of reinforced concrete, are located at major branch points for easy service and repair access. The wells contain valves, expansion compensators, flexible expansion hose connections and, most often, pipe anchor points. The wells are fitted with access covers on top to provide entrance for maintenance and inspection crews.

A distribution network pipe is commonly designed to withstand 10 to 25 bar internal pressure (1.0 to 2.5 MPa). The network is so designed that the lowest pressure at the intake to houses at maximum load is never less than 0.15 to 0.2 MPa. Another criterion is that the radiator system in connected buildings may not be able to withstand pressures in excess of 7 to 8 bars (0.7 to 0.8 MPa). This means that in districts where there is a large difference between the elevation of the highest and lowest buildings, the distribution system must be divided into a number of separate sections, each one serving buildings within a given range of elevation. In open systems, where the geothermal return water is discharged directly to drain, the water pressure in the house system is lower. Typically, pressure-sustaining valves in the discharge pipe are used to maintain house system pressure of 1 to 4 bars (0.1 to 0.4 MPa), while the supply pressure upstream of a differential pressure-regulating valve in the house intake is 2 to 7 bars (0.2 to 0.7 MPa). Normally, about 1 bar (0.1 MPa) differential pressure is sufficient for the house heating system.

Districts that are extensive in area and thus have large associated pressure losses require more than one pumping station to keep the maximum pressure within the acceptable pressure limits given above. During construction, the distribution network is typically tested at 12 bar (1.2 MPa) pressure.

House lines and house heating systems

The parts of the distribution system that connect consumers with the distribution system are the so-called 'house lines'. House lines differ in size depending on the size of the building and heating loads. The most common sizes are 20 and 25 mm nominal diameter, usually carbon steel pipes insulated with polyurethane foam and covered by a protective polyethylene pipe. Larger buildings need larger pipes – anywhere from 32 to 100 mm in nominal diameter.

The design of a house heating system is aimed at maximum utilization of the geothermal energy. Geothermal heating systems are, therefore, so-called 'once-through systems', that is, the hot water passes through the heating system only once and is returned to the district heating network, discharged to waste or re-injected. The typical house heating arrangement adopted for boiler plant-powered district heating systems, where a large quantity of heating water is circulated through the system and the temperature difference between the supply and return is kept low, is not economically viable for geothermal heating.

In order to maximize the utilization of the geothermal energy, the temperature of the heating water leaving the house (return) must be made as low as possible. A prerequisite for this maxim is that the heating equipment, most popularly radiators, has sufficient heating surface to ensure the maximum practicable temperature drop across the house (see Figure 3.7). Alternatively, the heating water may be passed through radiators and heating panels (floor or ceiling) in sequence, a type of cascaded use (see Figures 3.5c and 3.5d).

Figure 3.8 shows a typical house heating system, with radiators, a domestic water heater, pressure and temperature controls, as well as a hot water metering device, which is owned and maintained by the heating company.

Pipeline costs

In order to assess which system to use for a specific project one must evaluate collectively several parameters such as price, quality, reliability, construction period and density of traffic.

Heat loss from pipelines

The district heating system pipelines, whether above ground or buried underground, constitute one of the main sources of heat loss in a district heating system, and thus a loss in revenue. To minimize this, it is of utmost importance that the pipe insulation does not become damp, since this greatly reduces its thermal resistance and may increase heat losses to

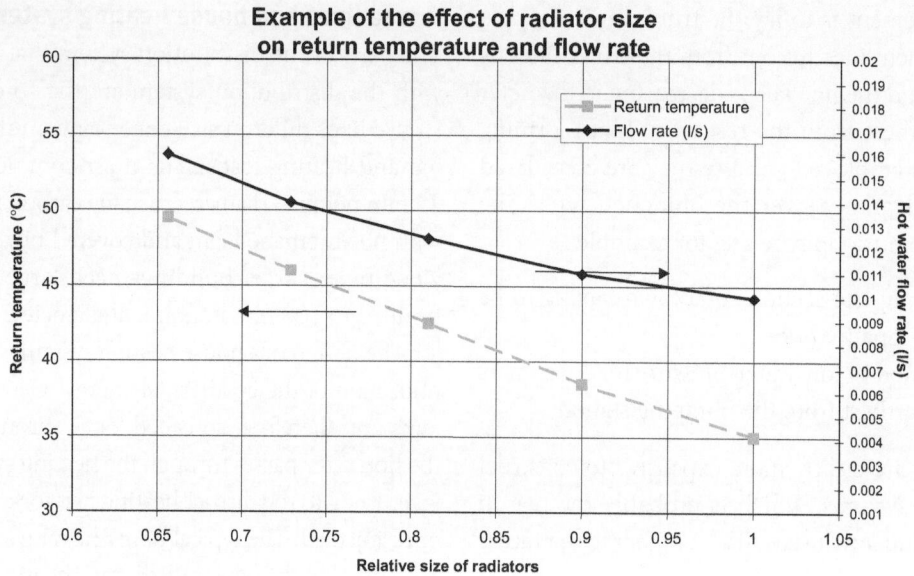

Figure 3.7 Effects of radiator size

such a level as to seriously impair the capacity of the district heating system. A damp layer of insulation on a steel pipe will furthermore induce rapid corrosion of the pipe.

The result of an evaluation of temperature losses in miscellaneous types of pipelines is depicted in Figure 3.3. The cooling of fluid flowing in pipelines made, for example, from polypropylene pipes installed with mineral wool is similar in range to that of buried carbon steel pipe, as long as the pipe insulation remains dry.

3.5 EQUIPMENT SELECTION

Most equipment for geothermal direct use can generally be described as off-the-shelf merchandise. Certain characteristics of the geothermal environment must, however, be borne in mind in the selection of this equipment. The most important of these factors are:

1. There is a high potential of scaling and corrosion.
2. Geothermal direct-use applications are typically capital-intensive and generate slow payback of investment.
3. The primary fluid side is generally compatible with high-technology electronic control components.
4. The geothermal resource is essentially non-renewable and typically best suited for base-load operation.
5. Environmental issues should be addressed. Geothermal energy is as a rule benign but notable exceptions exist.

The major equipment for geothermal direct-use applications will be described in the following sections.

3.5.1 Down-hole pumps

Most geothermal reservoirs are not artesian (self-flowing), especially low-temperature ones, and require pumping. In selecting a pumping unit for a geothermal well, the following aspects specific to geothermal utilization must be considered, in addition to the ones considered for normal pumping duties:

• The water temperature may range between 35 and 150 °C.
• The water chemistry is commonly one that promotes scaling and corrosion in the pump and associated equipment.
• The pump drawdown and setting depths associated with geothermal utilization are generally large, typically ranging between 50 and 300 m.
• Sand problems are relatively common in geothermal wells.

Shaft-driven deep-well pumps are less sensitive to fluid temperature than submersible ones, and are therefore usually selected where the water temperature is in excess of about 80 °C and the required setting depth less than about 230 m.

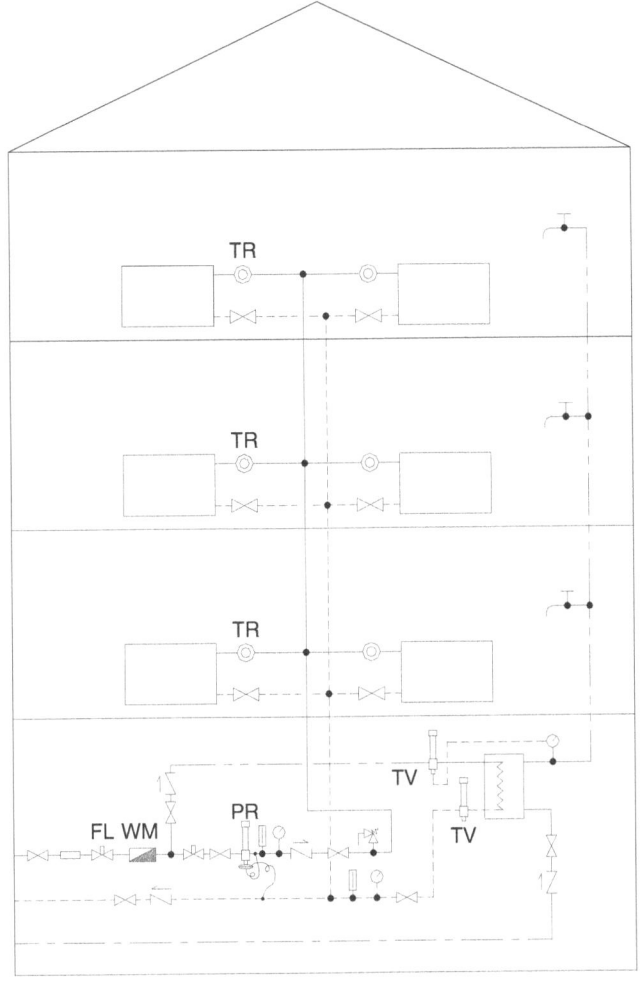

PR = Pressure regulator
TV = Thermostatic valve
TR = Thermostatic radiator valve
FL = Flow limiter
WM = Water meter

Figure 3.8 Typical house heating system

Submersible pumps are advantageous for water temperatures lower than about 80 °C. However, technical developments in the submersible pump sector have, in recent years, increased their temperature tolerance level to about 120 °C. Such pumps are expensive, and are most advantageous at setting depths greater than 200 m.

3.5.2 De-gassing tanks

To a varying extent geofluid contains non-condensable gas, the quantity and composition of which depends upon temperature, geological environment, age and origin of the fluid. It is important to remove this gas simply and efficiently in surface equipment without overly upsetting the chemical equilibrium of the fluid or introducing into it gases harmful to operations, notably atmospheric oxygen.

The de-gassers mostly used in the geothermal industry consist basically of vertical or horizontal cylindrical tanks made of carbon steel and suitably baffled to promote separation of gas and liquid. The fluid is introduced into the tank at a slight drop in pressure to flash out the gas. The de-gasser is fitted with a suitable gas relief valve, and commonly kept at a slightly super-atmospheric pressure to hinder the ingress of atmospheric oxygen.

3.5.3 Heat exchangers

The geothermal industry uses most of the standard-type heat exchangers commonly used in the chemical industry, namely plate and/or shell-and-tube type exchangers such as are commercially readily available. The only exception to this rule is the down-hole heat exchanger, used where the fluid chemistry is very problematic and the geothermal aquifer characteristics permit its use. Normally only used in small, low-cost heating projects powered by a single well, these exchangers can be found in the United States, New Zealand and Turkey.

The basic differences in the requirements imposed by geothermal applications, especially in the primary fluid cycle, are those due to scaling, corrosion and low fluid-temperatures.

The high scaling rates commonly characteristic of geothermal brines require that the heat exchangers be of a design promoting easy cleaning, such as the plate type, or a self-cleaning type, such as the fluidized-bed type. Where the shell-and-tube type heat exchanger is used, the brine flow is typically directed through the tubes instead of the shell, for the same reason.

The high corrosion potential often present in the geofluid requires that great care be exercised in the selection of heat exchanger materials. Exotic materials such as very special stainless steels and titanium are not uncommonly selected for the exchanger tubes or plates.

The low temperatures available require a large heat-transfer area, which commonly precludes selection of anything other than plate-exchangers for reasons of cost, flexibility and space requirements.

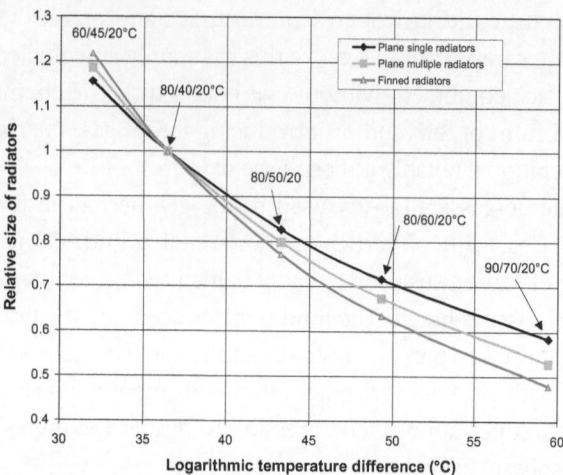

Figure 3.9 Effect of water temperature on the size of radiators

3.5.4 Radiators

Radiators are the water-to-air heat exchangers needed to heat the air at the user end of the district heating system. The heat is transferred in two ways, by convection and radiation. The relatively low water-temperature therefore requires fairly large heat-transfer areas in order to ensure the large temperature drop across the radiator, which is a prerequisite for an efficient geothermal energy conversion. The lower the water supply temperature, the larger the effective radiator area needs to be. The background to this adage is illustrated in Figure 3.9.

The radiator selected must, in addition to the above performance criteria, meet those of aesthetics and long life. Material plays an important role in both of these, especially the latter, and commonly cast-iron radiators are selected where the fluid is suspected of containing dissolved oxygen ($O_2 > 50$ μg/kg), especially in the presence of chloride ($Cl^- > 100$ mg/kg).

3.5.5 Control equipment

The type and refinement of the control equipment selected for geothermal district heating systems depend on local specifics, such as:

- size and sophistication of market and type of heating tariff structure
- geofluid characteristics, both physical and chemical
- geothermal reservoir characteristics and type.

The minimum control requirements may be summarized as follows:

- temperature, flow and pressure control at the well(s)
- flow, pressure and temperature control at the distribution centre(s)
- discrete, collective or quasi-individual sales metering (for example, a single meter for an apartment house having several individual apartments)
- individual radiator or apartment temperature control.

Control equipment for the well(s) is dictated by two basic requirements:

- the need to monitor the reservoir reaction to utilization and its long- and short-term behaviour
- the necessity to regulate water withdrawal relative to heating demand so as to satisfy the market while minimizing energy wastage.

The most important control function at the well(s) is to control pump discharge. Continuous speed control of the well pump(s) using electrical or mechanical devices, such as a thyristor or a fluid-coupling speed control, is usually the most desirable for flow and pressure control. Other methods are also used, such as combining the de-gasser/storage function in one tank fitted with a liquid-level control switch, which effects flow control by turning the well pump(s) on and off. Discharge totalization, regular drawdown and temperature measurements are also important for reservoir monitoring.

Pressure and flow control at the distribution centre(s) is, for a large heating system, effected by storage tanks, which, at a minimum, are sized to take the maximum twenty-four hour demand fluctuations of the heating system. In some cases they are sized to provide as much as a week's demand under severe cold spells. In some cases it may be necessary to control the supply temperature to the user by mixing in some return water via temperature-actuated valving or injection pumps and/or by selective pumping from geothermal fields of different temperatures.

In small district heating systems based upon artesian well(s), it may be a cost-effective alternative to sales metering to use a flow restriction to ration the flow to each house, apartment or high-rise building, and base the heating charges on the restriction setting. For larger district heating systems based on pumped reservoirs, energy conservation as well as cost-efficiency and ease of regulation considerations require sales metering to be of the volumetric or energy type.

At the user end the minimum requirement is efficient pressure balancing of the house system, combined with a thermostatic regulation between the hot water inlet and outlet of the house. Modern district heating systems have adopted individual radiator temperature regulation using thermostatic valves fitted at either the inlet to, or outlet from, each radiator. In addition the pressure across each house, apartment or apartment building is regulated using a membrane-type pressure-balancing valve. In some instances outdoor temperature compensating features are included.

3.5.6 Heat pumps

Heat pumps are often used in district heating systems, both alone and as a part of a bigger scheme. Heat pumps can be used for heating and for cooling, and for both purposes in combination, for example, cooling in the summer and heating during the winter.

The heat pump technique makes it possible to extract thermal energy from a heat source with a low temperature level and make it available as useful thermal energy at a higher temperature. The heat source can be of different types, for instance, outside air, groundwater or waste heat from industry. Geothermal water at low temperature, say 20 to 40 °C, which is too low for direct application in space heating, is an ideal heat source for heat pumps in a district heating plant because the economics of a heat pump installation are closely related to the temperature of the heat source.

A heat pump works like a refrigerator, where the working fluid is circulated in a close circuit, removing heat from inside the freezer and discharging it to the surroundings. In the heat pump, the working fluid extracts heat from the heat source through evaporation, and discharges it by condensation to the district heating water. To do this work external energy input is required, and the most commonly used type is a compressor driven by an electric motor, but chemical absorption, gas compression and other methods are available.

The ratio of the output energy to operating energy input is the basic measure of the effectiveness of a heat pump, and is very important to the economics of the heat pump operation, as referred to above. This ratio is known as the coefficient of performance (COP). For heat sources with a temperature in the range 20 to 40 °C the COP is very favourable. A likely COP factor in the case of a 30 °C geothermal resource cooled down to 20 °C to produce hot water

for space heating at 55 °C is, for example, about 4. This means that the resulting heat output for space heating is about four times the energy input to the compressor motor.

3.6 ECONOMIC CONSIDERATIONS

At the earliest planning stages of a district heating system, an evaluation must be carried out to compare the economic and technical viability of available energy resources and energy conversion systems in order to select the one most suitable to the specifics in question. The main factors in estimating the cost of the district heating system are as follows.

3.6.1 Cost of drilling

The cost of a geothermal well can vary between US$500,000 and US$2 million (1997 prices). The wells are typically drilled to a depth of 1,000 to 2,000 m. Conventional oil well and water well drilling equipment, materials and procedures are used with minor modifications to suit the geothermal specifics. The well casing sizes are normally larger than for oil wells, because far greater outputs are required for economical exploitation of geothermal wells. In many cases the well output is limited by the size of the deep well pump that can be accommodated within the well casing. Knowledge is required of the reservoir from which the geothermal fluid is to be produced, in order to estimate the cost of drilling. The following parameters need to be defined:

- geology of the reservoir
- depth of the wells to be drilled
- expected rate of penetration.

It usually takes one to two months to drill a 2,000 m deep geothermal well, and it is important to plan the work and material procurement thoroughly because time charges make up the bulk of the well costs. Time logs show that only 30 per cent of the time is typically spent on drilling 'on bottom'. The balance of the drilling time is spent on cementing, running casing, logging, moving equipment and so on. Figure 3.10 depicts the average costs of low-temperature drilling in Europe.

3.6.2 Cost of pipeline

The least expensive pipeline material available is carbon steel. As a result of corrosion problems, the steel pipes always

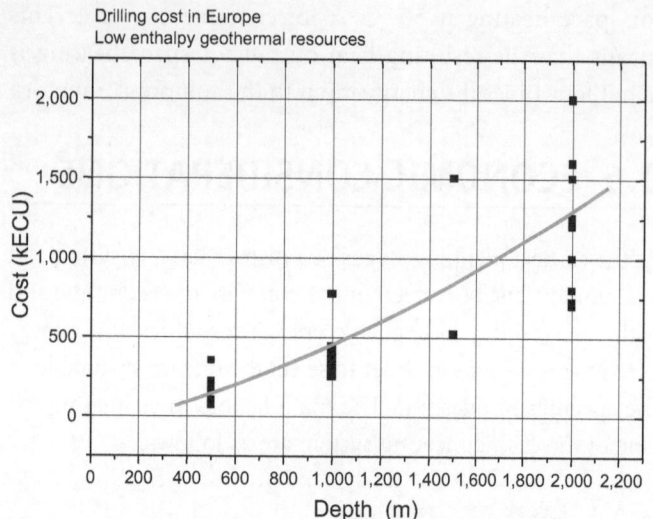

Drilling cost in Europe
Low enthalpy geothermal resources

Figure 3.10 Cost of low-temperature well drilling in Europe

have to be well protected against contact with water from the outside. Such protection of a hot pipeline is usually expensive. In many cases where the conditions are favourable (dry soil), it may therefore be economically advantageous to use another material, such as polyethylene, polypropylene or polybutylene. These are sometimes used without insulation, especially for short transmission lines.

High heat losses are associated with pipelines with no thermal insulation. Non-insulated pipes are, therefore, only used when the cost of obtaining and pumping the hot water is low, and high flow-rates can be maintained in the pipelines at all times in order to achieve acceptable temperature levels at the consumer end.

Asbestos-cement pipes, for example, buried under a layer of turf and without insulation or partially insulated, present a solution where an inexpensive transmission pipeline is of paramount importance to the viability of an undertaking. The above conditions also apply with regard to availability, cost and quantity of the hot water.

For many years, carbon steel piping pre-insulated with polyurethane foam and covered with a polyethylene jacket has been the most popular piping for distribution networks. Such pipes are also frequently used for transmission piping. The most common sizes used for single-pipe mains and networks are 20 to 350 mm and up to 800 mm nominal diameter. Pre-insulated carbon steel pipe selected for double-pipe systems is used most commonly in sizes of 20 to 600 mm nominal diameter.

Insulated carbon steel pipe carried in a concrete culvert provides a viable alternative to the popular pre-insulated steel pipe, for both transmission and distribution networks in regions where salaries are low relative to the cost of industrial products.

Surface pipes of carbon steel are used in open terrain as transmission mains and in geothermal fields as collection mains. These are less expensive than the pre-insulated pipe, as a single-pipe configuration in sizes above 300 mm nominal diameter. They should be used where permitted by planning, environmental and other site-specific conditions. Special care should be taken to ventilate the space between the pipe and cladding to avoid accumulation of moisture on the pipe surface.

High saline groundwater levels may, however, dictate the use of surface pipelines for distribution systems. Examples of this can be found in Russia and the People's Republic of China.

3.6.3 Capital investment

Prior to a decision on the establishment of a geothermal district heating system a large number of parameters must be taken into consideration. A feasibility study must be carried out taking into account all the factors involved. An acceptable rate of interest on the required investment must be agreed on, and the heating system must be the one closest to complying with the set criteria. In the Scandinavian countries, municipal energy companies typically require a 6 to 8 per cent rate of interest on their investments, whereas private companies normally demand somewhat higher values.

The establishment of a district heating system is usually highly capital-intensive. The economic analysis and finance scheduling should be based on long-term considerations. The energy planning underlying district heating systems should form a part of national energy schemes and be based on a fifteen- to twenty-five-year planning period. It is fairly normal to fix the heating tariff on the basis of a long payback period (for example, fifteen to twenty years) and minimal profit margins by making the total revenue roughly balance finance and operating costs. For this purpose long-term loans are advantageous.

Financing of geothermal district heating systems is commonly obtained through international financing institutions requiring government guarantees. In some cases, where the stand-alone economics of the geothermal system are marginal in relation to fossil-fuelled alternatives, government subsidies and loans may be made available for energy political reasons.

The principal capital investment cost items are as follows:

- geothermal well(s) (exploration, drilling, pumping, plant, utilization rights, land lease or purchase)
- main transmission pipeline (pumping plant, storage tanks)
- distribution network (heat exchanger plant, pumping plant, automatic control, sales metering equipment).

Capital investment costs differ greatly from location to location depending on the availability of geothermal energy, distance from the resource to the market and population density. The typical capital investment cost of geothermal district heating systems ranges from 450 to 700 US$/kW (1989 prices), where geothermal wells and the plant account for approximately 30 per cent, main pipelines 20 per cent and the distribution network 50 per cent.

For society as a whole it is a question of assessing the overall net economic benefit that can be achieved by using geothermal district heating rather than another type of heat source; it is important that this assessment also includes the fact that geothermal resources are indigenous, not imported from abroad, and that there is a consequent saving in foreign currency. Society as a whole is more inclined to adopt a long-term point of view, while individual consumers are more inclined to adopt the short-term one. It is, however, internationally recognized that long-term benefits are best served by designing the tariff structure so as to provide energy-saving incentives both to the consumer and the heating service companies alike.

3.6.4 Operating cost

Along with estimating the capital investment cost, it is necessary to estimate the annual operating cost of the system. The annual operating cost comprises the following:

- cost of capital
- operation and maintenance
- ancillary energy cost (pumping/peak load).

For new systems, the largest part of the annual operating cost is commonly the cost of capital, that is, loan and interest pay-

ments. Typical figures for the division of the annual cost for new systems can be 75 per cent to the cost of capital, 15 per cent to operation and maintenance and 10 per cent to ancillary energy provision. For older systems, when loans have been repaid, plant replacement and maintenance become the major items in the annual cost.

3.6.5 Cost of improving heat efficiency of buildings

There are numerous ways in which buildings may be improved to make them more heat efficient. Such improvements will clearly result in reduced heating requirements irrespective of the method of heating employed. The following improvements merit consideration:

- Improved *thermal insulation* of the buildings: The potential savings in heating requirements attainable by improving thermal insulation are large, as shown in Table 3.3.
- Installation of *thermal windows* (double or triple glass): Single-pane windows are typically one of the largest single sources of heat loss in buildings. When they are replaced by double or even triple glass windows, heat losses are drastically reduced.
- Reduced *air infiltration*: Detailed attention to the proper finishing of buildings is needed to ensure their being wind- and weatherproof. Improvements to walls and weather-stripping of window frames, which bring about a decrease in outdoor air leakage into the building, can be very effective in reducing heat losses.
- A more efficient *heating system*: It is usually possible to make improvements in existing heating systems to improve the utilization of the thermal energy carried by the district heating fluid. In cases where the heat is transmitted to the building by means of water radiators, significant improvements may be achieved by increasing the size of the radiators and/or improved thermal and pressure-control features. This will result in added cooling of the water flowing through the building and will thus achieve better utilization efficiency.

3.6.6 Cost of alternative thermal energy sources

It is evident that substantial investment capital is needed to build a geothermal space heating system. The location of the

Table 3.3 Effect of insulation on typical house heating demand

Insulation criteria	Typical heating demand KWh/year	Insulation			Glazing type
		Wall	Roof	Floor	
Insignificant insulation	>100 000	1	1	—	Single
Poor insulation	55 300	2.5	5.0	2.5	Single
Icelandic building code	30 000	10.0	20.0	7.5	Double
Swedish building code	26 800	14.0	20.0	63.0	Triple
Super insulation	22 500	20.0	25.0	10.0	Quadruple

thermal plants is more or less fixed by the location of the geothermal field, and cannot be chosen to best suit the market. This calls for transmission of the energy from the resource to the user, and a distribution system within the market area. The system installations have a long economic lifespan, and operating supplies and labour requirements are minimal.

The annual operating cost, which determines the cost of the energy to the consumer, consists of fixed and variable costs. The fixed costs comprise fixed charges such as interest, depreciation, maintenance, overhead, taxes and profits. The variable costs, on the other hand, depend on the quantity of energy produced and delivered, and are typically less important.

This feature of the geothermal space heating systems is quite different from conventional fuel-fired heating plants. Such systems are characterized by a much lower investment cost per installed kilowatt than the geothermal system. This results in much lower capital charges, which are offset by the high fuel costs. It is estimated that approximately 60 to 70 per cent of the annual operating cost for the fossil-fuelled plant comes from the cost of fuel.

It may be advantageous to build a combined geothermal/fossil-fuelled space heating system, where use is made of the different features of the two systems. The geothermal system would then be designed with a capacity such as to meet the energy requirements dictated by a set outdoor temperature equal to T_0. The resulting load factor will be far higher than would be achieved if the plant was designed to meet the energy requirements of an outdoor temperature lower than T_0 (which may occur in only brief periods of the year). The fossil-fuelled system would then come into operation when the outdoor temperature falls below T_0 in order to provide the necessary back-up to the geothermal system during peak load periods. The energy produced over the year by the fossil-fuelled plant is thus only a small fraction of the total energy production.

3.7 TARIFFS

3.7.1 Sales policies and metering

A district heating company can be either a municipal company or a private company. In Iceland it is mostly municipally or state-owned, or a company owned by the consumers. Where possible, it is considered preferable that the company be given a monopoly on distributing hot water or heating energy within the district. This point is very important because of the influence of population density on the heating system economics. But often this is not possible, for political reasons or other considerations, such as ensuring competition with another energy-producing company in the same district.

Most companies use a tariff system with three components:

- A *connection charge* paid once and for all. The connection charge covers in part the investment of the company and in part the cost of establishing supply to the individual consumer.
- A *fixed charge* based on the heated volume of the house (building).
- A *variable charge* paid per heat-unit consumed, based either on the quantity of water used or the energy consumption.

There are many parameters to be considered when selecting a sales policy. Geothermal fluid can have a very different energy content from one area to another. The useful energy content in one cubic metre of water with a temperature of 80 °C is not equal to that of the thermally equivalent quantity (1.33 m^3) at 60 °C. In large systems individual customers may receive water at different inlet temperatures. This is caused by differences in piping and water flow rates at different locations in the system. Geothermal district heating systems in Iceland are generally not double-pipe systems, that is, the spent discharge is discarded into the sewage system or is injected back into the reservoir (re-injection).

The utilization of the water can also be very different from one consumer to another. The temperatures of the discharge from individual customers are different; it is often around 35 to 40 °C, but it can vary greatly because of differences in the general condition of individual heating systems, the lifestyle of the inhabitants and the insulation of the houses. Some consumers may also use the water for other purposes than space heating, such as the melting of snow, greenhouses, Jacuzzis and swimming pools.

There are three methods available for measuring sales of geothermal water to individual customers: volumetric measurement of the quantity of water used, maximum flow restriction, and energy metering. The one most commonly used is volumetric metering. In the first two methods the temperatures of the water received and returned are not measured, so the actual amount of energy used is unknown. Of course other methods can be and are applied for sales purposes, for instance, basing the charges on the cubic metre content or square metre floor area of the heated house.

It must be pointed out that meters used as a basis for charging the end user must receive accreditation from special officially appointed agenc(y)ies. The normal accreditation period is five years. The difference in metering-equipment prices is quite large. Volumetric meters for individual houses cost US$40 to US$50 (1998 prices) and their guaranteed lifespan is at least five years. Maximum flow restrictors cost US$150 (1998 prices) each, and the energy meter around US$350 to US$400 (1998 prices), with a guaranteed lifespan of up to ten years.

The flow restriction method is mostly used in rural areas in Iceland and for very small district heating systems. The user decides what maximum flow rate will be needed (in litres per minute) and the flow rate is restricted to that amount. The distribution of water consumption over the year is more uniform using this method than with the other two. Although this method limits the maximum flow, it does not encourage energy conservation. Experience in Iceland has shown that hot water consumption decreases by 20 to 30 per cent when metering is changed from restructure to volumetric metering. Maximum flow metering can in some cases be justified, for example where long distances between the geothermal field and the consumer require excess water flow through the pipelines to keep the temperature to the user at an acceptable level. The advantage of maximum flow metering is also that the peak load effect is restricted.

In all cases the geothermal reservoir is limited and energy savings must be considered. Water metering is thus the most cost-effective method on which to base sales policy. If the consumers are to pay for the energy they consume, energy metering is the only method by which that goal can be achieved. The drawback of energy metering is that this method will not encourage customers to use the water down to low-temperature values to save water, since most geothermal systems are once-through systems and the spent return water is discharged directly to drains. This effect can be countered by fixing the return temperature, whereby a suitable return temperature is decided upon and built into the energy meter. The consumers may demand this metering method if the energy price is sufficiently high and the temperature in the system varies greatly.

Sales policy depends largely on the amount and cost of the hot water available. If the geothermal reservoir is limited, the sales policy must provide energy savings incentives.

3.8 INTEGRATED USES

It is important to recognize that each individual geothermal resource (geothermal field, reservoir) should be considered finite as regards viable utilization. In most cases energy is taken from the resource in the form of hot water, which is disposed of as effluent, after the thermal energy has been extracted. It is therefore important, from the point of view of economics and efficient energy conversion, to ensure that the thermal energy content in the effluent is as low as is practicable.

Disparate direct-use applications require dissimilar temperature ranges for economic operation. These requirements are not necessarily compatible with the maxim of high conversion efficiency for a particular selected reservoir/application combination. An important concept for improving the conversion efficiency, and thus the economics of a particular combination, is the so-called 'integrated' or 'cascaded' use. This scheme consists of integrating several direct-use applications suitable for stepwise-reduced temperature levels into one system in parallel and in cascade (series), so ensuring that minimum practicable energy is thrown away with the effluent. Cogeneration and combined heat and power (CHP) systems are variations of the concept.

To cite an example, a reservoir yielding a 150 °C hot brine unsuitable for use, and requiring re-injection back into the reservoir for environmental and pressure maintenance reasons, could supply primary energy for the following applications using a suitably paralleled and cascaded system of heat exchangers:

- Electricity production via a binary system: primary inlet temperature 150 °C; outlet temperature 80 °C.
- District heating system: supply temperature 80 °C, return temperature 40 °C.
- Balneological unit comprising public baths and swimming pools: supply temperature 50 °C, return temperature 25 °C.
- De-icing service for parking lots and pavements: supply temperature 30 °C, return temperature 10 °C.

The overall temperature drop attainable in the primary fluid might be as high as 125 to 130 °C.

A type of cascading in common use is the following. Relatively high-temperature water of, say, 70 to 80 °C is passed through buildings equipped with radiator heating systems. The return water from these systems (by then cooled to about 35 to 40 °C) is then piped to buildings fitted with radiant panel heating systems. A cascaded system of this type is, for example, in use in the city of Melun in France. With this type of use, the utilization efficiency of the geothermal energy is extremely high.

Typical direct-use applications, which merit consideration for cascaded use, are space heating, diverse balneological uses, greenhouse and soil heating, de-icing, production of domestic hot water, air conditioning, heat pump applications, assorted low-temperature washing applications, low-temperature drying, and fish farming.

3.9 ENVIRONMENTAL CONSIDERATIONS

3.9.1 Chemical pollution

Geothermal energy is relatively free from pollution problems. Even power stations employing high-temperature steam discharge less carbon dioxide to the atmosphere than fossil-fuelled power stations, and such emissions are negligible when low-temperature fluids are utilized. The gases that may conceivably cause problems in low-temperature applications are hydrogen sulfide and, in specific anomalous situations, ammonia and mercury. Attention has also been drawn to the possible ill effects of radiation from radon, although others have suggested that this may be beneficial in the small doses normally encountered.

Some chemical constituents of geothermal water may need to be monitored in case their concentrations exceed permitted pollution limits, but limits for water may vary according to the uses to which it is being put. Boron is, for instance, extremely harmful to vegetation and has to be avoided at all costs in irrigation water. Trace metals such as mercury, which are poisonous to organisms, are all the more insidious since they accumulate in tissue, and thus cause damage over the passage of time, and concentrate upwards in food chains. Direct removal of hazardous substances from the water has proved to be difficult, especially for boron, and such methods are generally not in use, although recent claims suggest that artificial wetlands may remove boron quite effectively. The only effective way of getting rid of hazardous chemicals is re-injection.

3.9.2 Thermal pollution

In difficult situations, geothermal effluent at a temperature of 35 to 40 °C may have to be discharged into streams, rivers or lakes. Many organisms are highly sensitive to temperature changes, and permanent changes of 1 °C or less may cause drastic changes in existing ecosystems. In cold countries this may occasionally be used to advantage, but more frequently geothermal energy producers are required to dispose of their spent fluids in a different manner, for example, by pre-cooling in ponds, by re-injection, and in some instances by constructing trenches or pipelines to the ocean.

3.9.3 Physical effects

Geothermal projects cause the same kind of disruption as other civil engineering projects of the same size and complexity. The locations of excavations, and the siting of bore holes and roads will have to be taken into account. Soil and plant erosion, which may cause changes in ecosystems, has to be watched. More specifically for geothermal projects, subsidence and possibly induced seismicity are potential effects that have to be watched. There is considerable noise involved during drilling and especially during discharge of high-temperature wells. Protection has to be ensured for people, and some permanent noise-reduction measures may need to be taken, such as the use of silencers during discharge. Last, but not least, geothermal projects may cause permanent changes in surface features, such as the disappearance of hot springs and the appearance of fumaroles, which, apart from aesthetic considerations, may affect the local tourist industry.

3.9.4 Social and economic considerations

The construction of geothermal installations often involves a temporary increase in employment and the import of an outside workforce calling for various services. The building of new roads may 'open up' areas and possibly increase tourism, especially if natural geothermal manifestations are left intact. Such changes may permanently affect traditional industries.

RECOMMENDED LITERATURE

ASHARE. 2000. *Handbook of HVAC Systems and Equipment, IP Edition: Chapter 11: District Heating and Cooling,* Atlanta, Ga., American Society of Heating, Refrigeration and Air Conditioning Engineers.

EUROPEAN INSTITUTE OF ENVIRONMENTAL ENERGY (EIEE). 1999. *District Heating Handbook,* Denmark, Herning. (eiee@eiee.dk). (Available in Chinese, English, Estonian, German, Latvian, Lithuanian, Polish, Russian.)

INTERNATIONAL DISTRICT HEATING ASSOCIATION. 1983. *District Heating Handbook,* 4th ed. International District Heating Association.

LUND, J.; LIENAU, P.; LUNIS, B. 1998. *Geothermal Direct-Use Engineering and Design Guidebook.* 3rd ed. Geo-Heat Center, Oregon Institute of Technology.

PIATTI, A.; PIEMONTE, C.; SZEGÖ, E. 1992. *Planning of Geothermal District Heating Systems.* Dordrecht, Kluwer Academic Publishers.

SELF-ASSESSMENT QUESTIONS

1. What is the minimum temperature needed to supply a heating system that uses radiators, without having to resort to booster facilities or heat pumps?
(a) 40 °C; (b) 60 °C; (c) 80 °C

2. Comment on the useful temperature range of geothermal fluids suitable for use in space or district heating.

3. There is sometimes considerable distance between the site of extraction of a hot geothermal fluid and the location where it is used. This distance is in some cases as much as 20 to 30 km. Beyond a certain distance the transmission of hot water is no longer economically viable. What are the main parameters that govern the economic transmission distance? What is the maximum acceptable distance for a large direct-use geothermal heating system and again for a similar small one?

4. The plastic piping (polypropylene and polybutylene) used for transporting hot water costs considerably less than carbon steel pipes. Comment on the drawbacks of using these plastic pipes.

5. In order to extract as much heat as possible from the geothermal fluid, the temperature of the heating water leaving the houses (return temperature) must be made as low as possible. Assuming that the water enters the house at 80 °C and the heating is by radiators, at what temperature should the water ideally leave the house? At what temperature will it leave the system if a cascaded scheme of the radiator/radiant-panel-type heating is used?

6. A few years ago the Reykjavik (Iceland) District Heating Service recommended that new radiator heating systems be of an 80/40/20 design (see explanation in discussion of Figure 3.9). Because of an expansion of the service area the mean supply temperature of the geothermal water is somewhat lower than before. To meet this the District Heating Service is now recommending the installation of somewhat larger radiators than before, defined by a 75/35/20 design. Discuss the performance of a radiator of this design in an area where the geothermal water supply temperature is 80 °C.

7. Water at 100 °C is passed through a pipe of 18 in diameter insulated with 2-in thick plastic-sheathed urethane insulation. The mean flow velocity is 0.5 m/s. Because of heat losses, the water temperature on leaving the pipe has dropped to about 84 °C. Comment on the effect on temperature loss of raising the flow velocity of the water.

8. On leaving a cascade geothermal heating the temperature may be as low as 20 °C. If the total dissolved chemical content of the water is low, as in Iceland, this water could be used again, for instance in irrigation. What is the maximum permissible concentration of chemicals for use in irrigation?
(a) 5 mg/kg
(b) 50 mg/kg
(c) 1,000 mg/kg
(d) 10,000 mg/kg

9. Comment on what can be deduced from an annual load duration curve.

10. A well produces water under pressure at a temperature of 105 °C, which contains a small quantity of dissolved chemicals (40 ppm of chlorides and 100 ppm of silica), as well as a small quantity of oxygen. The decision is taken to install a geothermal space-heating system. Which of the following two solutions should be adopted and why?
(a) The geothermal water is piped directly from the well to the houses, without undergoing chemical treatment or going through any intermediate equipment.
(b) The geothermal water is circulated only in a primary system; a secondary system heats the houses; in between the two are heat exchangers.

11. In order to maintain production in a hot water well that is not artesian (self-flowing), a pump must be used. Comment on the types of pumps normally used and the main differences between them.

12. What payback period is typically envisaged for the capital investment in a geothermal heating system?
(a) Five to ten years
(b) Ten to fifteen years
(c) Fifteen to twenty years
(d) Twenty to thirty years

13. What are the elements of the operating costs of a geothermal heating system? Specify approximately what percentage each of these elements represents in the total cost of the system.

14. A geothermal district heating system has to be gradually amplified to incorporate new apartments requiring heating. Consequently there is an increase in the demand for heat. The flow rate from the geothermal wells, however, cannot be increased, as this would damage the reservoir. If it is not planned to resort to conventional fuel-fired plants, what steps should be taken to meet, at least in part, the additional heat demand?

15. A district heating company has four potential geothermal areas (**I**, **II**, **III** and **IV** in Table 3.4) available for the construction of a district heating system; three are low-temperature areas and one a high-temperature area. The short-term (twenty to thirty years) and long-term (>fifty years) capacities of the areas have been estimated. These and the estimated lengths of pipelines from the source to the distribution system needed are also shown in Table 3.4.

Table 3.4 Geothermal areas: short- and long-term capacities

Area	I	II	III	IV
Nature	Low-temp.	Low-temp.	Low-temp.	High-temp.
Short-term capacity l/s	200	150	1 050	
MW$_t$	73	30	196	400
Long-term capacity l/s	150	75	500	
MW$_t$	44	15	93	300
Pipeline, km	<1	<1	15	27

The chemical analyses of the fluids from the geothermal areas and of the cold groundwater (**V**) used in the case of a heat-exchange utilization are presented in Table 3.5.

Table 3.5 Chemical analyses of fluids from geothermal areas (in mg/kg except for pH)

No.	I	II	III	IV	V
Temp °C	130	100	81	290	4
Water phase					
pH	9.4	9.6	9.4	7.1	7.3
SiO_2	167	130	71	846	19
B	0.05	0.03	0.04	1.8	0.03
Na	61	48	42	77	7.4
K	2.6	1.2	1.1	15	0.8
Ca	2.6	1.5	2.1	0.2	6.1
Mg	1.1	0.2	0.06	0.01	4.5
HCO_3	13	55	18	40	61
CO_3	28	20	19	1	2
SO_4	19	14	17	3.8	8.2
H_2S	0.4	0.3	0.3	126	0
Cl	35	19	11	66	8.6
F	1.1	0.4	0.7	0.9	0.1
TDS	328	250	211	1,178	104
Steam phase					
CO_2	—	—	—	1,790	—
H_2S	—	—	—	1,538	—
H_2	—	—	—	115	—
CH_4	—	—	—	0.3	—
N_2	—	—	—	9.2	—

(a) In view of the chemical composition, how would you utilize the fluid from the different areas (directly or indirectly with heat exchangers)?
(b) There already exists a distribution system for a large part of the area. The low-temperature areas are being utilized to their short-term capacities. You are ready to reduce their production and start producing with the aid of the high-temperature area, **IV**. How would you then organize your distribution system?

16. In two geothermal district systems, 'a' and 'b', the chloride concentration has developed as shown in Figures 3.11 and 3.12, respectively.

System 'a' is inland close to a lake, whereas system 'b' is on a peninsula close to the sea. Suggest reasons for the chloride variations, discuss potential problems and suggest ways of dealing with the situation in each case.

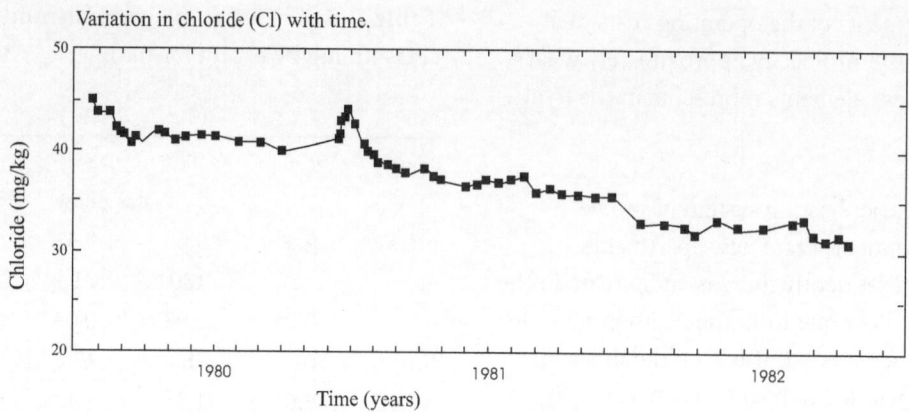

Figure 3.11 Chloride variation in system 'a' with time

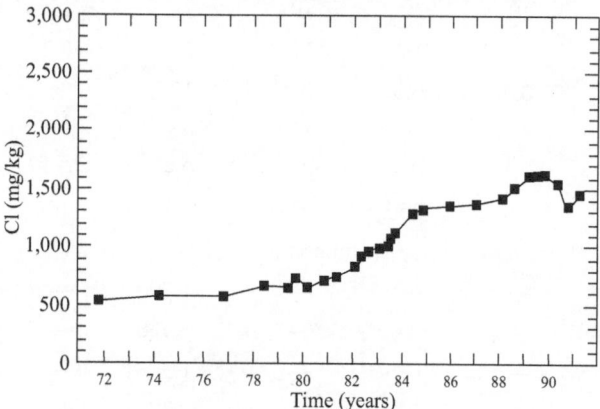

Figure 3.12 Chlorine variation in system 'b' with time

17. Refer to the example given in self-assessment question 15, Table 3.4.

Area **I** has already been exploited and a further 25 MWt are needed for the district heating system. Four alternatives are to be considered.

(a) Exploit area **II** to its full short-term capacity.

(b) Start exploiting area **III** but leave area **II** intact for the time being.

(c) Exploit area **II** to its full long-term capacity and start exploiting area **III**.

(d) Start exploiting area **IV** but leave areas **II** and **III** intact.

Area **II** is situated in the middle of a suburb of a fairly large town, which is the main beneficiary of the enlargement of the heating system.

Area **III** is located in a small town about 15 km away from the large town. It is a separate community and has already exploited a small part of the geothermal area for its own benefit and therefore does not stand to gain from the enlargement of the system.

Area **IV** is in a remote place with no neighbouring community that is immediately affected by the engineering activity.

Use this information and that in the example given in self-assessment question 15 to prepare an environmental impact statement where the pros and cons of the four alternatives are discussed.

ANSWERS

1. (b) Roughly 60 °C.

2. Geothermal fluids ranging in temperature from 125 to 35/40 °C are generally used for this purpose. With adequate equipment we can of course also utilize fluids at a higher temperature, but it is usually more convenient to exploit the latter in other applications.

3. The main parameters governing the economic transmission radius of a geothermal fluid are:
(a) The global cost of the entire transmission system, that is, all material, equipment and instrumentation from the point of extraction of the fluid to its entry into the utilization plant.
(b) The operating costs, including thermal energy loss due to heat losses in the pipeline.
  The economic transmission radius may be of the order of magnitude of 30 km for a large plant and 2 to 3 km for a small plant.

4. The plastic pipes cannot be used to transport water with a temperature above 90 °C. They also permit the ingress of atmospheric oxygen by diffusion, and the oxygen can cause corrosion in the equipment.

5. The temperature of the water leaving the radiator system will be no lower than about 40 °C. If we add radiant panels to create a cascade scheme we will recover a further roughly 20 °C. Figure 3.5 shows a number of possible heating schemes.

6. A supply temperature higher than the radiator design value results in a higher initial rate of heat transfer from the radiator to the surrounding air than called for by the design. This has the effect that the room temperature control (see Section 3.5.5) reduces the water flow rate below the design value. The result is added cooling of the geothermal water, so that it leaves the radiator at a temperature below 35 °C.

7. If we increase the flow rate, the velocity of the water in the pipes will also increase, and we will thus have a minor loss of heat from the pipes. For example, at a velocity of 1.5 m/s, the temperature will drop to about 95 °C after 25 km.

8. (c) Roughly 1,000 mg/kg.

9. The curve shows that, to effect the required heating wholly by geothermal means, we would have to install a heating system of a high enough capacity to meet the heat requirements of the coldest days. These are very few; a system of such a high capacity would thus operate at a very low annual load factor. The curve also shows that the normally short duration of the load-peak permits the economical use of fossil-fuelled plants in combination with geothermal to improve the load factor.
  An annual load duration curve shows the statistical distribution of daily temperature in a particular climate and the number of days that have a mean daily temperature lower than any given temperature value. The area below the curve is a measure of the heat necessary to ensure comfort.

10. Solution (a) cannot be adopted because of the high temperature of the water. The correct solution is (b). Alternatively, the temperature of the geothermal water could be decreased (to about 80 °C) by mixing it with the water returning from the house radiators and the mixed water piped directly to the houses (see the scheme in Figure 3.5).

11. Shaft-driven and submersible pumps are commonly used in geothermal applications. The shaft-driven pumps have a depth limit because of torsional stress on the shaft and difficulty in axial alignment of pump impellers due to thermal expansion and load-induced elongation of the shaft. The submersible pumps are sensitive to high temperatures.

12. (c) Usually the payback period varies from fifteen to twenty years.

13. The operating costs of a geothermal heating system comprise:
(a) cost of capital (75 per cent);
(b) operation and maintenance (15 per cent)
(c) ancillary energy costs (for pumping/peak load) (10 per cent)

14. We could either:
(a) increase the efficiency of the heating system, or
(b) improve the thermal insulation in the buildings that are being heated.

15. The following answer suggests itself:
(a) Fluids from areas **I**, **II** and **III** are innocuous and can be piped directly into the system. There should be no danger of deposition and the small H_2S concentration should prevent O_2 from causing corrosion. The fluid from area **IV** is unsuitable for direct utilization and cannot be piped directly into the heating system. It needs to be utilized via heat exchangers in which cold groundwater is heated and piped into the system. As there is always a danger of oxygen entering and causing corrosion, it is a good idea to dose the heated groundwater with a small amount of H_2S from the geothermal gas to prevent this. Special care should be taken that there is no mixing of geothermal water and cold groundwater across the exchangers as this can cause magnesium silicate deposition.

(b) The silica concentration of fluids from areas **I**, **II** and **III** is relatively high as is to be expected from geothermal water, but the magnesium concentration of the cold groundwater (**V**) is relatively high. There is therefore a danger of the formation of magnesium silicate deposits if the two are mixed. Thus, it is sensible to organize the distribution in such a way that the two do not mix, i.e. by distributing the heated groundwater and the piped geothermal water to separate areas.

16. In both cases there are clear signs of mixing of geothermal water with a different and, most probably, cooler water. The accompanying cooling of the systems may not show immediately because the hot rock may heat up the intruding water. Therefore the chemical changes may give an early warning that cooling is imminent. In geothermal district system 'a' this could be lake water with a low chloride concentration (5 to 10 ppm), but in district system 'b' it is most probably seawater with chloride concentration ≈19,000 ppm.

The change in chloride concentration in system 'a' upon exploitation is drastic. Assuming an initial chloride concentration of 50 ppm for the geothermal water and 5 to 10 ppm for the lake water, the portion of geothermal water in the mixture is drawn down to between one-third and a quarter of the total flow after about three years' production. Thus the reservoir cannot withstand this level of production and it has to stop or be drastically reduced.
New geothermal aquifers need be found and exploited. If this does not happen, alternative ways of space heating will have to be considered.

The changes in system 'b' can be considered as warning signs. Assuming an initial chloride concentration of 500 ppm for the geothermal water and 19,000 ppm for seawater, the increase of salinity represents about 5 per cent gradual increase in the portion of seawater over twenty years. Thus drastic cooling is not likely to occur, but careful monitoring is essential and it would not be advisable to drill additional wells close to this area if it can be avoided. Another consequence of such mixing is a possible increase in the corrosion potential of the produced fluid. This has to be watched and appropriate measures taken.

17. The relative advantages and disadvantages for the three areas, considered from the environmental point of view, are listed in Table 3.6.

In any environmental statement, the alternatives – that is, no additional heating, the use of other forms of heating, and the question as to whether the estimated need for

Table 3.6 Advantages and disadvantages for the three areas considered

Area no.	Relative advantages	Relative disadvantages
II	Short pipeline, no gas emissions. Little noise. In large community and work not likely to affect community socially. No surplus effluent.	Small field, danger of over-exploitation with possible disappearance of natural manifestations and mixing of hot and cold aquifers. Situated in the middle of the community and construction would interfere physically.
III	No gas emissions. Little noise. No surplus effluent. Construction not likely to interfere much physically. No danger of over-exploitation.	Fairly long pipeline. Small local community might be significantly affected socially.
IV	No local community to interfere with. No danger of over-exploitation.	Long pipeline. Gas emissions. Noise. Surplus effluent needs to be disposed of. Possible interference with undisturbed nature.

additional heating is realistic – would have to be considered very carefully. Leaving out economic factors, the choice would be between alternatives (b) and (c), but the fact that alternative (b) involves construction in one place only would favour alternative (b) because the disadvantages connected with production from area **III** would combine with those connected with production from area **II**. If, on the other hand, the study of the alternatives reveals that there was a real need for 10 to 15 MW_t then alternative (a), with a production of that order of magnitude, would seem attractive.

Alternative (c) would in all cases come last.

CHAPTER 4

Space Cooling

Kevin D. Rafferty
Geo-Heat Center, Klamath Falls, Oregon, USA

AIMS

1. To show that the heat content of geothermal fluids can be exploited as a source of energy for air conditioning and industrial refrigeration plants; that the existing technology for absorption machines (lithium bromide/water and water/ammonia cycles) can be utilized in geothermal applications; and that machines of this type are available commercially for conversion to the geothermal sector.

2. To provide the basic configurations of the absorption equipment and explain their operating principles in simple terms.

3. To provide the elements for comparing geothermal and electrical plants, and for analysing the economic feasibility of utilizing a geothermal resource in space cooling. Also, to show that the efficiency of geothermal plants are highly dependent on the temperature of the resource.

4. To emphasize that the additional costs required for adapting the equipment to a geothermal resource are often high, that the payback period is long, and consequently that careful and accurate economic analyses must be made beforehand.

OBJECTIVES

When you have completed this chapter, you should be able to:

1. Describe the main features of the absorption cycles used for air conditioning and industrial refrigeration in geothermal applications (lithium bromide/water and water/ammonia machines).

2. Discuss the performance, in diverse operating conditions, of the absorption machines that use geothermal fluids.

3. Discuss the economic feasibility of geothermal absorption machines, comparing them with electric cooling machines.

4. Give a brief description of the machines available commercially.

5. Discuss the effects of the geothermal resource temperature on the operating efficiency of the machines.

This material was originally published by the author as part of the *Geothermal Direct Use Engineering and Design Guidebook* by the Geo-Heat Center, Klamath Falls, Oregon, USA.

4.1 INTRODUCTION

Space cooling can be accomplished from geothermal energy by means of a process (absorption cycle) in which the refrigeration effect is achieved through the use of two fluids and some quantity of heat input, rather than electrical input as in the more familiar vapour compression cycle. The vapour compression and absorption refrigeration cycles both accomplish the removal of heat through the evaporation of a refrigerant at a low pressure and the rejection of heat through the condensation of the refrigerant at a higher pressure. The method of creating the pressure difference and circulating the refrigerant is the primary difference between the two cycles. The vapour compression cycle employs a mechanical compressor to create the pressure differences necessary to circulate the refrigerant. In the absorption system, a secondary fluid or absorbent is used to circulate the refrigerant. Because the temperature requirements for the cycle fall into the low-to-moderate temperature range,

and there is significant potential for electrical energy savings, absorption would seem to be a good prospect for geothermal application.

Absorption machines are commercially available today in two basic configurations. For applications above 0 °C (primarily air conditioning), the cycle uses lithium bromide as the absorbent and water as the refrigerant. For applications below 0 °C, a water/ammonia cycle is employed with water as the absorbent and ammonia as the refrigerant.

4.2 AIR CONDITIONING

4.2.1 Lithium bromide/water cycle machines

Figure 4.1 shows a diagram of a typical lithium bromide/water machine (Li Br/H$_2$O). The process occurs in two vessels or shells. The upper shell contains the generator and condenser; the lower shell, the absorber and evaporator.

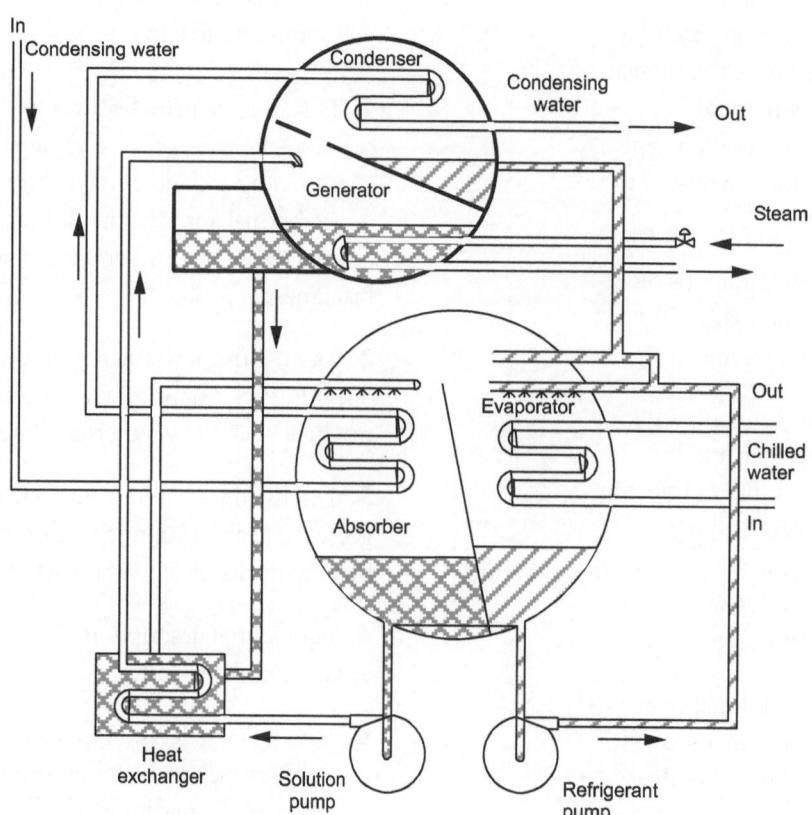

Figure 4.1 Diagram of two-shell lithium bromide cycle water chiller
Source: ASHRAE, 1983.

Heat supplied in the generator section is added to a solution of Li Br/H_2O. This heat causes the refrigerant, in this case water, to be boiled out of the solution in a distillation process. The water vapour that results passes into the condenser section where a cooling medium is used to condense the vapour back to a liquid state. The water then flows down to the evaporator section where it passes over tubes containing the fluid to be cooled. By maintaining a very low pressure in the absorber-evaporator shell, the water boils at a very low temperature. This boiling causes the water to absorb heat from the medium to be cooled, thus lowering its temperature. Evaporated water then passes into the absorber section where it is mixed with a Li Br/H_2O solution that is very low in water content. This strong solution (strong in Li Br) tends to absorb the vapour from the evaporator section to form a weaker solution. This is the absorption process that gives the cycle its name. The weak solution is then pumped to the generator section to repeat the cycle.

As shown in Figure 4.1, there are three fluid circuits that have external connections:

- generator heat input
- cooling water
- chilled water

Associated with each of these circuits is a specific temperature at which the machines are rated. For single-stage units, these temperatures are: 82.7 kPa steam (or equivalent hot water) entering the generator, 29 °C cooling water, and 7 °C leaving chilled water (ASHRAE, 1983). Under these conditions, a coefficient of performance (COP) of approximately 0.65 to 0.70 could be expected (ASHRAE, 1983). The COP can be thought of as a sort of index of the efficiency of the machine. It is calculated by dividing the cooling output by the required heat input. For example, a 1.76 MW absorption chiller operating at a COP of 0.70 would require: (1.76 × 3.60 × 10^3 MJ/h/MW) divided by 0.70 = 9,050 MJ/h heat input. This heat input suggests a flow of 4,100 kg/h of 82.7 kPa steam, or 63.6 l/s of 116 °C water with a 9.5 °C ΔT. (Note that US equipment is usually sized in tons of cooling capacity [1 ton = 3.52 kW].)

Two-stage machines with significantly higher COPs are available (ASHRAE, 1983). However, temperature requirements for these are well into the power generation temperature range (177 °C). As a result, two-stage machines would probably not be applied to geothermal applications.

4.2.2 Performance

Based on equations that have been developed (Christen, 1977) to describe the performance of a single-stage absorption machine, Figure 4.2 shows the effect on COP and capacity (cooling output) versus input hot-water temperature. Entering hot water temperatures of less than 104 °C result in a substantial reduction in equipment capacity. The steep drop-off in capacity with temperature is related to the nature of the heat input to the absorption cycle. In the

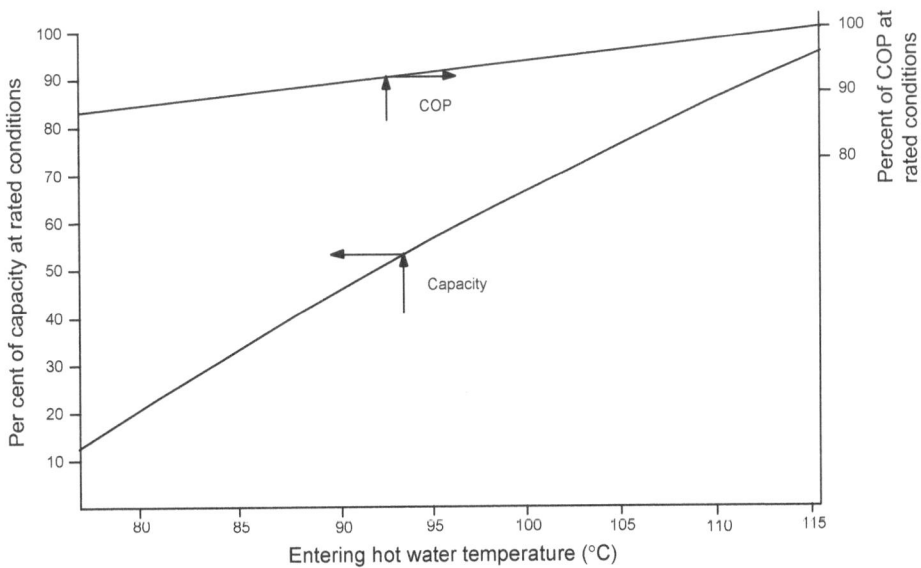

Figure 4.2 Capacity of lithium bromide absorption chiller
Source: Christen, 1977.

generator, heat input causes boiling to occur in the absorbent/refrigerant mixture. Because the pressure is fairly constant in the generator, this fixes the boiling temperature. As a result, a reduction in the entering hot water temperature causes a reduction in the temperature difference between the hot fluid and the boiling mixture. Because heat transfer varies directly with temperature difference, there is a nearly linear drop-off in absorption refrigeration capacity with entering hot water temperature. In the past few years, one manufacturer (Yazaki, undated) has modified small capacity units (7 to 35 kW) for increased performance at lower inlet temperature. However, low-temperature modified machines are not yet available in large outputs, which would be applicable to institutional- and industrial-type projects. Although COP and capacity are also affected by other variables such as condenser and chilled water temperatures and flow rates, generator heat input conditions have the largest impact on performance. This is a particularly important consideration with regard to geothermal applications.

Because many geothermal resources in the 116 °C and above temperature range are being investigated for power generation using organic Rankine cycle (ORC) schemes, it is likely that space-conditioning applications would see temperatures below this value. As a result, chillers operating in the 82 to 110 °C range would (according to Figure 4.2) have to be (depending on resource temperature) between 20 and 400 per cent oversized, respectively, for a particular application. This would tend to increase capital cost and decrease payback when compared to a conventional system.

An additional increase in capital cost would arise from the larger cooling tower costs that result from the low COP of absorption equipment. The COP of single-effect equipment is approximately 0.7. The COP of a vapour compression machine under the same conditions may be 3.0 or higher. As a result, for each unit of refrigeration, a vapour compression system would have to reject 1.33 units of heat at the cooling tower. For an absorption system, at a COP of 0.7, 2.43 units of heat must be rejected at the cooling tower. This results in a significant cost penalty for the absorption system with regard to the cooling tower and accessories.

In order to maintain good heat transfer in the generator section, only small ΔTs can be tolerated in the hot-water flow stream. This is because the machines were originally designed for steam input to the generator. Since heat transfer from the condensing steam is a constant-temperature process, in order to have equal performance the entering hot

water temperature would have to be above the saturated temperature corresponding to the inlet steam pressure at rated conditions, so as to allow for some ΔT in the hot-water flow circuit. In boiler-coupled operation, this is of little consequence to operating cost, but because ΔT directly affects flow rate, and thus pumping energy, this is a major consideration in geothermal applications.

For example, assuming a COP of 0.54 and 8 °C ΔT on the geothermal fluid, 750 kPa pump head and 65 per cent wire-to-water efficiency at the well pump, approximately 60 W/kW pumping power would be required. This compares to approximately 140 to 170 W/kW for a large centrifugal machine (compressor consumption only).

The small ΔT and high flow rates also point out another consideration with regard to absorption chiller use in space-conditioning applications. Assume a geothermal system is to be designed for heating and cooling a new building. Because the heating system can be designed for rather large ΔTs in comparison to the chiller, the incremental cost of the absorption approach would have to include the higher well and/or pump costs to accommodate its requirements. A second approach would be to design the well for space-heating requirements and use a smaller absorption machine for base load duty. In this approach, a second electric chiller would be used for peaking. In either case, capital cost would be increased.

4.2.3 Large tonnage equipment costs

Figure 4.3 presents some more general cost information on large size (>350 kW) cooling equipment for space-conditioning applications. The plot shows the installed costs

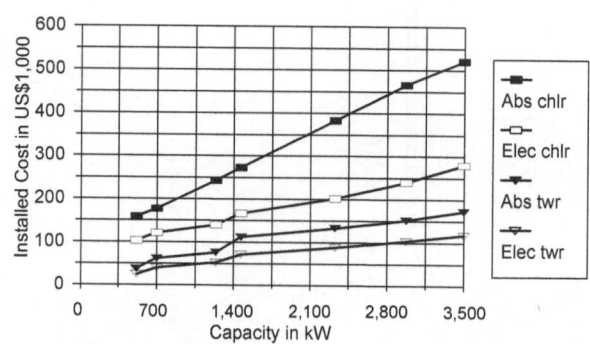

Figure 4.3 Chiller and auxiliary equipment costs: electric and absorption
Source: Means, 1996.

for both absorption chillers (Abs chlr), centrifugal chillers (Elec chlr), and auxiliary condenser equipment (cooling tower, cooling water pumps and cooling water piping) for both absorption chillers (Abs twr) and centrifugal chillers (Elec twr). As shown, both the chiller itself and its auxiliary condenser equipment costs are much higher for the absorption design than for electric-driven chillers. These are the primary capital cost differences that a geothermal operation would have to compensate for in savings.

4.2.4 Small tonnage equipment

To our knowledge, there is only one company (Yazaki, undated) currently manufacturing small size (<70 kW) lithium bromide refrigeration equipment. This firm, located in Japan, produces equipment primarily for solar applications. Currently, units are available in 4.6, 7.0, 10.5, 17.6, 26.3 and 35.2 kW capacities. These units can be manifolded together to provide capacities of up to 175 kW.

Because the units are water-cooled chillers, they require considerably more mechanical equipment for a given capacity than the conventional electric vapour compression equipment usually applied in this size range. In addition to the absorption chiller itself, a cooling tower is required. The cooling tower, which is installed outside, requires interconnecting piping and a circulation pump. Because the absorption machine produces chilled water, a cooling coil and fan are required to deliver the cooling capacity to the space. Insulated piping is required to connect the machine to the cooling coil. Another circulating pump is required for the chilled water circuit. Finally, hot water must be supplied to the absorption machine. This requires a third piping loop.

Figure 4.4 was developed to evaluate the economic merit of small absorption equipment compared to conventional electric cooling. This plot compares the savings achieved through the use of the absorption equipment to its incremental capital costs over a conventional cooling system. Specifically, the figure plots cost of electricity against simple payback in years for the five different-sized units. In each case, the annual electric cost savings of the absorption system (at 2,000 full-load hours per year) are compared with the incremental capital cost of the system to arrive at a simple payback value. The conventional system to which absorption is compared in this case is a rooftop package unit. This is the least expensive conventional system available. A comparison of the absorption approach to more sophisticated cooling systems (VAV, four-pipe chilled water, etc.) would yield much more attractive payback periods.

The plot is based on the availability of geothermal fluid of sufficient temperature to allow operation at rated capacity (88 °C or above). In addition, other than piping, no costs for

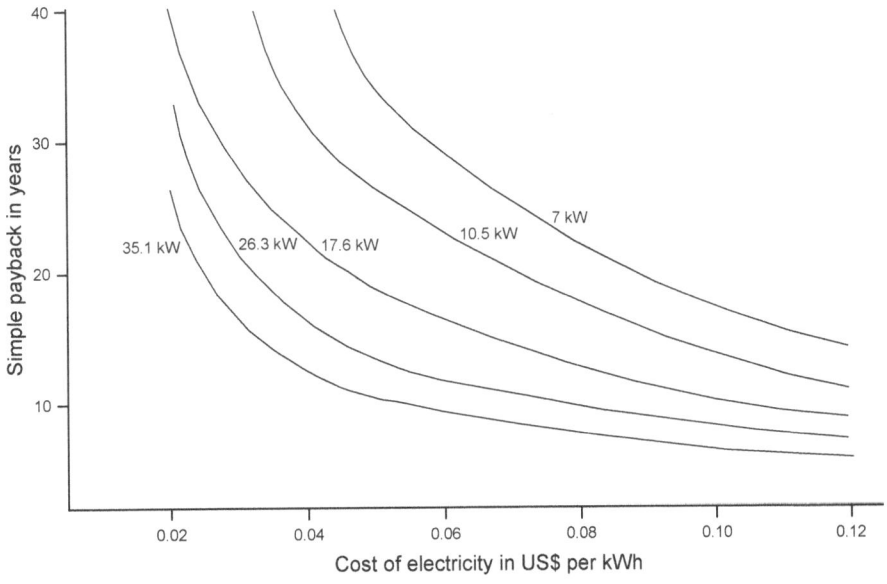

Figure 4.4 Simple payback on small absorption equipment compared to conventional rooftop equipment

geothermal well or pumping are incorporated. Only cooling-equipment-related costs are considered. As a result, the payback values in Figure 4.4 are valid only for a situation in which a geothermal resource has already been developed for some other purpose (such as for space heating or aquaculture), and the only decision at hand is that of choosing between electric and absorption cooling options.

Figure 4.4 also shows that the economics of small-size absorption cooling are attractive only in cases of 17.6 to 35.2 kW capacity requirements and more than US$0.10 kW/h electrical costs. Figure 4.4 is based on an annual cooling requirement of 2,000 full-load hours per year. This is on the upper end of requirements for most geographical areas. To adjust for other annual cooling requirements, simply multiply the simple payback from Figure 4.4 by actual full-load hours and divide by 2,000.

The performance of the absorption cooling machine was based on nominal conditions in order to develop Figure 4.4. It should be noted that, as with the larger machines, performance is heavily dependent upon entering hot water temperature and entering cooling water temperature. Ratings are based on 88 °C entering hot water, 29 °C entering cooling water and 9 °C leaving chilled water. Flow rates for all three loops are based upon a 5 °C ΔT.

Figure 4.5 illustrates the effect of entering hot water temperature and entering cooling water temperature on small machine performance. At entering hot water temperatures of less than 82 °C, substantial de-rating is necessary. For a preliminary evaluation, the 29 °C cooling water curve should be employed.

4.3 COMMERCIAL REFRIGERATION

4.3.1 Water/ammonia cycle machines

Most commercial and industrial refrigeration applications involve process temperatures of less than 0 °C and many are −18 °C. As a result, the lithium bromide/water cycle is no longer able to meet the requirements, because water is used for the refrigerant, and a fluid that is not subject to freezing at these temperatures is required. The most common type of absorption cycle employed for these applications is the water/ammonia cycle, in which water is the absorbent and ammonia is the refrigerant.

Use of water/ammonia equipment in conjunction with geothermal resources for commercial refrigeration applications is influenced by some of the same considerations as space cooling applications. Figure 4.6 illustrates the most important of these. As refrigeration temperature is reduced, the required hot water input temperature is increased. Because most commercial and industrial refrigeration applications occur at temperatures below 0 °C, required heat input temperatures must be at least 110 °C. It should also be remembered that the required evaporation temperature is 6 to 8 °C below the process temperature. For example, for a −7 °C cold storage application, an evaporation temperature of −15 °C would be required.

Figure 4.6 suggests that a minimum hot water temperature of 135 °C would be required, but there are not many

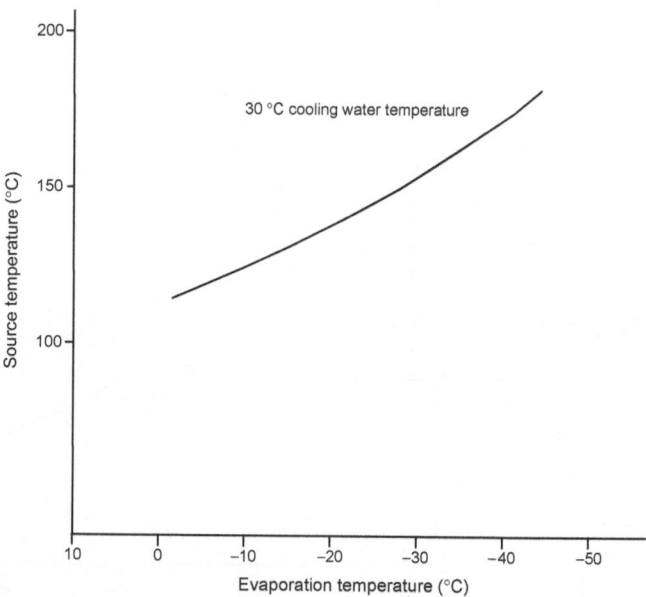

Figure 4.6 Required resource temperatures for water/ammonia absorption equipment
Source: Hirai, 1982.

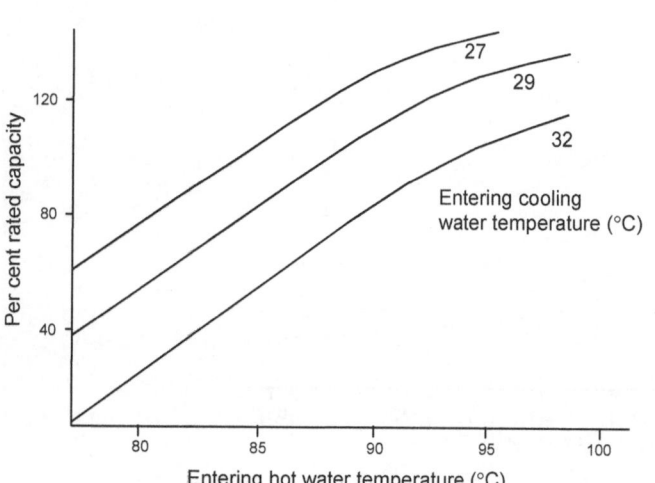

Figure 4.5 Small tonnage absorption equipment performance

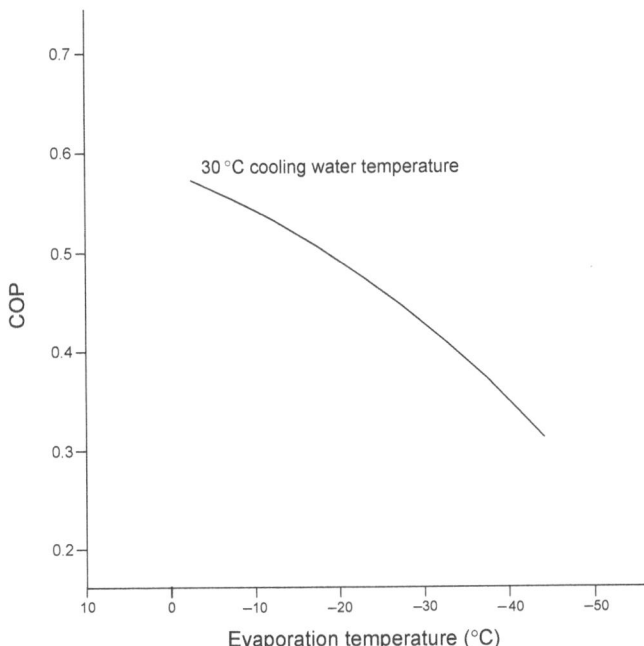

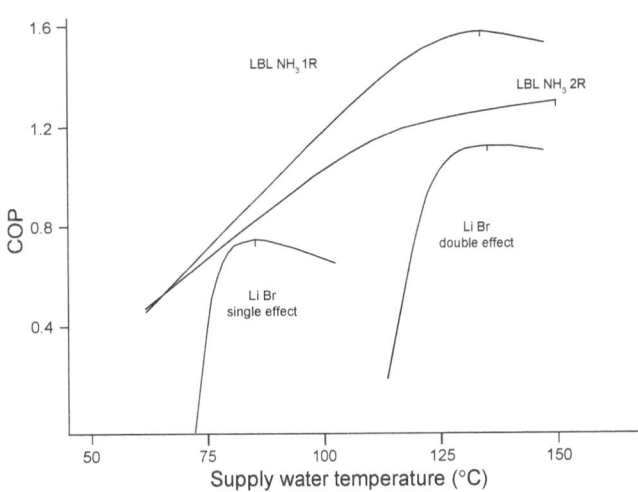

Figure 4.8 Water/ammonia single and double effect regenerative cycle performance
Source: Wahlig, 1984.

Figure 4.7 The COP for water/ammonia absorption equipment in refrigeration applications
Source: Hirai, 1982.

geothermal resources at this temperature, and where they do exist they are more likely to be utilized for small-scale power generation, unless cascade uses are envisaged.

Figure 4.7 indicates another consideration for refrigeration applications. The COP for most applications is likely to be less than 0.55, with the result that hot-water flow requirements are substantial. In addition, the cooling tower requirements, as discussed above, are much larger than for equivalently sized vapour compression equipment.

4.4 ABSORPTION RESEARCH

Studies at the Lawrence Berkeley Laboratory (LBL) (Wahlig, 1984) resulted in significantly improved absorption cycle performance. Researchers at LBL, working to improve absorption cycle performance for solar application, developed two advanced versions of the water/ammonia machine. Water/ammonia was chosen as the working fluid pair in order to allow the use of an air-cooled condenser for potential heat-pump operation.

The two cycles (1R and 2R) were developed for a single-effect regenerative cycle and double-effect regenerative cycle, respectively. As shown in Figure 4.8, these two cycles show substantially higher COP, over a much broader range of

generator input temperatures, than the conventional lithium bromide cycles. The superior performance is achieved by operating the chiller input stage at constant temperature, rather than constant pressure as in conventional systems. This has the effect of reducing the thermodynamic irreversibilities in the absorption cycle (Wahlig, 1984).

It is not known to what extent the major manufacturers of indirect fired absorption equipment have incorporated this technology.

4.5 MATERIALS

The generator section is the only portion of the absorption machine that is likely to be exposed to the geothermal fluid. In this section, the heating medium is passed through a tube bundle to provide heat to the refrigerant/absorbent mixture located in the shell.

The generator tube bundle is generally constructed of copper or a copper alloy (90/10 cupro nickel). These alloys are not compatible with most geothermal resources, particularly if hydrogen sulfide (H_2S), ammonia (NH_3) or oxygen is present. Because most resources contain some or all of these dissolved gases, exposure of standard construction chillers to these fluids is not recommended. Two available options are:

- special-order chiller with corrosion resistant tubes
- an isolation heat exchanger and clean-water loop.

Conversations with at least one major large-size absorption machine manufacturer indicate that the first option may be the most cost-effective (Todd, 1987). Although a 316 stainless steel tube would appear to be the most cost-effective, the manufacturer suggests the use of titanium. Because titanium tubes are more generally available in the enhanced surface configurations necessary for this application, their cost is very competitive with the stainless steel tubes. In addition, the use of unenhanced stainless steel tubes would (according to the manufacturer) result in a large de-rating of the chiller because of less-effective heat transfer.

The incremental capital cost for this type of construction (titanium generator tubes) would amount to approximately 10 to 15 per cent of the basic machine cost. In most cases, this would be far less than the cost associated with the heat exchanger, circulating pump, piping, and controls necessary for an isolation loop. An additional advantage is that the alternate generator construction avoids the losses associated with the heat exchanger.

4.6 CONCLUSIONS

In conclusion, it is necessary to evaluate the following factors when considering a geothermal/absorption cooling application for space conditioning.

- *Resource temperature*: substantial de-rating factors must be applied to equipment at temperatures less than 104 °C. Very-high resource temperatures or two stages are required for low-temperature refrigeration.
- *Absorption machine hot-water requirements compared to space-heating flow requirements*: incremental well and pumping costs should be applied to the absorption machine.
- *Refrigeration capacity required*: larger machines have lower incremental capital costs on a US$/kW basis. Coupled with the larger displaced energy, this results in a more positive economic picture.

- *Annual cooling load for space conditioning, in full-load hours or for process cooling, in terms of load factor*: obviously, higher utilization of the equipment results in more rapid payout.
- *Pumping power for resources with unusually low static-water levels or drawdowns*: pumping power may approach 50 per cent of high efficiency electric chiller consumption.
- *Utility rates*: as with any conservation project, high utility-rates for both consumption and demand result in better system economics.

REFERENCES

Ashrae (American Society of Heating, Refrigerating and Air-Conditioning Engineers). 1983. *ASHRAE Handbook: Fundamentals*. Atlanta, Ga., ASHRAE, pp. 14.1–14.8.

Christen, J. E. 1977. *Central cooling: absorptive chillers*. Oak Ridge, Oak Ridge National Laboratory.

Hirai, W. A. 1982. *Feasibility Study on an Ice-Making and Cold Storage Facility Using Geothermal Waste Heat*. Klamath Falls, Geo-Heat Center.

Means, R. S. 1996. *1996 Means Mechanical Cost Data*. Kingston, R.S. Means Inc.

Todd, M. 1987. (Airefco Inc., Portland). Personal communication.

Wahlig, M. 1984. (Lawrence Berkeley Lab.). Personal communication.

Yazaki Corp. Undated. *Yazaki Gas and Solar Air Conditioning Equipment: Cat. No. 15.3 AME*. Tokyo, Yazaki Corp.

SELF-ASSESSMENT QUESTIONS

1. What are the two basic configurations available commercially for absorption machines? For what temperature applications can they be utilized?

2. List the three fluid circuits that have external connections in a geothermal water chiller and the approximate temperatures of the entering cooling water and leaving chilled water using 83 kPa steam (or equivalent hot water).

3. For which temperature of geothermal fluids are the absorption machines most efficient?
(a) <82 °C; (b) between 82 °C and 116 °C; (c) >116 °C

4. Figure 4.3 shows that the added costs for the cooling tower, pumps and pipelines are higher for the absorption machines than for the vapour compression machines. Why?

5. List the main factors to be considered when comparing the economic feasibility of the absorption cycle with that of the vapour compression cycle.

ANSWERS

1. The two basic configurations are lithium bromide/water and water/ammonia cycles. The first can be used for temperature applications of more than 0 °C and the second for temperatures of less than 0 °C.

2. The fluid circuits are (1) generator heat input, (2) cooling water and (3) chilled water. The temperatures are 29 °C entering cooling water and 7 °C leaving chilled water.

3. (c) Figure 4.2 shows that the higher the temperature, the higher the coefficient of performance (COP), and the higher the per cent of capacity at rated conditions.

4. Because the absorption machines have a much lower coefficient of performance (COP) than the vapour compression machines and so have to reject more heat to the cooling towers. The latter must, therefore, be of greater capacity.

5. The key factors are cost of electricity, nominal capacity (investment costs), temperature of the geothermal fluid, annual cooling requirements, temperature of the cooling water, and design temperature required in the space to be cooled.

CHAPTER 5

Greenhouse Heating

Kiril Popovski

Faculty of Technical Sciences, St Kliment Ohridski University,
Bitola, Republic of Macedonia

AIMS

1. To describe the energy aspects of protected crop cultivation and their influence on the production rate and quality of vegetables, flowers and other crops.

2. To illustrate the feasibility in technological and economic terms of using geothermal energy to meet the heat requirements of greenhouses in different climates and conditions.

3. To present commercially viable types of installation for heating greenhouses with low-temperature fluids; to describe the transmission systems for the geothermal fluids from their source to the installations.

4. To illustrate the influence of different factors on the choice of heating installation.

5. To identify the problems that could arise when using geothermal energy in greenhouse heating.

6. To describe how to conduct an economic analysis of geothermal greenhouse heating in a real situation: how to select the optimal technical-economic solution.

OBJECTIVES

When you have completed this lesson you should be able to:

1. Discuss the factors influencing greenhouse climate.

2. Define the heat requirements of a greenhouse, depending on the type of greenhouse and crop.

3. Describe the characteristics of different low-temperature heating installations and how these characteristics influence cultivation techniques and the economics of production.

4. Discuss the transmission systems for carrying fluids from the source to the heating installation.

5. Analyse the economic aspects of utilizing geothermal energy for greenhouse heating.

5.1 INTRODUCTION

Geothermal energy has been used most extensively in agriculture in greenhouse heating. Many European and other countries (Table 5.1) are experimenting (Popovski and Popovska-Vasilevska, 1998) but also regularly using geothermal energy for commercial out-of-season production of vegetables, flowers and fruits.

Table 5.1 Geothermal greenhouses in the world

Country	Total area (ha)
United States	183.12
Hungary	130.38
China	115.92
Rep. Macedonia	62.46
China (Taiwan)	60
Italy	50.5
Russia	34
France	24.3
Spain	20
Iceland	18
Greece	17.95
Bulgaria	17.6
Slovakia	17.36
Georgia	16.5
Romania	13
Japan	12
Yugoslavia	10.13
New Zealand	10
Turkey	7.3
Slovenia	6
Israel	3
Bosnia/Herzegovina	2
Algeria	0.72
Germany	0.3
Croatia	0.25
Portugal	0.22
Belgium	0.05
Poland	0.05
TOTAL	833.11

Source: Popovski, 1998.

Although not in order of importance, the reasons for choosing geothermal energy in this sector are:

1. Good correlation between the sites of greenhouse production areas and low-enthalpy geothermal reservoirs.
2. The fact that greenhouses are one of the largest low-enthalpy energy consumers in agriculture.
3. Geothermal energy requires relatively simple heating installations, but advanced computerized installations can later be added for total conditioning of the inside climate in the greenhouses.
4. The economic competitiveness of geothermal energy for greenhouse heating in many situations.
5. The strategic importance of energy sources that are locally available for food production.

As with other uses of geothermal energy, it is not possible to make a general statement that greenhouse heating is the optimal form of application. Each situation must be evaluated separately, and local factors play a decisive role in any decision making (Popovski, 1988*b*).

5.2 ENERGY ASPECTS OF PROTECTED CROP CULTIVATION

5.2.1 Why protected crop cultivation?

Each plant can be considered a chemical factory. Sunlight converts the carbon dioxide (CO_2) from the air and the water (H_2O) from the air and soil into plant material such as sugars. The 'production technology' involved in this conversion is called *photosynthesis*. Photosynthesis is a reversible process: the 'free' energy of the surrounding environment is captured and stored in the form of plant material, which may be transformed back into CO_2 and H_2O by the process of plant respiration.

The energy stored in the form of plant material is used by the plant as 'building material' for its growth. By creating optimum growth conditions we can accelerate such life processes of the plant.

In order to complete a full life cycle, each type of plant needs a specific quantity of energy (that is, heat). The duration of its development and the quantity and the quality of the crop are governed by the quality and density of the energy supplied to the plant, which are in their turn correlated to the intensity of light available. The ideal natural conditions for supplying the energy required for plant development are available during only a part of the climatic year. These conditions will also be different for different plants, depending on their origin and specific needs. The objective of protected crop cultivation is therefore to guarantee these optimum growth conditions, independently of climatic conditions outside the greenhouse, so as to increase the quantity and improve the quality of the crop. The main problem is thus to identify the factors influencing plant development and the techniques and technologies necessary to maintain optimum growth conditions, within the framework of the prevailing social and economic situation.

5.2.2 Greenhouse climate

A greenhouse is a space bounded by transparent partitions, in which we can maintain a desired 'climate' that is different from the climate outside the greenhouse. There are four physical phenomena that play a major role in creating these differences in climate:

1. Solar radiation, in particular short-wave radiation, penetrates the glass or plastic film covering the greenhouse with practically no loss of energy. On reaching the soil surface, the plant canopy, and the heating and other installations, this radiation changes to long wave. Since long waves are unable to penetrate the walls of the greenhouse as easily as short waves, the energy is trapped within the enclosed space.
2. The air within the greenhouse is stagnant.
3. The concentration of plant mass in a protected space is much higher than outside, so that mass transfer is different.
4. The presence of heating and other types of installations changes some of the energy factors of greenhouse climate.

All the above phenomena involve physical parameters that control the plant growth process. We will now discuss these parameters.

Light

This is the most important parameter in plant growth. All the other parameters depend directly on light intensity. It should be stressed at this point that only radiation with wavelengths between 400 and 700 nm influences the intensity of plant life processes.

Temperature

The transfer of energy from the environment to the greenhouse is governed by greenhouse temperature. This transfer of energy is influenced mostly by the temperature of the surrounding air, but also by the temperature of the soil and of other elements of the environment (greenhouse construction, installations and so on).

Optimum development of plant life processes depends directly on the level of plant temperature, which depends on the intensity of light available: that is, the higher the intensity of light, the higher the plant temperature.

CO_2 concentration

The CO_2 in the atmosphere surrounding the plant is used to form the plant 'building' materials, such as sugars. Its transformation is governed by light intensity and plant temperature. The higher the intensity of the light and the temperature, the higher must be the concentration of CO_2 in the air.

Air movement

Air movement in a greenhouse influences the transfer of heat between the plant and the air, and the exchange of water between them. Different plant species require different types of air movement for optimum growth.

Water transport

Water is also an important element in the production of plant 'building' materials. The plant takes in water from the surrounding air and from the soil around its root system. Optimum conditions of air and soil humidity will, however, depend on the type of plant and its stage of development.

Heating installation(s)

The heating system affects air and soil temperature, and also influences the type and velocity of inside air movement. It therefore plays an active role in the plant energy balance.

Cooling installation(s)

Cooling installations are also an active element in creating a greenhouse climate, as they influence air temperature and humidity. There are other elements that also affect greenhouse climate, such as the type of construction of the greenhouse, the materials used for the transparent walls, the type of crop and its stage of development.

The optimum greenhouse climate, that is, the climate that will permit optimum plant growth, depending on the available light intensity and quality, is the result of a compromise between the optimal values of each of the above parameters or elements. These values are interdependent and at times may be contradictory (Popovski, 1991).

5.3 CHARACTERISTICS OF HEAT CONSUMPTION

A greenhouse is a construction aimed at creating a protected space for plant cultivation in a controlled environment, even during climatically unfavourable periods. The importance of light in the life processes of the plants entails the use of transparent materials such as glass, plastic films, and plates, fibreglass and so on, which also exploit solar energy to raise the inside temperature conditions. However, this is not enough to maintain optimum growing conditions during periods when solar radiation is not strong enough and during the night (Figure 5.1). An additional source of heat is required that can be regulated. The amount of extra heat required depends on the local climate, plant requirements and the type of greenhouse construction. Over a twelve-month period, it mainly depends on changes in the outside air temperature and in the intensity of solar radiation (Figure 5.2).

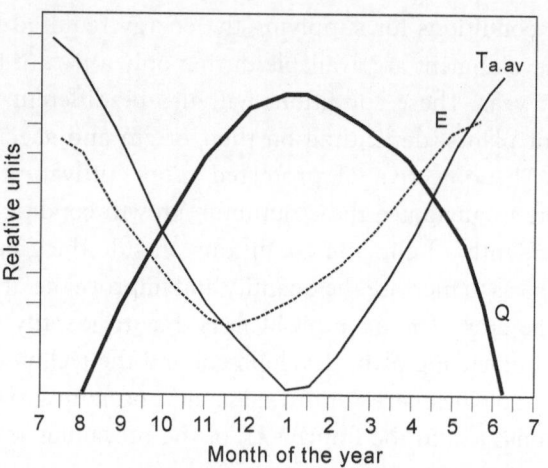

Figure 5.2 Heat requirement fluctuations in a greenhouse over a typical year in Gevgelia, Republic of Macedonia. E = solar radiation energy flux (Wh/m^2); $T_{a.av}$ = monthly average outside air temperature (°C); Q = greenhouse heat requirements (W)
Source: Popovski, 1984.

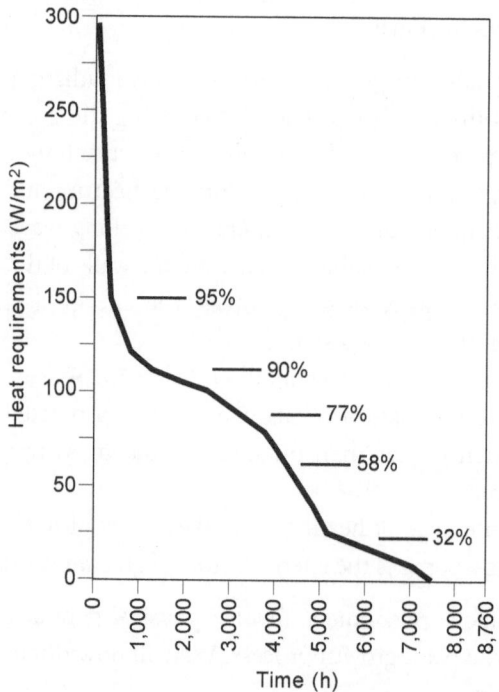

Figure 5.3 Percentage coverage of annual heat demand for greenhouse heating by alternative energy sources relative to Naaldwijk, Netherlands
Source: van den Braak and Knies, 1985.

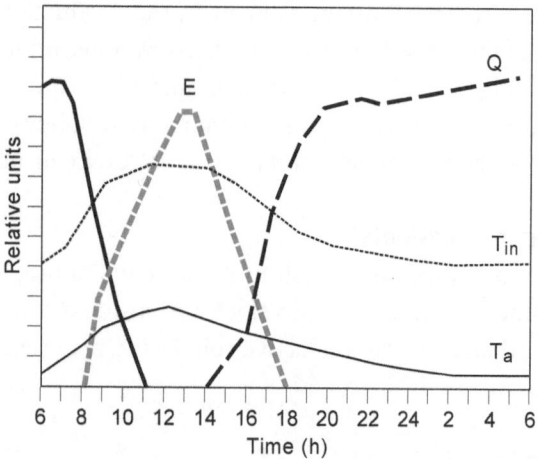

Figure 5.1 Heat requirement fluctuations in a greenhouse; average conditions in January in a greenhouse in Gevgelia, Republic of Macedonia. E = solar radiation energy flux (Wh/m^2); T_a = outside air temperature (°C); T_{in} = optimal inside air temperature (°C); Q = greenhouse heat requirements (W)
Source: Popovski, 1984.

In Figure 5.3, which shows the heat requirements of a greenhouse in the Netherlands, we can note that peak demands are of very short duration. Even in the conditions found in north-west Europe, only 50 per cent of the maximum heat capacity is necessary to cover 95 per cent of the total annual heat demand. Similarly, 40 per cent of the maximum heat capacity can cover 90 per cent of the annual heat demand, 25 per cent covers 77 per cent, and so on.

We can therefore conclude that heat consumption varies on a daily and yearly basis, with rather short periods of maximum heat demand. It is not justified to invest in expensive installations to cover the total heat requirements of greenhouses. Peak load should be avoided by selective production or met by heat sources that require low investments (Popovski, 1988c).

5.4 TECHNICAL SOLUTIONS FOR GEOTHERMAL GREENHOUSE HEATING

5.4.1 Factors influencing the choice of technological solution

The technological solution adopted for the greenhouses depends on:

1. The type of greenhouse production, that is, whether the plants require a controlled climate throughout the year or only during some months of the cold season.

2. The role of the heating installation, that is, to improve or to totally control the inside temperature conditions.
3. The type of well: that is, artesian or producing by means of electric-driven pumps.
4. Chemical characteristics of the geothermal fluids.
5. Limitations dictated by eventual other uses of the geothermal water.

5.4.2 Hot-water transmission systems

The type of system adopted for transmitting the geothermal water from its source to the utilization plant depends on the type of well (artesian or non-artesian), the chemical characteristics of the water, and investment and operation costs.

Direct connections

Low-temperature water with a low mineral content permits us to adopt simple technical solutions (Figure 5.4). The water is collected in an open de-aeration tank installed above ground level so that it can flow by gravity through the transmission pipeline and the heating installation. The heat is regulated by means of one or more hand valves.

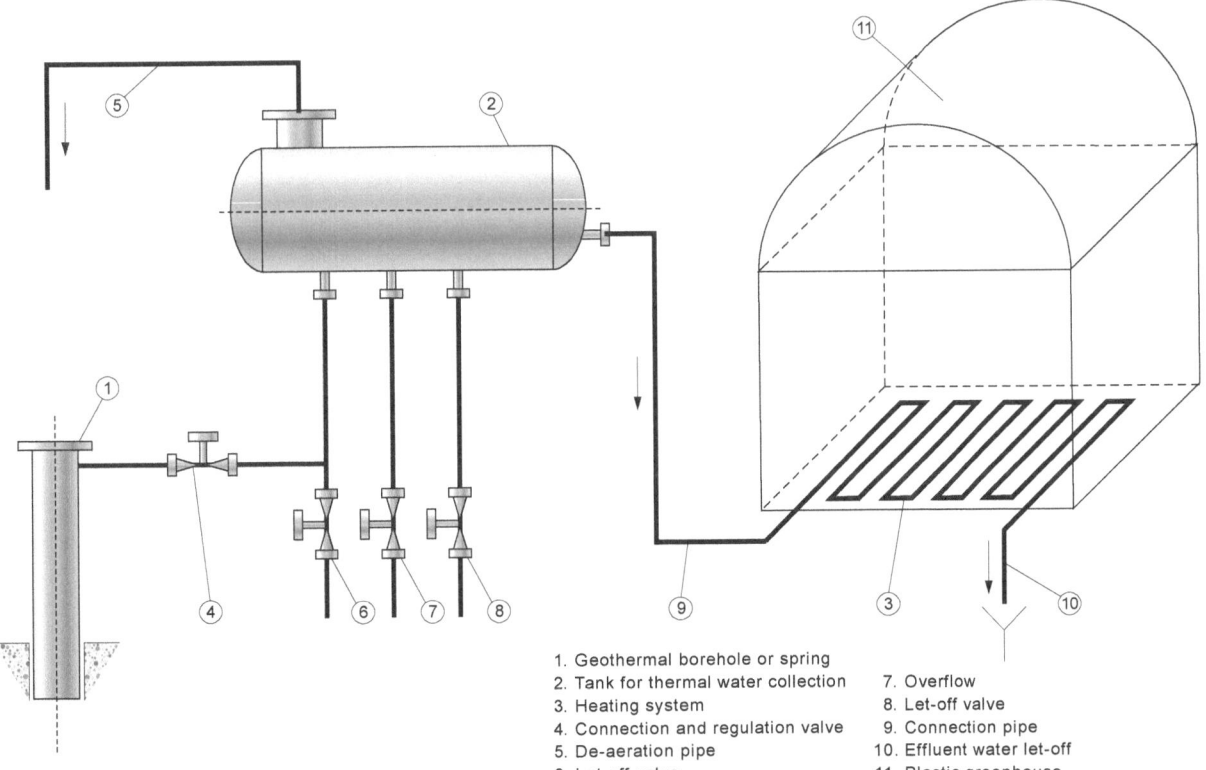

1. Geothermal borehole or spring
2. Tank for thermal water collection
3. Heating system
4. Connection and regulation valve
5. De-aeration pipe
6. Let-off valve
7. Overflow
8. Let-off valve
9. Connection pipe
10. Effluent water let-off
11. Plastic greenhouse

Figure 5.4 Simple direct connection to the geothermal spring or well

This type of solution is very popular in Mediterranean countries, where simple plastic-covered greenhouses are used for an earlier spring crop. It is also a viable option when the water is highly corrosive, provided that only plastic materials are used for the transmission pipes, the heat exchangers and other construction parts that come into contact with the water.

Indirect connections

If the mineral content of the water is very low and the plants under cultivation permit higher investments, a simple transmission from the well can be connected to a more sophisticated heating system (Figure 5.5), which includes a heat pump (4, 5 and 6 in the figure). The transmission system is therefore connected indirectly to the heating system. Environmental constraints and waters with high mineral content entail more complicated water transmission systems, including de-aeration equipment, regulators of CO_2 content and pH, and corrosion preventers.

The problems arising with indirect connections (primary and secondary circuits) are mainly tied to the type of heat exchanger used. Plate heat exchangers have proved to be the best solution for geothermal waters with a high salt content. There are several advantages with this model. Their large heat-exchanging surfaces are contained within a small space. Different temperature regimes can be combined in one single heat-exchanger system. Cross-fluid flows between the plate sections provide high heat-transfer coefficients, and the shape of the plates is such that they can readily be assembled and dismantled. The latter is particularly important where geothermal fluids with a high mineral content are concerned, as the equipment has to be dismantled and cleaned frequently to remove scaling deposits.

5.4.3 Combined uses

One of the major disadvantages of using geothermal energy in greenhouses is the rather high investment cost for the well and the transmission and regulation systems, which may be used for part of the year only (Figure 5.3). Consequently, the price of the heat used will be high.

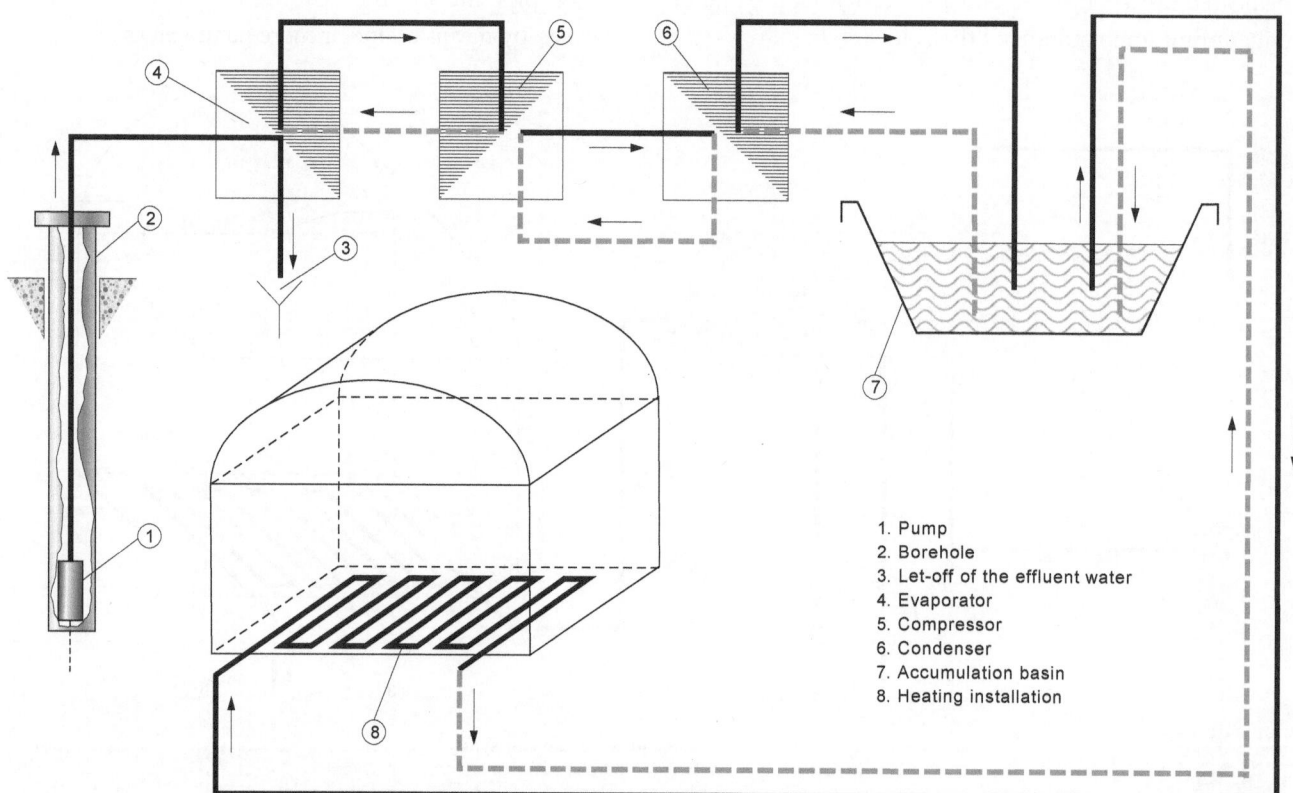

1. Pump
2. Borehole
3. Let-off of the effluent water
4. Evaporator
5. Compressor
6. Condenser
7. Accumulation basin
8. Heating installation

Figure 5.5 Heat pump installation with simple connection to the well
Source: Agence pour les Économies d'Énergie, 1982.

This problem can be overcome by finding other potential users of the heat. Ideally, the annual and daily heat requirements of these potential users should be different; that is, when one user needs maximum heat the other should need a minimum. For example, the price of heat in Bansko (Macedonia) decreased from 6.25 US cents/kWh for greenhouse heating only, to 2.89 US cents/kWh for the integrated system shown in Figure 5.6.

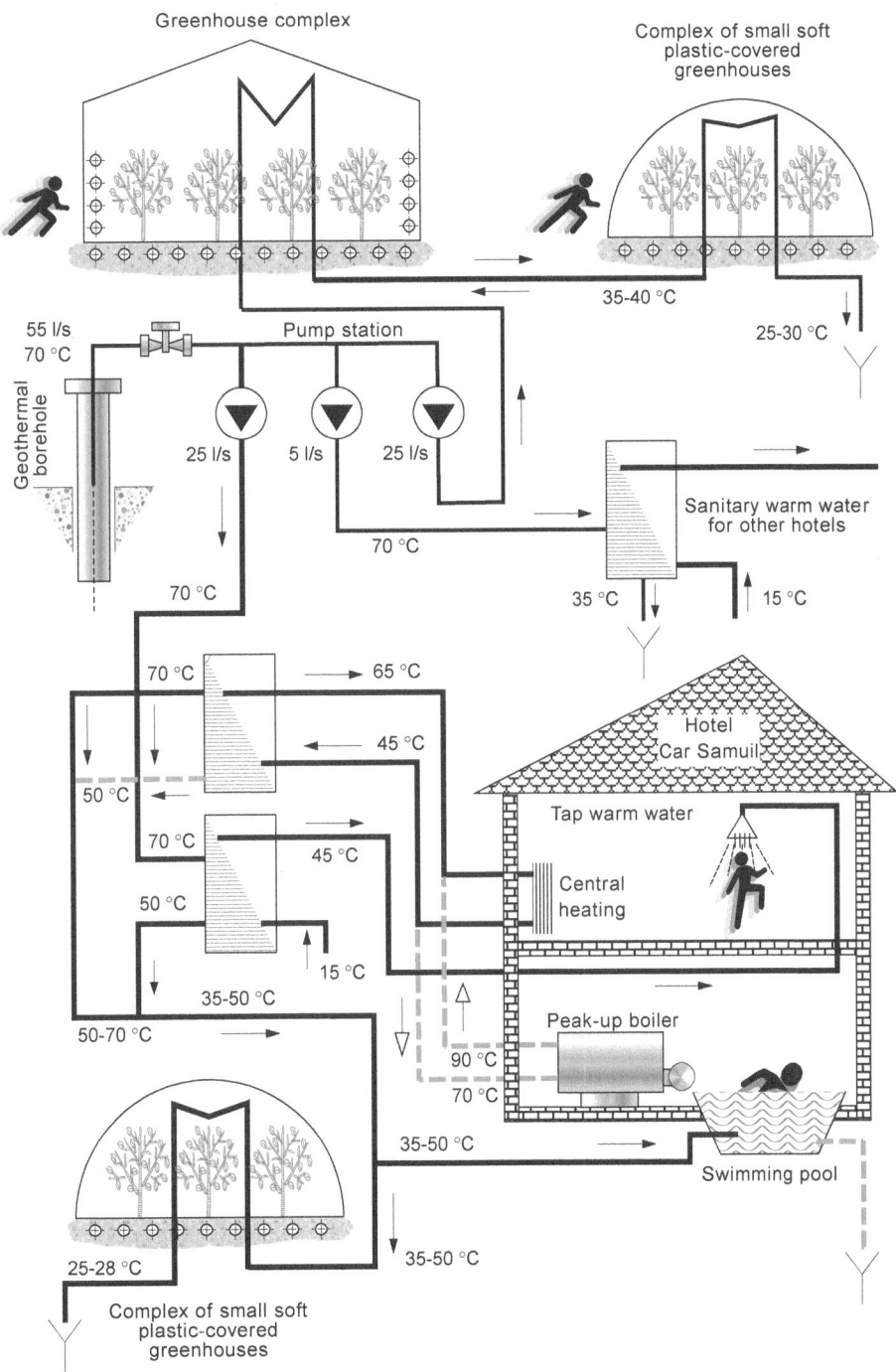

Figure 5.6 Simplified flow diagram of the Bansko integrated geothermal system (Republic of Macedonia), consisting of greenhouse heating and different heat users of a hotel-spa complex

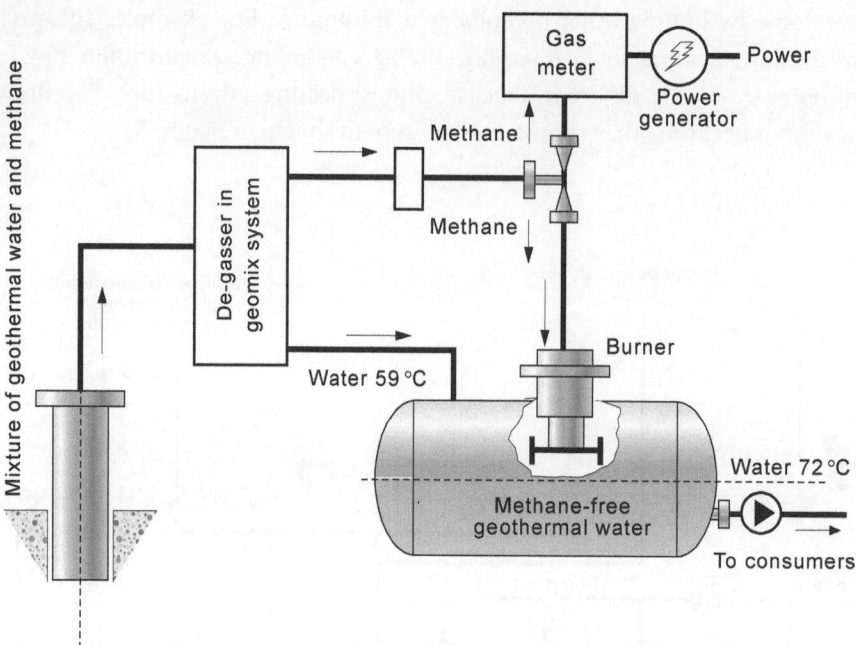

Figure 5.7 Technically improved geothermal installation in Srbobran (Yugoslavia), with extraction of the CH$_4$ from the geothermal water for electricity generation and direct heat use for peak demand in the greenhouse

In particular situations, it may be more convenient to add cheap auxiliary heating equipment to the system in order to meet peak heat loads of short duration (Figure 5.7).

5.5 GEOTHERMAL GREENHOUSE HEATING INSTALLATIONS

5.5.1 Classification

Heating installation is the term commonly used for the heat exchanger providing supplementary heat in greenhouses. The temperature of the heating medium, that is, the geothermal water, and the particular requirements of the plants in the greenhouse, dictate its design, location, material, means of regulation and so on. The level of sophistication depends on the technological level of production, greenhouse construction, climate, and technical and economic factors.

There are two extreme technological solutions, encompassing a wide spectrum of intermediate schemes. The two extremes are described below:

1. A simple heating installation made of plastic materials that connect the heat source directly with the greenhouse, and manual regulation of the heat supply.

The aim of this type of installation is only to improve inside temperature conditions all year round, in mild winter climatic conditions, or during early spring and late autumn in more severe climates (Campiotti et al., 1985). It is in use in cheap plastic-house constructions which are not technologically suitable for intensive production. Usually there is no provision for water treatment. Pumping of the water is not always economically justified, but depends on the plants grown, production levels and marketing conditions.

2. Sophisticated heating installations for total air conditioning, with automatic regulation of the heat supply. The factors influencing heat requirements are indoors and outdoor conditions, plant growth and production schedules. These installations are economically justified in expensive glasshouse or rigid plastic constructions, equipped with the technology for intensive production. If the chemical characteristics of the thermal water are unfavourable, the well can be indirectly connected to the greenhouse installations.

These two technological solutions use different types of heat exchanger, depending on heat requirement, chemical characteristics of the geothermal water, greenhouse construction and economic factors.

Aerial heat exchangers can be classified according to the type of heat transfer and location of the heating elements (von Zabeltitz, 1986) (Figure 5.8). From the technical viewpoint this approach is incomplete, since it does not include some typical low-temperature installations that are not limited to heating the air only. Nowadays, the combinations that are most commonly used are those developed between 1986 and 1988 by Bailey, von Elsner and Popovski, because these combinations have been accommodated to the particular needs of the growers and are therefore much more

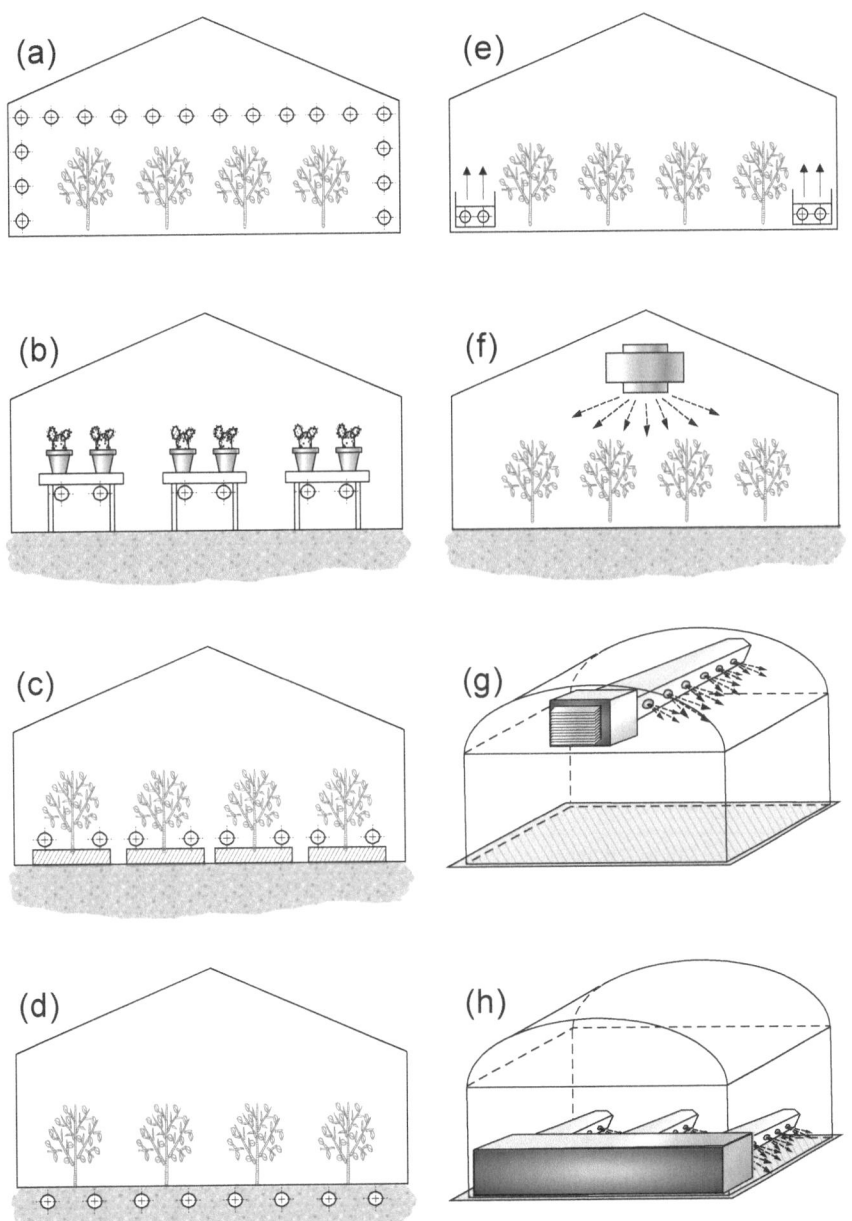

Figure 5.8 Classification of low-temperature heating systems. Heating installations with natural air movement (natural convection): (a) aerial pipe heating; (b) bench heating; (c) low-position heating pipes for aerial heating; (d) soil heating. Heating installations with forced-air movement (forced convection): (e) lateral position; (f) aerial fan; (g) high-position ducts; (h) low-position ducts
Source: von Zabeltitz, 1986.

practical and easy to understand. They are as follows:

- heating systems with heat exchangers within the soil or cultivation base
- heating systems with the exchangers laid on the ground surface or on benches
- aerial pipe heating systems
- fan-assisted convector air heating systems

- non-standard heating systems
- combined heating systems.

5.5.2 Soil heating installations

The objective of soil heating installations is to heat the plant root system (Figure 5.9). They can cover only a part of the

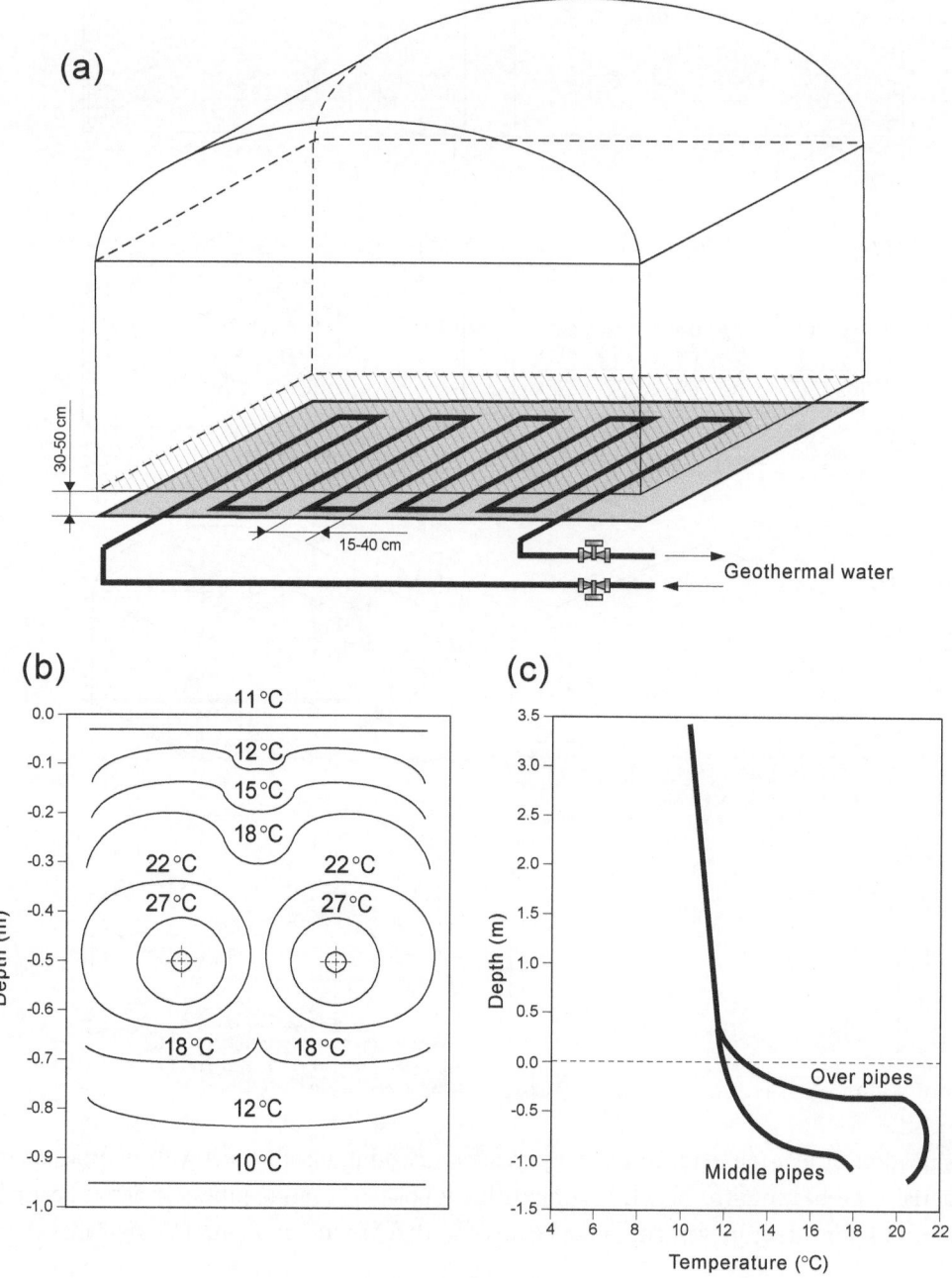

Figure 5.9 Installation for heating the soil in greenhouses: (a) position of heating pipes; (b) temperature profile of the heated soil; (c) vertical temperature profile in the greenhouse

total heat demand, and can be used without other heating systems only in very mild climatic conditions. They are economically feasible only for cultures that require precise temperature regulation of the root system, or in combination with other types of heating installation for temperature control of the greenhouse environment.

5.5.3 Soil–air heating installations

These installations consist of a system of heating elements positioned on the ground surface (Figure 5.10). With this layout the upper layer of the soil and the air are heated, which is very convenient for many cultures. There are

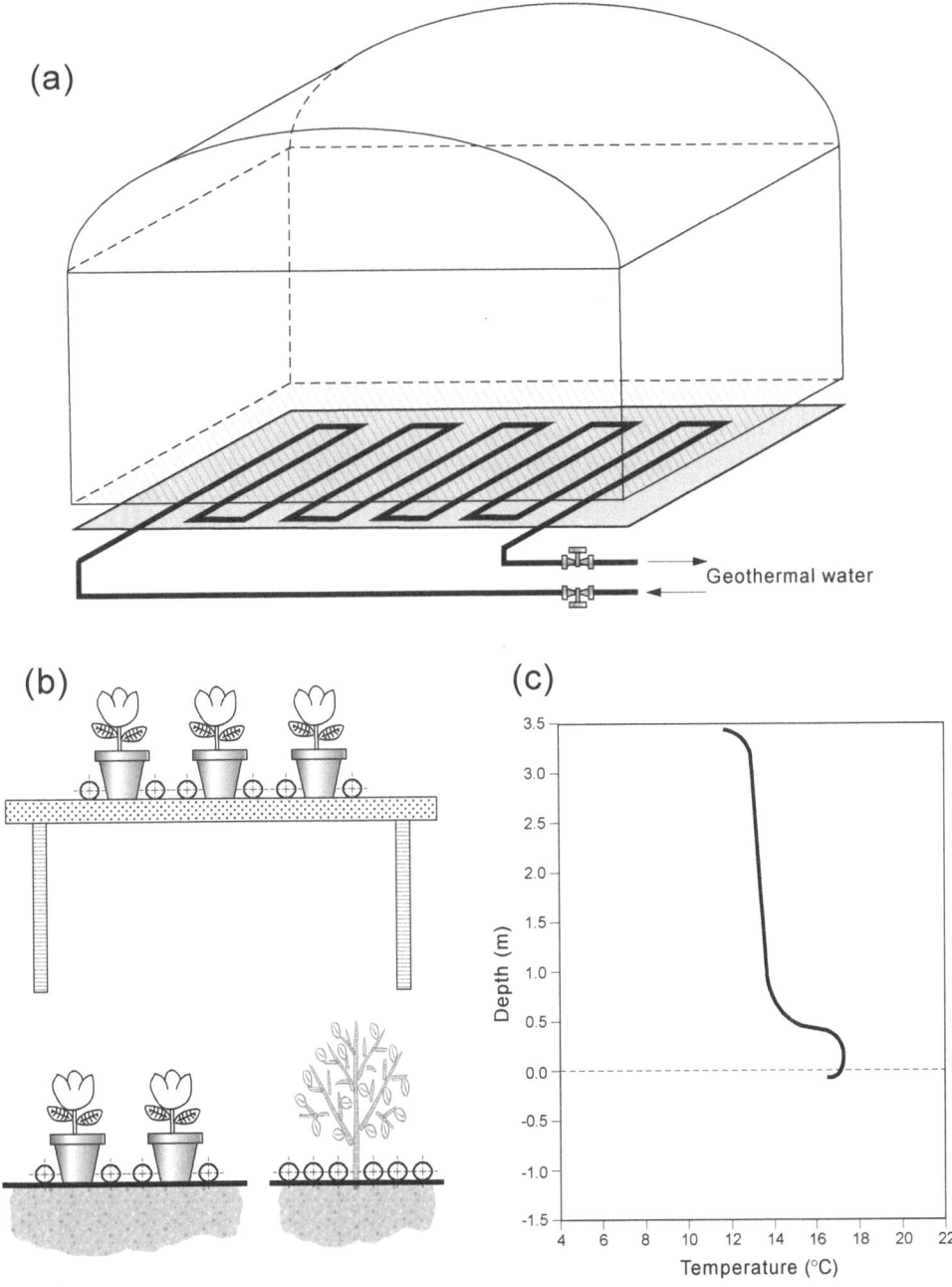

Figure 5.10 Installation for heating the air and soil in greenhouses (heating pipes placed on the soil surface). (a) Position of heating pipes; (b) different solutions for allocation of heating pipes (cultivation on benches, in pots on soil surface and in soil); (c) vertical temperature profile in greenhouse

different types of heating elements commercially available for this purpose:

- thin metallic pipes, smooth and corrugated thin plastic pipes, as for soil heating
- transparent plastic sleeves (tubes of large diameter)
- polytube plates of soft and rigid plastic
- plastic plates.

Proper positioning of the heating elements (Figure 5.10) permits an optimum transfer of heat to the plants and minimum heat loss to the environment. It is an excellent solution for covering total heat demand in milder climates, or the base demand in moderate and rigid climates.

5.5.4 Aerial pipe heating installations and convectors

This group of heating installations consists of metallic pipes, finned metallic pipes or convectors, positioned above the ground surface (Figure 5.11). The advantage with these installations is that they permit rapid and precise regulation of temperature, and can be used on their own even in moderate and rigid climates. The drawback is that the heat-transfer coefficient for low-temperature heating fluids is very low, which means that the heating surfaces must be very large, and may thus reduce light diffusion in the greenhouse and jeopardize working conditions.

The vertical temperature profiles are rather uneven for the pipe heating elements, but not for the convectors. However, convectors are unsuitable for low-temperature thermal waters.

5.5.5 Fan-assisted convectors

Significant improvements to the heat-transfer coefficient can be achieved by introducing fan assistance, even for rather low-temperature heating fluids. Compact fan-assisted air heating units can be used as the heating element, blowing warm air directly into the protected environment; alternatively, these can be combined with one or more soft-plastic air distribution tubes positioned over or between the plant rows (Figure 5.12). A good technical solution is to position the fan-assisted convectors longitudinally.

The quality of heating achieved depends very much on the technical solution adopted. Vertical temperature profiles

are fairly uniform when the air distribution pipes are positioned in the middle of the plants, but very uneven when they are located over them. This type of installation is very popular in the Mediterranean countries and the United States as it is simple, cheap, and suitable for automatic regulation by means of cheap equipment, and guarantees a fast response to outside temperature changes.

5.5.6 Other types of heating installation

Some ingenious heating installations have been designed for particular local conditions. For example, an air–water heat exchanger has been developed in Greece (Figure 5.13) to overcome the scaling problems caused by the geothermal water. The pipe carrying the water is positioned inside the air distribution tube. The latter is made of cheap polyethylene material, which can be replaced after one or two productive seasons.

Evaporative heating units are also adopted in some cases.

5.6 FACTORS INFLUENCING THE CHOICE OF HEATING INSTALLATION

One of the most difficult design problems in a geothermal greenhouse project is the choice of a technically, technologically and commercially feasible heating installation. Each case must be judged on its own merits, taking into consideration all the influencing factors in order to reach the optimal solution.

5.6.1 Temperature profiles in the greenhouse

The primary objective of the greenhouse heating installation is to maintain the temperature of the environment at values near to optimal. However, none of the installations that we have just described will guarantee a totally uniform vertical (Figure 5.14) or horizontal (Figure 5.15) temperature profile inside the greenhouse. For a reference air temperature 1.5 m above ground level, the difference at ground level and below the greenhouse roof could be as high as 5 to 6 °C (Figure 5.16). It is obvious that such differences can be of crucial importance for the choice of heating installation.

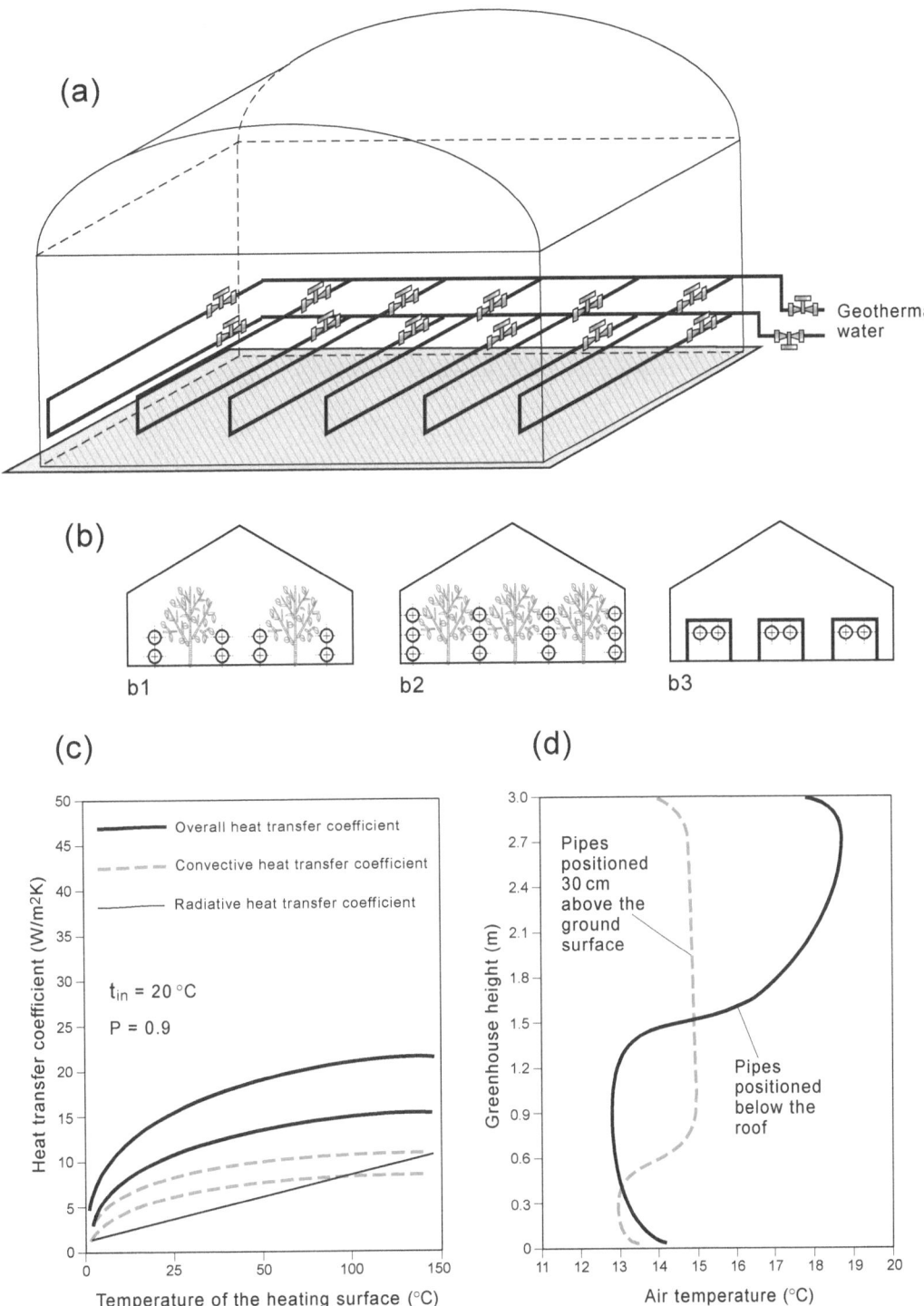

Figure 5.11 Aerial heating installations made of smooth or finned steel pipes. (a) and (b) Position of aerial pipe heat exchangers for low-temperature heating fluids: (b1) along the plant rows; (b2) in the plant canopy; (b3) below the benches; (c) heat transfer coefficient for the greenhouse interior based on pipe diameter and temperature of the heating fluid; (d) vertical air temperature profile in a greenhouse heated by aerial pipe heating installations

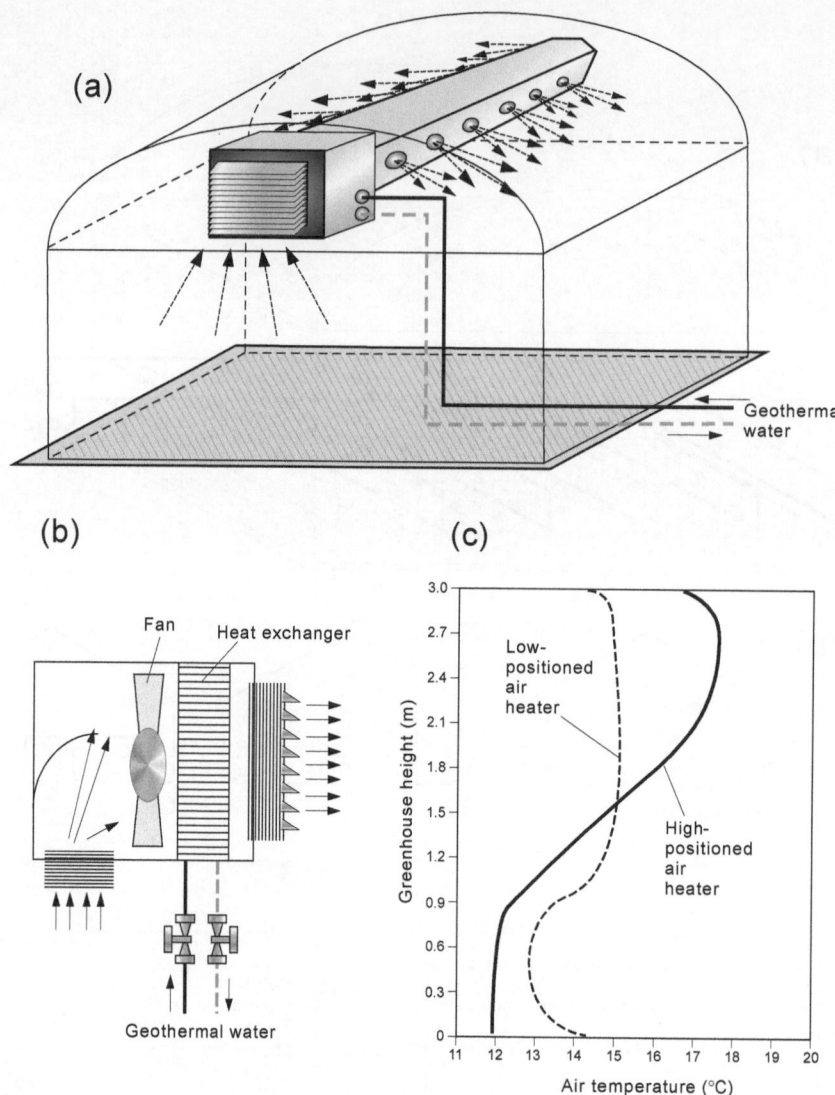

Figure 5.12 (a) Position of the 'fan-jet' heating system in a greenhouse; (b) layout of a fan-assisted convector unit; (c) vertical temperature profiles in a greenhouse heated with fan convector units at different heights

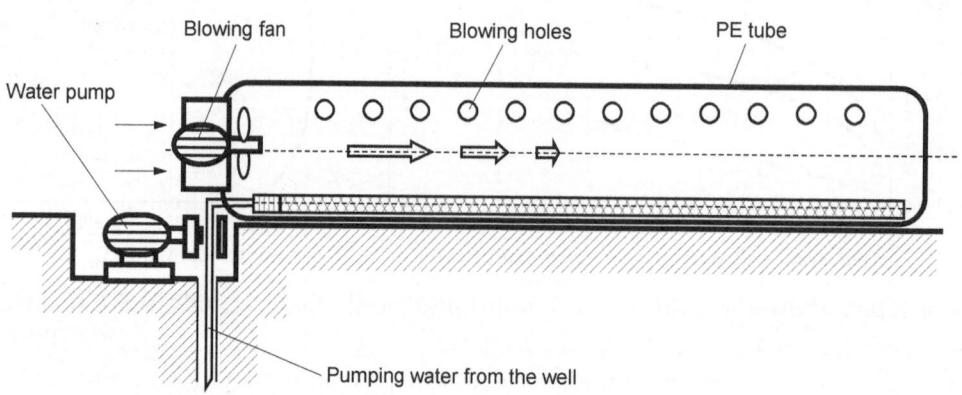

Figure 5.13 Water/air heat exchanger in air distribution tube
Source: Grafiadellis, 1986.

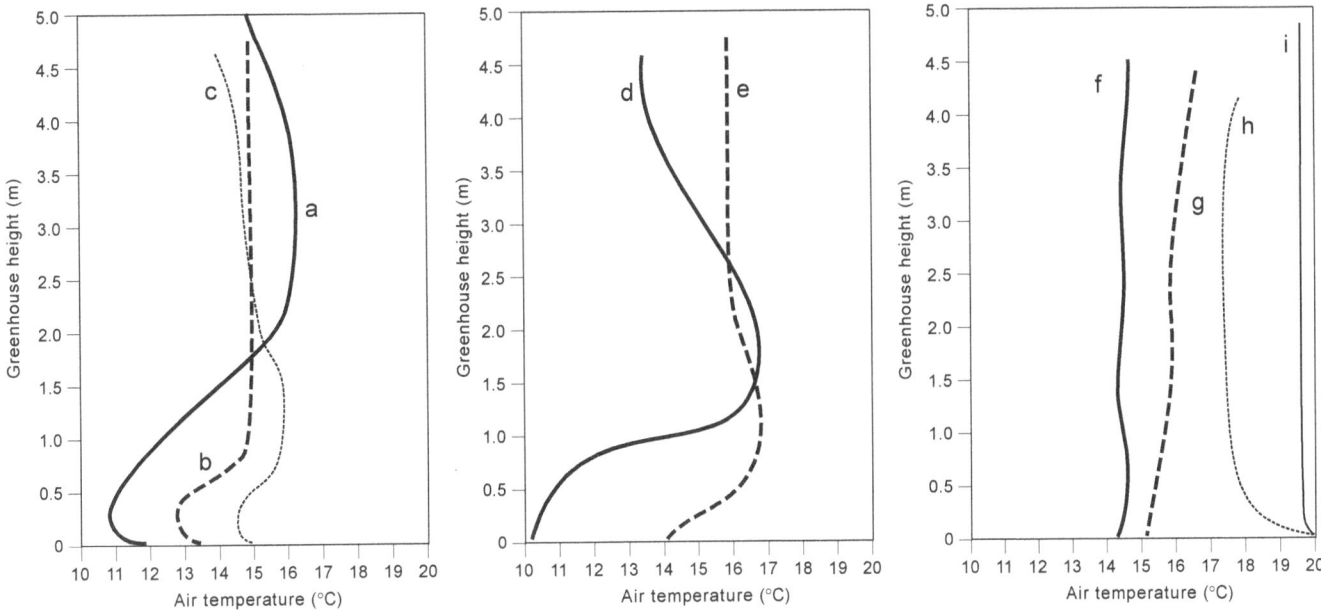

Figure 5.14 Vertical temperature profiles in a greenhouse, depending on the type and location of the heating installation. (a) High aerial pipes; (b) high pipes; (c) low pipes; (d) high-positioned air heaters; (e) 'fan-jet' system 2 m above the crop; (f) high-speed air heating; (g) convectors; (h) 'fan-jet' between the plants; (i) 'fan-jet' below the benches
Source: von Zabeltitz, 1986.

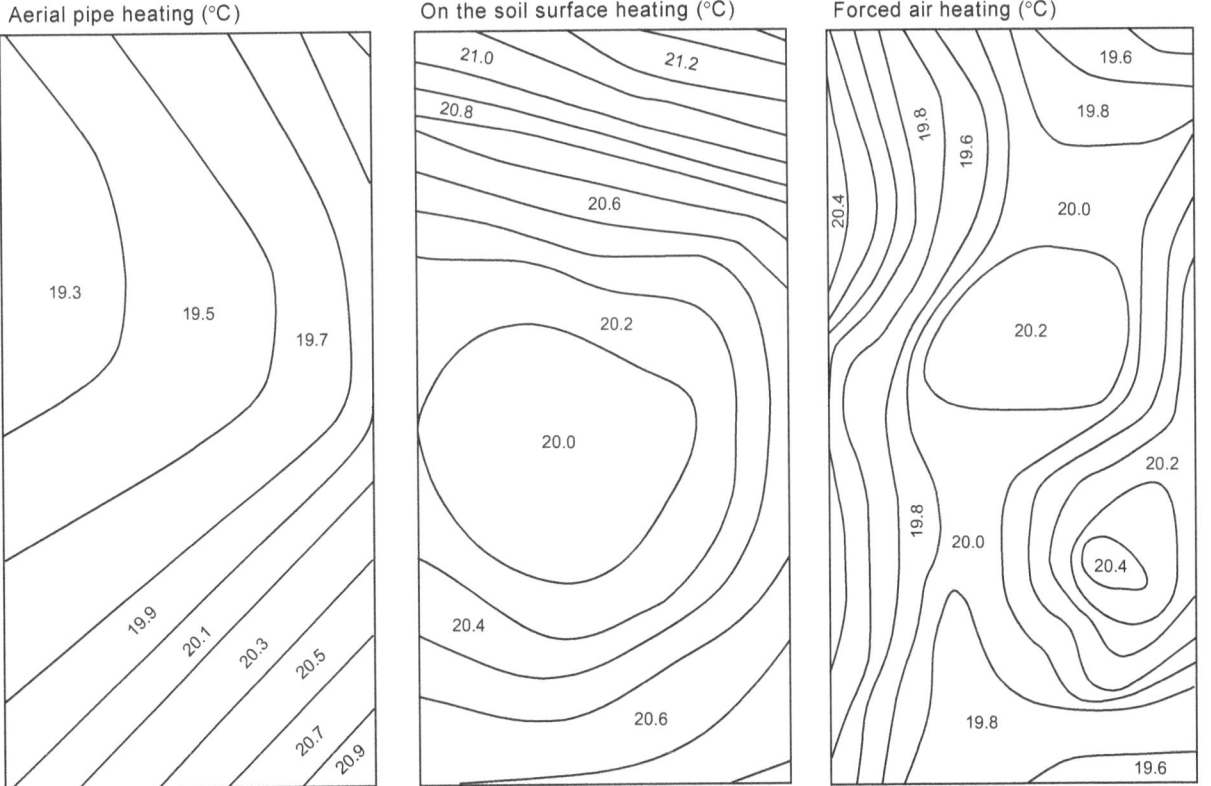

Figure 5.15 Horizontal temperature distribution (°C) in a greenhouse heated by different types of heating installation, under the same outside climatic conditions

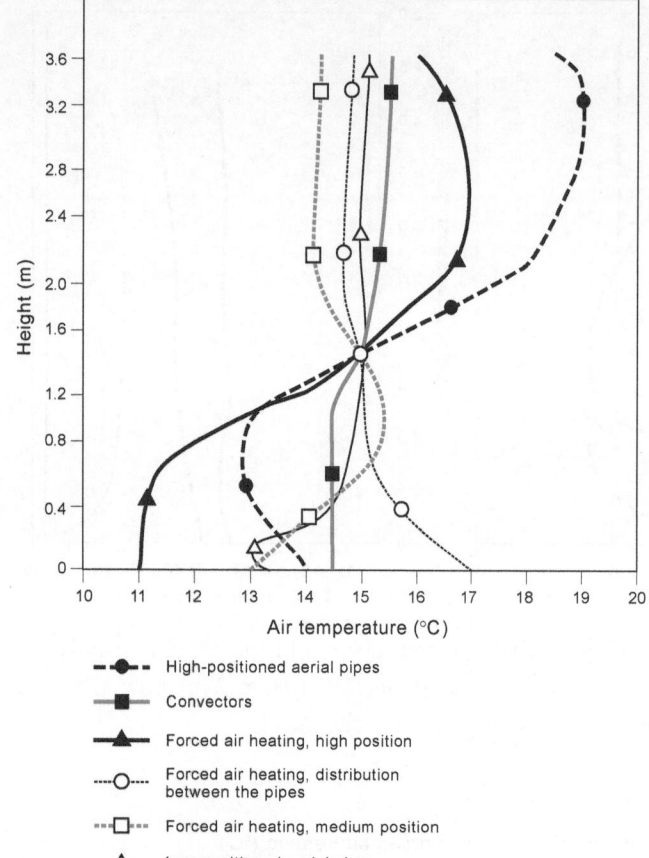

- —●— High-positioned aerial pipes
- —■— Convectors
- —▲— Forced air heating, high position
- ---○--- Forced air heating, distribution between the pipes
- ---□--- Forced air heating, medium position
- —△— Low-positioned aerial pipes

Figure 5.16 Vertical air temperatures in a greenhouse, heated by different types of heating installation

5.6.2 Economic aspects

The commercial feasibility of geothermal greenhouse heating depends on a number of factors, such as capital investment costs, operating costs, cost of energy with respect to the value of the product and market available for the product. If we leave out all the factors that are common to any energy source, it is obvious that the crucial factor for estimating the commercial feasibility and competitiveness of geothermal energy is the price of the heat *used*. The final price of heat depends on factors such as capital investment required, credit conditions, maintenance and utilization costs, insurance and taxes, labour costs, fuel prices, plant efficiency and utilization coefficient. Depending on the type of energy utilized and local conditions, characteristic curves can be plotted (Figure 5.17) to show the dependence of the used heat price on the annual heat load factor (hours of utilization of the installed heat capacity) (Popovski, 1988*a*).

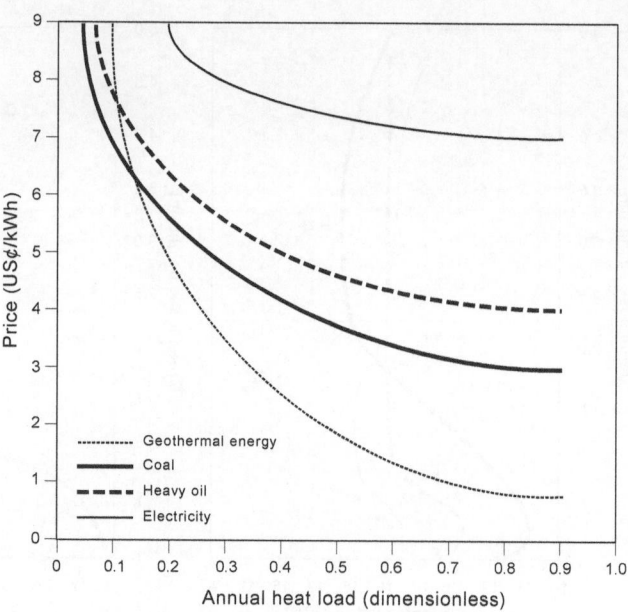

Figure 5.17 Heat cost of geothermal energy, heavy oil, coal and electricity, as a function of annual load factor. Annual load factor L = (Q/P) 8,760 × 3,600, where Q is the heat consumption during the year (joules) and P is the heat capacity of the geothermal well (watts). Example of greenhouse heating in Bansko, Macedonia
Source: Popovski and Popova, 1987.

These curves will differ from case to case, but generally they show that geothermal energy is not economical for short utilization periods. Real benefits can be attained if the available thermal power is used over long periods of the year; that is, it is very convenient for covering base-heat demands. It is more economical to meet peak demands with some other energy that does not require high capital investments.

As an example, Figure 5.17 reports an analysis of economic data for geothermal greenhouses located in Macedonia (Popovski, 1988*b*). The greenhouses cover 3 ha, with 6 MW$_t$ of installed capacity; the geothermal well produces 50 l/s and the temperature regime for the heating installations is 70/40 °C. The price of heavy oil is 200 US$/t, of 12,000 J/kg coal it is US$75 and of electricity US$0.065/kWh. Credit conditions are: eight years repayment, two years grace period, 10 per cent rate of interest. The analysis was made by determining the price of the used kWh of thermal energy for different loads during the year, for each of the listed fuels and geothermal energy.

The optimal ratio of the different energy sources covering total heat consumption can be defined by additional analysis, which again accounts for local factors. In the case described above, if total heat consumption is met by geothermal energy, the price of the used heat would be about US$0.07/kWh. If 50 per cent of the thermal capacity is designed to be supplied by geothermal energy and 50 per cent by a heavy-oil boiler plant, then the price would drop to about US$0.04/kWh, which is more economical than using a single form of energy. Ideally the project should envisage a combination of different heat users, with different utilization factors.

For instance, if 6 MW_t of the total capacity is utilized for greenhouse heating, 6 MW_t for space heating and sanitary warm water, and 0.6 MW_t for crop drying (from July to October), all connected to the same well mentioned above and to a heavy-oil boiler plant of 6 MW_t, it would be possible to attain a utilization factor of 59 per cent for the well (heat load L = 0.59). The price per kWh would then drop to only US$0.023/kWh, which is far more competitive than other energy sources.

5.6.3 Operating problems

Operating problems are mainly related to the nature of the geothermal resources. The problems begin at the spring or well-head. If it is a natural artesian spring then few difficulties will be encountered, but if it is a borehole where pumping is necessary, two important problems may arise: overexploitation could exhaust the reservoir, and sand could damage the pumps. To avoid the first problem, the reservoir characteristics must be thoroughly investigated before starting commercial use, to define the optimal resource management. The problem of sand can be solved only by filtering the water before it reaches the pumps and by avoiding abrupt start-up of the pumps.

Another problem is corrosion, mainly caused by oxygen (O_2), carbon dioxide (CO_2), chlorine (Cl) and hydrogen sulfide (H_2S) in the geothermal water. Three solutions are now being adopted: non-corrosive materials, modification of the chemical composition by de-aeration and separation or neutralization of aggressive chemical components, and indirect connection of the heating installations to the well. A good compromise between technical requirements and economic results has still to be reached. Simple solutions will lead to frequent maintenance, whereas sophisticated solutions are too expensive and rarely economically justified.

A similar problem occurs with scaling. Several methods for combating scaling have been developed, such as chemical inhibitors and treating the water with a magnetic field and ultrasonic waves. Again each situation must be assessed separately.

The problems with regard to the heat exchanger stem from the physical properties of the materials used and the regulation of the supply. The soft-plastic exchangers are liable to puncture, for example. Apart from impairing the supply of heat, this can also cause damage to the surrounding plants. There is still no real solution to these problems, except in the short term.

Heat-regulation problems are mainly related to the voluminous heat exchanger installations and their large inertia. For the moment, the best solution seems to be that of avoiding situations where precise regulation is required (simple installations only, aimed at improving the inside temperature characteristics) and using a combination of types of installation, some of which are easy to control.

5.6.4 Environmental aspects

Although considered a 'clean' form of energy for a long time, geothermal energy is in fact not so. At least two negative effects on the environment can be identified: thermal and chemical pollution, which are particularly important when large flows of thermal water are involved, as in the case of greenhouse heating.

Thermal pollution is the result of poor technical design of the geothermal heating installations, irregular modifications to the design in order to lower investment costs, and irregular use during routine operations. As a consequence, only the upper part of the available temperature interval is used, so that the wastewater is discharged at high temperatures, representing a possible risk to the surrounding environment.

Chemical pollution is a result of the chemical composition of the thermal waters, which often contain aggressive and toxic elements; chemical disequilibrium will also cause scaling. Both phenomena have negative effects on the environment.

Any measures adopted to protect the environment will have a direct influence on the economy of geothermal energy use, and positive results can only be achieved through legislation. Regulations must be enforced, prescribing the maximum temperature of the thermal effluents and maximum concentrations of harmful chemical constituents.

One way of tackling thermal pollution is to use low-temperature heating installations or a combination of installations that guarantees use of the entire temperature interval available.

The only real solution to chemical pollution is to re-inject the used water, but unfortunately this is also the most expensive one. Only a few countries have regulations for environmental protection during geothermal exploitation. In the others, and particularly in developing countries, this problem has been more or less neglected. This is unfortunate, as the negative consequences sooner or later will become evident.

5.6.5 Adaptation of technological solutions to local conditions

Changeable market conditions and requirements have a negative influence on the choice of geothermal energy for greenhouse heating. Geothermal energy requires higher capital investments than other types of energy, and therefore entails longer payback periods. Considering that each production necessitates a specific type of heating installation, very expensive additional investments are needed if we change the type of crop. It is for this reason that geothermal energy has been used so far for 'guaranteed sale' cultures, such as tomatoes, cucumbers, and carnations, and for cheap and simple greenhouse constructions. Nowadays, with such a large number of installations commercially available, this is not a major limitation, but it is still not possible to make drastic changes; that is, good results can only be achieved with the same heating installation if we keep to similar cultures.

Low-position heating installations that provide natural air movement in the greenhouse offer, in principle, better vertical temperature profiles and lower heat consumption (lower air temperatures below the cold surfaces of the roof). However, this characteristic of these installations is also one of their limitations, as they do not protect the plants from cold radiation from outside, which is very important in colder regions and for some high-growing plants. Their use is thus limited to milder climates, when they are the only heating installation in use. An auxiliary installation has to be used in colder climates to cover peak heating demands and protect from cold radiation.

Another important factor is shading of the plant by the elements of the heating installation. Ideally there should be as little shading as possible of incoming solar radiation. For low-growing plants, medium- and high-positioned heating elements should be avoided; high-growing plants, on the other hand, should receive heat directly on the plant canopy from high-positioned elements.

Greenhouse construction can limit our choice of heating installation. Heavy heating elements cannot be hung from cheap, lightweight plastic constructions, and similarly for expensive highly sophisticated equipment.

Local climate can influence our choice in one or another direction. For example, although the optimal installation for the greenhouse and for the plants may be natural movement of heated air, we may have to resort to forced air movement because of local strong winds, so as to reduce their negative effects on horizontal temperature distribution.

A ready supply of materials is important when choosing the type of installation. Spare parts, materials and equipment must be easy to procure to ensure the smooth running of operations. The extremely small thermal inertia of the greenhouse construction will not permit long breaks for maintenance.

Local traditions should also not be neglected. It must be remembered that few greenhouse operators have had a technical education, and they are incapable of appreciating immediately all the advantages of a new type of heating installation. It is better to avoid drastic change and adopt a step-by-step approach of gradual small improvements, even though better technical solutions may actually be available.

5.7 FINAL CONSIDERATIONS

As we have already said, the choice of heating installation for a geothermal greenhouse is a complicated process because of the interdependence of various scientific and technical factors. It is impossible to define any precise methodology for approaching this problem, except for the iterative method of identifying the influencing factors and gradually eliminating those that are least important in the case in question. However, it is of paramount importance in all cases to carry out a careful and thorough study of the characteristics of the geothermal resource before proceeding with this type of application, so as to avoid expensive mistakes later. Local conditions will dictate the type and extent of the investigations that will be necessary, but in general the recommended approach is as shown in Figure 5.18.

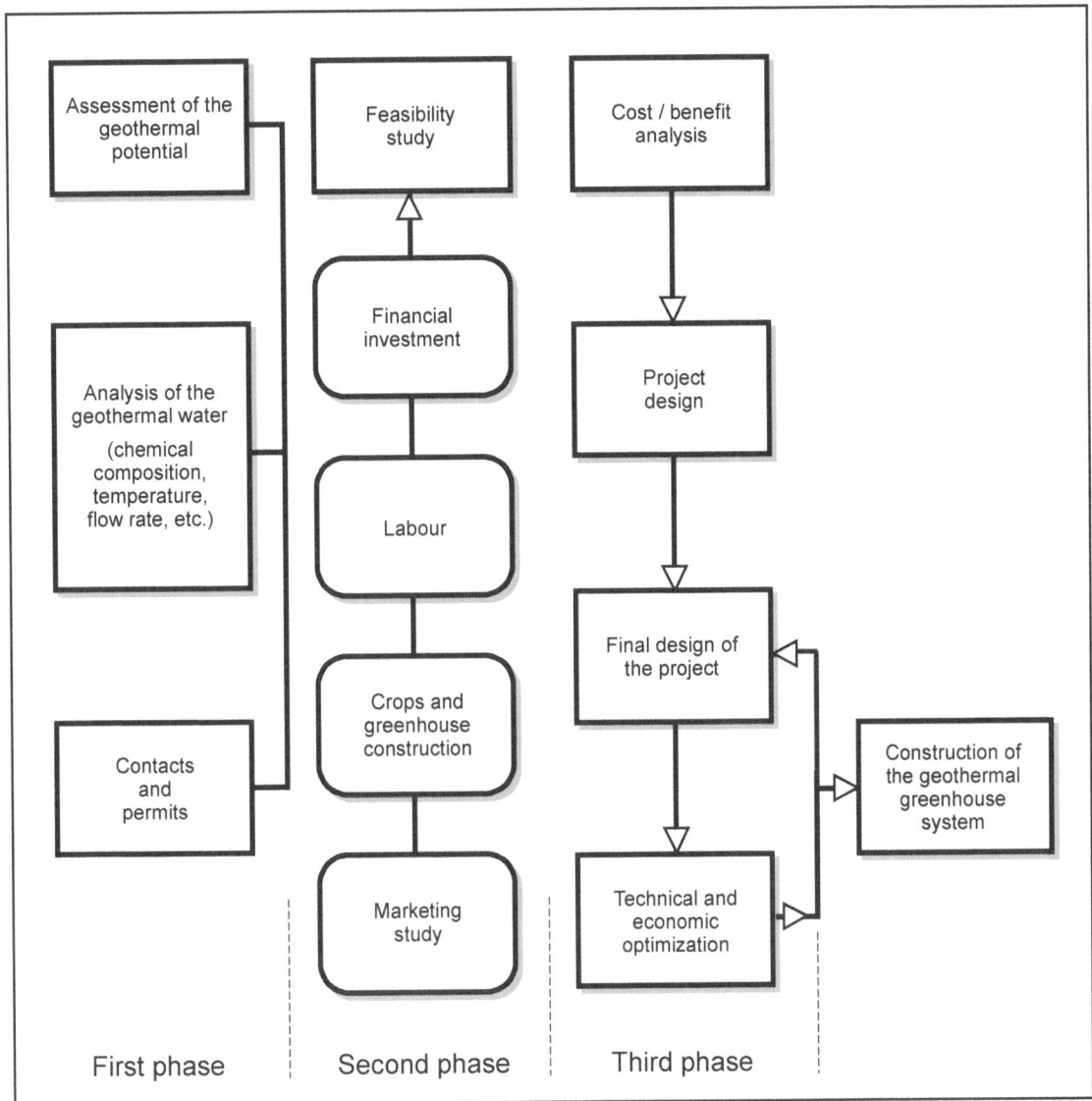

Figure 5.18 Block diagram of the development of a geothermal greenhouse heating project

REFERENCES

AGENCE POUR LES ECONOMIES D'ENERGIE. 1982. *Utilization des Energies de Substitution dans les Serres*, Paris, France.

CAMPIOTTI, C. A.; PICCIURRO, G.; ROMANAZZO, M.; TAGGI, R. 1985. Climatic and environmental data monitoring in greenhouses heated with geothermal water. *Acta Horticulturae*, Vol. 174, pp. 533–9.

GRAFIADELLIS, M. 1986. In: *Development of Simple Systems for Distribution of Geothermal Energy in Greenhouses*, FAO/CNRE First Workshop on Geothermal Energy Use in Agriculture, Skopje, Macedonia.

POPOVSKI, K. 1984. Use of solar energy and heat accumulator in the soil for greenhouse heating. Ph.D. dissertation. Ljubljana, Slovenia, Faculty of Mechanical Sciences, 'Edward Kardelj' University.

POPOVSKI, K. (ed.). 1988*a*. *Geothermal Energy Resources and Their Use in European Agriculture*. Rome, Italy, REUR Technical Series 4, CNRE Study 2.

POPOVSKI, K. 1988*b*. *Present status of geothermal energy use in agriculture in Europe*. In: FAO/CNRE Workshop on Solar and Geothermal Energy Use for Greenhouses Heating, Adana.

POPOVSKI, K. 1988*c*. *Simple heating methods for mild Mediterranean climate conditions*. In: ISHS Symp. on Simple Methods for Heating and Ventilating Greenhouses in Mild Climate Conditions, Djerba-Tozeur.

POPOVSKI, K. (ed.). 1991. *Engineering aspects of geothermal energy use in agriculture, guideline and textbook*. In: International Summer School on Direct Application of Geothermal Energy, Skopje, Macedonia.

POPOVSKI, K. 1998. *Geothermally heated greenhouses in the world*. Guideline and Proc. International Workshop on Heating Greenhouses with Geothermal Energy, Ponta Delgada, Azores, pp. 42–8.

POPOVSKI, K.; POPOVA, N. 1987. *Economy of geothermal energy use in agriculture*. In: First FAO/CNRE Workshop on Geothermal Energy Use in Agriculture, Skopje, Macedonia.

POPOVSKI, K.; POPOVSKA-VASILEVSKA, S. (eds.). 1998. *Heating Greenhouses with Geothermal Energy in the World*. Guideline and Proc. International Workshop on Heating Greenhouses with Geothermal Energy, Ponta Delgada, Azores.

VAN DEN BRAAK, N. J.; KNIES, P. 1985. *Waste heat for greenhouse heating in The Netherlands*. In: First FAO/CNRE Technical Consultations on Geothermal Energy and Industrial Thermal Effluents Use in Agriculture, Rome.

VON ZABELTITZ, C. 1986. *Gewachshauser, Handbuch des Erwerbgartners*. Stuttgart, EV Vertag Eugen Ulmer.

SELF-ASSESSMENT QUESTIONS

1. Why do we grow plants in an environment protected from the outside ambient environment, that is, in greenhouses?

2. There are many greenhouses of very simple construction, with no heating installations. Usually these greenhouses are quite small. What are their main characteristics of operation?

3. The amount of thermal energy required by a greenhouse depends on three main elements. Name them.

4. Which of the following parameters have a direct influence on the growth processes of greenhouse plants: air temperature in the greenhouse, CO_2 concentration, soil temperature, wind velocity, air and soil humidity, light?

5. The connections between the geothermal well and the installations of the heat user(s) can be either direct or indirect. Our choice will be based on which of the following?
(a) The chemical composition of the geothermal water.
(b) The type of heat exchanger or the type of heating installation adopted.
(c) The chemical composition of the geothermal water and the materials used for the installations of the heat user.

6. The heat exchanger is the part of the heating installation that extracts the heat from the geothermal water and passes it to the space or material requiring heat. On which of the following do we base our classification of the heat exchangers used in greenhouse heating:
(a) The type of heat transfer involved.
(b) The location of the heating elements with respect to culture allocation and their arrangement in the protected space.
(c) Both (a) and (b), because of the different heat requirements of the different plants and types of cultivation.

7. Referring to the discussion of environmental aspects in this Chapter (and also to Chapter 8), what are the possible effects on the environment of setting up geothermal greenhouses? What steps can be taken to reduce these effects?

8. A borehole, sited at an elevation of 600 m above sea level, produces water at 60 °C and a pressure slightly above atmospheric. The decision is taken to use this water to heat a greenhouse located further downhill, at an elevation of about 300 m above sea level. The water does not contain any chemicals that could cause scaling or corrosion, so that, from the chemical point of view, it could be carried directly into the greenhouse heating installation. However, a double circuit is adopted: in the primary circuit the geothermal water circulates between the borehole and heat exchangers, sited near the greenhouse; in the secondary (closed) circuit water circulates in the heating installation inside the greenhouse. Between these two circuits are the heat exchangers. Why was this double circuit solution adopted, considering it is more expensive than a single one?

9. Are plant life processes affected by solar radiation of any wavelength?

ANSWERS

1. The reason for constructing greenhouses is that each type of plant needs a specific quantity of energy in order to grow, that is, in order to transform the CO_2 in the air and the H_2O in the air and soil into organic matter (photosynthesis). The energy required by the plant is not always available in sufficient and continuous quantities, and in the localities desired. Greenhouses are therefore intended to provide the plant with the optimum quantity and quality of energy needed for photosynthesis in the localities deemed most convenient, independently of outdoor conditions, by either 'helping nature' or recreating these optimum conditions artificially.

2. The walls of these simple greenhouses are made of transparent material, such as glass or plastic film, which allow the short waves of solar radiation to penetrate to the inside of the protected area practically without any loss of energy. These waves change to long waves once they reach the soil surface and the plant canopy, and are trapped in the greenhouse because long waves cannot penetrate the walls as easily as short waves.

3. (a) The climate in the zone where we construct the greenhouse, (b) the type of greenhouse and (c) the plant species we want to cultivate.

4. All these parameters have a direct influence on plant growth, with the exception of wind velocity, which has no effect inside the greenhouse. However, wind does affect the amount of heat loss from the greenhouse walls and must therefore be taken into consideration.

5. (c) We must determine the effects of the chemical composition of the geothermal water on the materials used in the heat user's installation, in order to avoid corrosion problems. The possibility of scaling should also be considered. Indirect connections should be used only when it is impossible to avoid doing so, and only where it is economically feasible. For example, it may be more economic to adopt a direct connection and change the heat user's installation from time to time (as is the case with simple plastic installations), rather than using indirect connections.

6. (c).

7. The impact on the environment derives mainly from:
(a) Possible damage to animal and plant life and to the landscape, caused by the buildings, the pipelines for transporting the geothermal water, the boreholes and so on. This type of environmental impact can be reduced somewhat by adopting particular architectural solutions, but it certainly cannot be avoided altogether.
(b) Environmental pollution can also be caused by the temperature of the wastewater discharged from the plant, which may be higher than the ambient temperature. In this case measures should be taken to decrease the fluid temperature as much as possible.
(c) Finally, chemical pollution is another environmental hazard, caused by the content of the geothermal water. These chemicals are effectively impossible to eliminate. However, both thermal and chemical pollution can be avoided by re-injecting the wastewater back into the underground.

8. The double circuit had to be adopted because of the 300 m difference in elevation between the borehole and the utilization plant. This elevation difference corresponds to a pressure, at the end of the circuit, of about 300 kPa. This pressure would create severe problems in the heating installations in the greenhouse and would entail very complicated and expensive technical solutions.

9. No. Only solar radiation with wavelengths between 400 and 700 nm influences plant life processes. It is therefore important that the materials used in the construction of a greenhouse permit these radiation wavelengths to penetrate.

CHAPTER 6

Aquaculture

AIMS

1. To show that geothermal energy can readily be used in the raising of numerous aquatic species, both marine and freshwater; that the geothermal fluids needed for aquaculture have low temperatures and are, therefore, available in many countries.

2. To give specific examples of successful geothermal aquaculture projects.

3. To outline some simple methods for calculating heat loss from a body of water, and thus estimating the heat that has to be supplied to maintain the desired temperature conditions; to suggest various systems for reducing these heat losses.

OBJECTIVES

When you have completed this unit, you should be able to:

1. Estimate the optimal temperature required by several aquatic species to create the most favourable growth conditions.

2. Cite several examples of successful geothermal aquaculture projects and the typical species being raised.

3. Calculate, by simple equations, the heat loss from a body of water as a result of (a) evaporation, (b) convection, (c) radiation and (d) conduction.

4. Describe the main systems for reducing heat loss from bodies of water.

5. Evaluate the maximum surface area allowable for a pond as a function of the quantity of heat available.

6. Give examples of calculations of the amount of heat required by a pond used in aquaculture.

6.1 INTRODUCTION TO GEOTHERMAL AQUACULTURE

John W. Lund
Geo-Heat Center, Klamath Falls, Oregon, USA

6.1.1 Background

Aquaculture involves the raising of freshwater or marine organisms in a controlled environment to enhance production rates. The main species reared in this way are carp, catfish, bass, tilapia, frogs, mullet, eels, salmon, sturgeon, shrimp, lobster, crayfish, crabs, oysters, clams, scallops, mussels, abalone, tropical fish (cichlids), alligators and crocodiles.

It has been demonstrated that more fish can be produced in a shorter period of time if geothermal energy is used in aquaculture than if the water is solely dependent upon the Sun for its heat. When the water temperature falls below the optimal values, the fish lose their ability to feed because their basic body metabolism is affected (Johnson, 1981). A good supply of geothermal water, by virtue of its constant temperature, can therefore 'outperform' even a naturally mild climate.

Ambient temperature is generally more important for aquatic species than for land animals, which suggests that the potential of geothermal energy in aquaculture may be greater than in animal husbandry, such as pig and chicken rearing (Barbier and Fanelli, 1977). Figure 6.1 shows the growth trends for a few land and aquatic species. Land animals grow best in a wide temperature range, from just under 10 °C and up to about 20 °C. Aquatic species such as shrimp and catfish have a narrower range of optimum production at a higher temperature, approaching 30 °C. Trout and salmon, however, have a lower optimum temperature, around 15 °C.

A total of sixteen countries reported geothermal aquaculture installations at the World Geothermal Congress 2000 (Lund and Freeston, 2001). The leading countries are China, the United States, Turkey, Israel, Iceland, Japan and Georgia. Unfortunately very little information on pond sizes, the use of raceways, or kg of fish produced was presented in the country update reports. Thus, based on work in the United States, it was calculated that it requires 0.242 TJ/yr/tonne of fish (bass and tilapia) using geothermal water in ponds and 0.675 TJ/yr/tonne of fish in raceways. Using these approximate numbers, the 11,733 TJ/yr of energy reported for aquaculture should be equivalent to producing between

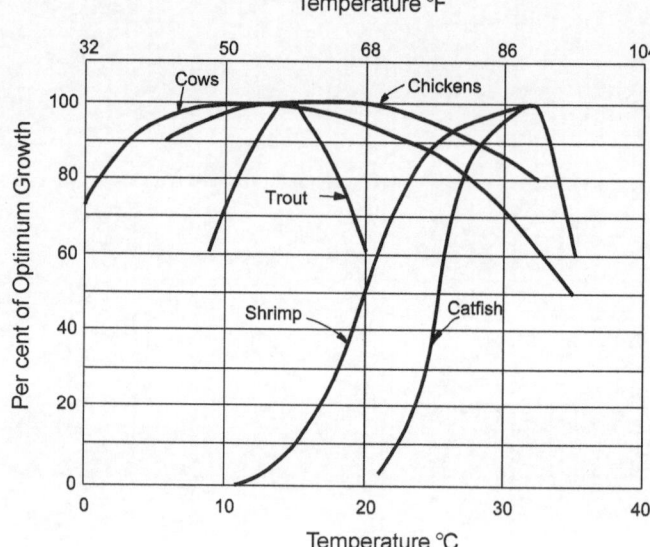

Figure 6.1 Optimum growing temperatures for selected animal and aquatic species
Source: Beall and Samuels, 1971, modified.

17,100 and 47,800 tonnes of fish per year. The reported 600 MW$_t$ of installed capacity gives a capacity factor for the international aquaculture industry of 0.62.

6.1.2 Examples of geothermal projects

Fish breeding is a successful business in Japan, where carp and eels are the most popular species raised. Eels are the most profitable, and are reared in earthenware pipes 25 cm in diameter and 0.9 m long. Water in the pipes is held at 23 °C by mixing hot spring water with river water. The adult eels weigh from 100 to 150 grams, with a total annual production of 3,800 kg. Alligators and crocodiles are also reared in geothermal water, but these reptiles are bred purely for tourism. In combination with greenhouses exhibiting tropical flora, alligator farms are becoming even more popular, making a significant contribution to the growth of the domestic tourist industry (JGEA, 1974). Alligators are now raised in the United States in conjunction with aquaculture operations in Idaho and Colorado (Clutter, 2001b). In Iceland, 610,000 salmon and trout fingerlings are raised annually in geothermal water in ten fish hatcheries, in a new and fast-growing industry (Hansen, 1981; Georgsson and Fridleifsson, 1996).

In the United States, aquaculture projects using geothermal water exist in Arizona, Idaho, Nevada, Oregon and California.

Fish Breeders of Idaho, Inc., located near Buhl, have been rearing channel catfish in high-density concrete raceways for almost thirty years. The water is supplied by artesian geothermal wells flowing at 380 l/s at 32 °C. Cold water from springs and streams is used to cool the hot water to 27 to 29 °C for the best production temperature. Normal stocking densities are from 80 to 160 kg of fish per cubic meter. The maximum recommended inventory for commercial production is about 1.6 to 2.4×10^5 kg per cubic meter per second of water. Yearly production will usually be three to four times the carrying capacity. Oxygen and ammonia are the principal factors limiting production (Ray, 1979). In addition to raising catfish they also raise tilapia, with an annual production of 227 and 45 tonnes, respectively. Rainbow trout and sturgeon are grown in cold water on adjacent property. In their operation, over 90 tonnes of waste are produced annually from processing the fish for market. To solve this disposal problem, they started raising alligators in 1994, which were fed the fish waste and are eventually harvested for their meat and hides (Clutter, 2001*b*). Since 1995, the operation has processed 3,500 alligators at an average length of over two metres.

Giant freshwater prawns (*Macro brachium rosenbergii*) were raised at Oregon Institute of Technology (OIT) from 1975 to 1988. Additional research in trout culture and mosquito fish (*Gambuzia affinis*) at this site demonstrated that these and the tropical crustaceans can be grown in a cold climate as low as −7 °C if the water temperature is maintained at the optimal growing temperature. Initially, two smaller outdoor ponds (1.2 m deep) were used, before another two of 0.2 ha each were built (Figure 6.2). A selected brood stock

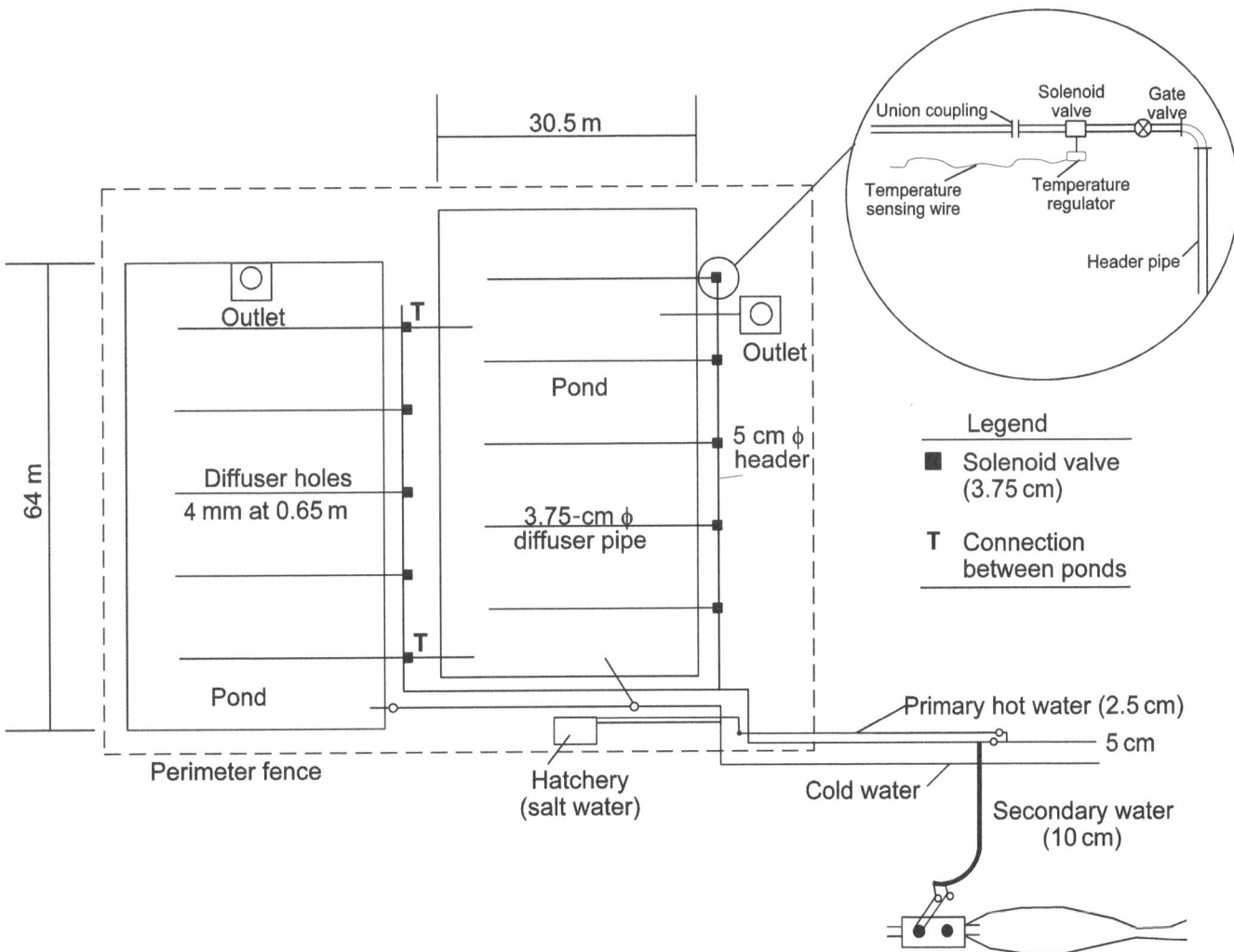

Figure 6.2 The geothermal aquaculture research project at Oregon Institute of Technology
Source: Smith, 1981.

of prawns was held in a small spawning building where larvae were hatched in artificial saltwater and reared to the post-larva stage, which made the facility self-supporting. Growth rates of 2 cm per month were maintained (twice that obtained in tropical climates) with a 900 cm^2 of surface area per animal maximum density. The plumbing system of the ponds consisted of perforated diffuser pipes, control valves and thermostats to maintain an optimum temperature in the ponds (27 to 30 °C, which provided an even distribution of geothermal energy throughout the ponds (Johnson, 1978 and 1981; Smith, 1981).

A very successful catfish raising operation was launched by the Native American community at Fort Bidwell (northeastern California). Geothermal well water at 40 °C is mixed with cold water to produce 27 °C water, which is then piped into raceways 7.6 m long × 2.4 m wide × 1.2 m deep. Two sets of parallel raceways use 57 to 63 l/s. A 0.3 m drop between raceways is used to aerate the water. The initial stock of 28 g fish at 3,000 per raceway produced a surviving 2,000 fish at 0.9 kg each in five months. The cost of construction of the raceways and well was US$100,000. The fish are sold live at the source for US$6.60 to US$8.80 per kg. The production cost at Fort Bidwell is approximately US$1.32/kg (Johnson, 1990).

In a tropical fish raising operation near Klamath Falls, Oregon (Lund, 1994; Clutter, 2001a), the effluent water from a greenhouse operation is used to heat thirty-seven shallow tropical fish ponds lined with diatomaceous Earth. These ponds are 30 m long and 4 m wide, and vary from 1.0 to 1.4 m in depth. They are kept at a constant 23 °C temperature, using 7.0 TJ per year of geothermal heat. At present, the owner raises eighty-five varieties of cichlid fish for pet stores on the west coast of the United States. Approximately 250,000 fish, 7.5 to 10 cm in length, are shipped annually from the local airport or sent by truck. The geothermal heat is a real advantage, as the greatest demand for the fish is during the winter months, and the owner estimates the operation saves US$100,000 annually using geothermal energy.

In 1987, one of the largest and most successful freshwater prawn farms was established on North Island, New Zealand, to take advantage of geothermal waste heat from the Wairakei power-generating field. At present, the farm has nineteen ponds varying in size from 0.2 to 0.35 ha and from 1.0 to 1.2 m in depth. The ponds are kept at a temperature of 24 °C, with a variation of 1 °C from one end of the pond to the other. The farm is currently capable of producing up to 30 tonnes of prawns per year. The adult prawns are harvested at about nine months, averaging thirty to forty per kg, and sold at US$37/kg wholesale, and US$60/kg retail. A restaurant on the property catering to about 25,000 tourists each year buys 90 per cent of the harvested prawns. In the near future, another 40 ha will be added on the other side of the Wairakei power plant, using waste cooling water from a proposed binary power generator. The farming operation could then well become the third-largest freshwater prawn producer in the world, at 400 tonnes per year, which would mean an income of more than US$6.7 million annually (Lund and Klein, 1995).

An eel farm was started in Slovakia near Turčianske Teplice in 1992 using geothermal waters (Lund, 1998). Well water at 42 °C is mixed with colder water and supplied at 25 °C to sixty circular tanks 4 m in diameter. Each tank will hold from 5,000 to 30,000 eels, weighing a total of about 500 to 1,500 kg, depending on their size. Eels (*Anguilla anguilla*) are initially provided to the farm weighing about 0.3 grams. They are then raised for eighteen to twenty months until they reach a weight of about 150 grams. The eels are harvested in the spring and autumn, mainly for export.

6.1.3 General design and considerations

Based on experience at the Oregon Institute of Technology, ponds for raising shrimp, gambusia and trout are best constructed with 0.1 ha of surface area. A size of 15 by 60 m is ideal for harvesting. A minimum-sized commercial operation should have 3 to 4 ha under development (water surface area), or about thirty to forty ponds. The maximum surface area that should be considered for a single pond is 0.2 ha. For tropical fish, smaller sizes in the 5 by 30 m range are used. Figure 6.2 illustrates the geothermal pond design of the Oregon Institute of Technology. Recent trends are to use circular holding tanks, constructed of either metal or fibreglass, of 6 to 10 m diameter. An example of this type of geothermal installation for tilapia is found in Imperial Valley, California (Rafferty, 1999). The optimum size and shape is a function of the species.

The most important items to consider are quality of the water and disease. If geothermal water is to be used directly, evaluation of heavy metals such as fluorides, chlorides and arsenic must be undertaken to determine whether the fish or prawns can survive. A small test programme is often a wise first step. An aeration pond preceding the stocked ponds will

Table 6.1 Crops that are good candidates for aquaculture

Species	Growth period (months)	Water temperature (°C)
Tropical fish	2–3	23–27
Catfish	4–6	27–29
Trout	4–6	13–18
Prawns	6–9	27–30
Tilapia	6–9	22–30

often solve the chemical problems. If necessary, a heat exchanger can be used to isolate the geothermal water from benign pond water in a secondary loop.

Crops that are a good candidate for aquaculture are listed in Table 6.1.

Tropical fish (cichlids) are generally the easiest to raise, and have a low investment and high yield. As indicated above, smaller ponds can also be used. Freshwater prawns generally have a high market value, with marketable sizes being thirty-five to forty-four tails to the kilogram. Channel catfish are also popular, especially as fillets. Production rates depend upon water quality and flow rates. Tilapia appears to be one of the fastest growing fish products in the United States, and is popular with geothermal operations.

Ponds require geothermal water of 38 to 66 °C and a peak flow of 10 to 25 l/s for 0.2 ha of uncovered surface, depending on the climate. The long axis of the pond should be constructed perpendicular to prevailing winds to minimize wave action and temperature loss. The ponds are normally constructed of excavated Earth and lined with clay or plastic where necessary to prevent seepage loss. Temperature loss can be reduced, thus reducing the required geothermal flow, by covering the pond with a plastic bubble. Construction cost, exclusive of geothermal wells and pipelines, will run to US$75,000 to US$125,000 per hectare.

6.1.4 Additional information

An engineering design guide on how to determine heating requirements, as well as various alternatives on how to prevent heat loss, such as surface covers, pond enclosures and using thermal mass for an aquaculture pond, is provided by Kevin Rafferty in the following section. His design guide calculates heat losses from evaporation, convection, radiation and conduction.

A geothermal *Aquaculture Information Package* (Boyd and Rafferty, 1999) is available from the Geo-Heat Center. This package includes material on market and price information, water quality issues in aquaculture, culture information, pond and raceway heat-loss calculations, aquaculture bibliography, and an aquaculture glossary. This package is available from the Geo-Heat Center website: http://geoheat.oit.edu/pdf/ aqua.pdf.

REFERENCES

Barbier, E.; Fanelli, M. 1977. Non-electrical uses of geothermal energy. *Progr. Energy Combustion Sci.*, Vol. 3, pp. 73–103.

Beall, S. E.; Samuels, G. 1971. The use of warm water for heating and cooling plant and animal enclosures. *Oak Ridge National Laboratory*, ORNL-TM-3381.

Boyd, T.; Rafferty, K. 1999. *Aquaculture Information Package*. Klamath Falls, Ore., Geo-Heat Center.

Clutter, T. 2001a. Out of Africa. *Geothermal Bulletin*, Vol. 30, No. 2, Davis, Calif., Geothermal Resources Council, pp. 74–6.

Clutter, T. 2001b. Gators in the sage. *Geothermal Bulletin*, Vol. 30, No. 6, Davis, Calif., Geothermal Resources Council, pp. 246–9.

Georgsson, L. S.; Fridleifsson, O. 1996. High technology in geothermal fish farming at Silfurstjarnan Ltd., N.E. Iceland. *Geo-Heat Center Quarterly Bulletin*, Vol. 17, No. 4, pp. 23–8.

Hansen, A. 1981. Growing under glass. *Iceland Rev.* Vol. 19, No. 4.

J.G.E.A. 1974. *Geothermal energy utilization in Japan*. Tokyo, Japan, Japan Geothermal Energy Association.

Johnson, W. C. 1990. Personal communication.

Johnson, W. C. 1981. The use of geothermal energy for aquaculture. *Proc. 1st Sino/U.S. Geothermal Resources Conference*, Tianjin, P.R. China.

JOHNSON, W. C. 1978. *Culture of Freshwater Prawns Using Geothermal Waste Water*. Klamath Falls, Ore., Geo-Heat Center.

LUND, J. W.; FREESTON, D. 2001. Worldwide direct uses of geothermal energy 2000. *Geothermics*, Vol. 30, pp. 29–68.

LUND, J. W. 1998. Geothermal eel farm in Slovakia. *Geo-Heat Center Quarterly Bulletin*, Vol. 19, No. 4 pp. 17–18.

LUND, J. W.; KLEIN, R. 1995. Prawn park: Taupo, New Zealand. *Geo-Heat Center Quarterly Bulletin*, Vol. 16, No. 4, pp. 27–9.

LUND, J. W. 1994. Agriculture and aquaculture cascading the geothermal way. *Geo-Heat Center Quarterly Bulletin*, Vol. 16, No. 1, pp. 7–9.

RAFFERTY, K. 1999. Aquaculture in the Imperial Valley: A geothermal success story. *Geo-Heat Center Quarterly Bulletin*, Vol. 20, No. 1, pp. 1–4.

RAY, L. 1979. Channel catfish (*Ictalurus punctatus*) production in geothermal water. *Geothermal Resources Council Special Report 5*, pp. 65–7.

SMITH, K. C. 1981. *A Layman's Guide to Geothermal Aquaculture*. Klamath Falls, Ore.

SELF-ASSESSMENT QUESTIONS

1. List at least three aquatic species that have been successfully raised in geothermal waters.

2. What is the main benefit, besides energy savings, of raising aquatic species using geothermal energy?

3. List the two most common types of enclosures used for raising aquatic species.

4. What is the approximate size (width, length and depth) of ponds used in commercial operations for prawns, trout and tilapia? What is the recommended total surface area for a commercial operation?

5. What are some of the possible problems using the geothermal water directly in aquaculture facilities? How might these problems be mitigated?

ANSWERS

1. Prawns, eels, alligators, crocodiles, catfish, trout, salmon, and tropical fish (cichlids).

2. Enhanced production rate (that is, more crops per year).

3. Ponds or tanks, and raceways.

4. Size: approximately 15 m by 60 m by 1.0 to 1.2 m. Recommended total surface area is 3 to 4 hectares.

5. The water temperatures may be too high, thus requiring mixing. Heavy metals such as fluorides, chlorides, arsenic, etc. may be present. An aeration pond, or using a heat exchanger and a secondary source of water, can remove or isolate these elements from the ponds or tanks.

6.2 AQUACULTURE TECHNOLOGY

Kevin D. Rafferty
Geo-Heat Center, Klamath Falls, Oregon, USA

6.2.1 Introduction

Aquatic species, particularly those produced in the commercial aquaculture industry, are cold-blooded animals. Maintaining them in a warm environment increases the metabolism, and allows the animal to feed, and thus grow, at increased rates. The increased growth rate permits the production of market-sized products in a shorter period of time, and enhances the commercial viability of an aquaculture operation. Key to this is the maintenance of optimum temperatures for the particular species being grown. Table 6.2 provides some data on culture temperatures for key aquaculture species.

Table 6.2 Temperature requirements and growth periods for selected aquaculture species

Species	Tolerable extremes (°C)	Optimum growth (°C)	Growth period to market size (months)
Oysters	0–36	24–26	24
Lobsters	0–31	22–24	24
Penaeid shrimp			
Kuruma	4–?	25–31	6 to 8 typically
Pink	11–40	22–29	6 to 8
Salmon (Pacific)	4–25	15	6 to 12
Freshwater prawns	24–32	27–30	6 to 12
Catfish	17–35	27–29	6
Eels	0–36	23–30	12 to 24
Tilapia	8–41	22–30	—
Carp	4–38	20–32	—
Trout	0–32	15	6 to 8
Yellow perch	0–30	22–28	10
Striped bass	? –30	16–19	6 to 8

Source: Behrends, 1978.

This material was originally published by the author as part of the *Geothermal Direct Use Engineering and Design Guidebook* by the Geo-Heat Center, Klamath Falls, Oregon, USA.

To accomplish the maintenance of temperatures similar to those indicated in Table 6.2, it may be necessary to supply substantial quantities of heat to the water in the culture tanks, ponds or raceways. For outdoor operations, it is frequently impractical to accomplish this using conventionally fuelled heating equipment. Low-temperature geothermal resources are uniquely suited to this application. In fact, for the last five years, aquaculture has been the largest and fastest growing geothermal direct-use application in the United States.

The focus of this chapter is the calculations specific to heat losses from bodies of water. Information on the culture methods and biological considerations for individual species has been published widely and may be found in Avault (1996), Wheaton (1985) and many others.

6.2.2 Heat exchange processes

For those involved in the initial planning of an aquaculture project, one of the first questions to be addressed relates to project size. In most geothermal applications, the maximum heat available from the resource restricts the maximum pond area that can be developed. This section will therefore deal briefly with the subject of heat loss from ponds (or pools), one of the main parameters that must be determined to make an informed evaluation of geothermal resources intended for aquaculture use.

A non-covered body of water, exposed to the elements, exchanges heat with the atmosphere by way of four mechanisms:

- evaporation
- convection
- radiation
- conduction.

Each of these is influenced by different parameters that are discussed separately in the following paragraphs.

Evaporative loss
Evaporation is generally the largest component of the total heat loss from the pond. Considering evaporation, the loss of volume generally comes to mind rather than the loss of heat. However, in order to boil water (and hence cause evaporation) heat must be added. The quantity of heat required to evaporate one kilogram of water varies with temperature and pressure, but under normal atmospheric conditions the

value is approximately 2,440 kJ. When water is evaporated from the surface of the pond, the heat is taken from the remaining water. As a result, as each kilogram of water evaporates from the surface, approximately 2,440 kJ are lost with escaping vapour.

Losses can occur by evaporation even when the water temperature is at or below the surrounding air temperature.

The rate at which evaporation occurs is a function of air velocity and the pressure difference between the pond water and the water vapour in the air (vapour pressure difference). As the temperature of the pond water is increased or the relative humidity of the air is decreased, the evaporation rate increases. The equation that describes the rate of evaporation is (ASHRAE, 1995):

$$W_p = (11.0 + 4.30 \ v) \times (P_w - P_a) \times A$$

where

  W_p = rate of evaporation (kg/h)
  A = pond surface area (m^2)
  v = air velocity (m/s)
  P_w = saturation vapour pressure of the pond water (bar-absolute)
  P_a = saturation pressure at the air dew point (bar-absolute)

For enclosed ponds or indoor swimming pools, this equation can be reduced to (ASHRAE, 1995):

$$W_p = 14.46 \times A \times (P_w - P_a)$$

where

  W_p = rate of evaporation (kg/h)
  A = pond area (m^2)
  P_w = saturation pressure of the pond water (bar-absolute)
  P_a = saturation pressure at air dew point (bar-absolute).

The following are some common values for P_w and P_a:
  For P_w: at 15 °C water, P_w = 0.0170 bar
      at 20 °C water, P_w = 0.0234 bar
      at 25 °C water, P_w = 0.0317 bar
      at 30 °C water, P_w = 0.0424 bar.

For outdoor locations with a design dry-bulb air temperature below 0 °C, P_a can be taken as 0.0061 bar. For indoor locations with a design of approximately 24 °C and 50 per cent relative humidity, P_a can be taken as 0.0145 bar.

For example, assume a pond with a surface area of 50 m^2 (5 m × 10 m) located outside in an area with design temperature of −10 °C. Wind velocity is 2.5 m/s and pond water is to be 25 °C.

$$\begin{aligned} W_p &= [11.0 + (4.30 \times 2.5)] \times (0.0317 - 0.0061) \times 50 \\ &= 27.8 \ \text{kg/h} \end{aligned}$$

To obtain the heat loss (q_{EV}) in kJ/h, simply multiply the kJ/h loss by the value of 2,440 kJ/kg:

$$q_{EV} = 27.8 \ \text{kg/h} \times 2,440 \ \text{kJ/kg}$$

$$q_{EV} = 67,900 \ \text{kJ/h}.$$

This is the peak or design heat loss. It is important to note that the example values given above are for the design (worst) case. At higher outdoor air temperatures and different relative humidities, this value would be less. As mentioned earlier, the rate of evaporation loss is influenced by the vapour pressure difference between the pond water and the water vapour in the air. Reduced water temperature would reduce the vapour pressure differences, and hence the rate of evaporation.

Wind speed over the surface of the water has a very substantial impact upon both evaporative and convective heat losses from ponds. When calculating the design heat loss for ponds, it is not necessary to use unrealistically high wind speeds. In general, the coldest outdoor temperatures are not accompanied by high wind-speed conditions. Furthermore, sustained high-wind conditions are generally not experienced for extended periods of time. This, coupled with the high thermal mass of the water, allows the pond or pool to sustain brief high-wind periods without significant water temperature drop. The mean wind speeds given in Chapter 1 of the US Department of Defense publication *Engineering Weather Data* (AFM 88-29, 1978) are appropriate values for these calculations.

Pond surface area can be influenced by surface disturbances caused by waves or the use of splash-type aeration devices. The calculations presented above are based upon a calm water surface. If surface disturbances exist, the pond surface area ('A' in the earlier equation) should be increased to reflect the departure from the calm water condition.

Convective loss

The next major mechanism of loss from the pond surface is convection. This is the mode associated with the heat loss caused by cold air passing over the pond surface. The two most important influences on the magnitude of convective heat loss are wind velocity and temperature difference between the pond surface and the air. This is evidenced in Wolf (1983):

$$q_{CV} = (9.045 \, v) \times A \times (T_w - T_a)$$

where

q_{CV} = convection heat loss (kJ/h)
v = air velocity (m/s)
A = pond area (m^2)
T_w = water temperature (°C)
T_a = air temperature (°C)

The shape of the pond and the direction of the prevailing wind influence the magnitude of the convective heat loss. The method used here is appropriate for pond dimensions up to approximately 30 m. For very large ponds, convective losses would be up to 25 per cent lower than the figure obtained with this method.

For an indoor pool, this equation would be (Lauer, undated):

$$q_{CV} = 22.6 \, (T_w - T_a)^{1.25} \times A$$

Using the example from above (−10 °C design temperature, 25 °C water and 2.5 m/s wind), the following convective heat loss can be calculated:

$$q_{CV} = (9.045 \times 2.5 \, \text{m/s}) \times 50 \, \text{m}^2 \times (25° + 10°)$$

$$q_{CV} = 39,600 \, \text{kJ/h}$$

Radiant loss

Radiant heat loss, the third-largest component of the total heat loss, is dependent primarily on the temperature difference between the pond surface temperature and surrounding air temperature. Under normal circumstances, radiant heat exchange is assumed to occur between solid bodies with little or no gain to the air in between the bodies. However, because of the evaporative losses near the pond surface, the air tends to contain a large quantity of water vapour. When this is the case, the pond surface radiates to the water vapour in the air, which is assumed to be at the temperature of the air itself. The equation describing this process is (Stoever, 1941):

$$q_{RD} = 1.836 \times 10^{-8} \times [(492 + 1.8 \, T_w)^4 - (492 + 1.8 \, T_a)^4] \times A$$

where

q_{RD} = radiant heat loss (kJ/h)
T_w = pond water temperature (°C)
T_a = air temperature (°C)
A = pond surface area (m^2)

Again, referring to the above example (−10 °C design temperature, 25 °C pond temperature), the following radiant heat loss is calculated:

$$q_{RD} = 1.836 \times 10^{-8} \times [(492 + 1.8 \times 25 \, °C)^4 - (492 - 1.8 \times 10)^4] \times 50$$

$$q_{RD} = 30,000 \, \text{kJ/h}$$

Conductive loss

The final mode of heat loss is conduction. This is the loss associated with the walls of the pond. Of the four losses, conduction is by far the smallest, and in many calculations is simply omitted. The following method (ASHRAE, 1985) is valid for a pond depth of 0.9 to 1.5 m:

$$q_{CD} = \{[(L + W) \times 12.45] + (L \times W \times 0.4084)\} \times [T_w - (T_a + 8.33)]$$

where

q_{CD} = conductive heat loss (kJ/h)
L = length of pond (m)
W = width of pond (m)
T_w = design water temperature (°C)
T_a = design outside air temperature (°C)

This calculation assumes the use of a lined pond construction. That is, there is no significant leakage of water from the walls or floor of the pond.

Using the previous example, the following conductive heat loss is calculated:

$$q_{CD} = \{[(5 + 10) \times 12.45] + (5 \times 10 \times 0.4084)\} \times [25 - (-10 + 8.33)]$$

$$q_{CD} = 5,500 \, \text{kJ/h}$$

Table 6.3 summarizes the results of the calculations performed for the example of a 50 m^2 pond.

Table 6.3 Summary of example of heat loss

Heat loss method	Loss (kJ/h)	Amount (%)
Evaporation	67 900	47
Convection	39 600	28
Radiation	30 000	21
Conduction	5 500	4
TOTAL	143 000	100

Table 6.4 Summary of example of heat loss using pool cover

Heat loss method	Loss (kJ/h)	Amount (per cent)
Evaporation	0	0
Convection	6 400	41
Radiation	3 500	23
Conduction	5 500	36
TOTAL	15 400	100

These losses are the peak or maximum heat loss. At any given time during the year, other than the design case, the heat loss would be less than this value. The annual heating requirement cannot be determined from simply multiplying the peak heating requirement by 8,760 h/yr. Because of the need to consider varying temperature, wind, humidity and solar heat gain, methods for calculating the annual heating requirement are beyond the scope of this chapter.

6.2.3 Reduction of heating requirements

Depending on the specifics of the application (surface area, water temperature) and the site (climate, wind velocity), heat losses from the facility may be excessive. It may be possible to employ strategies to reduce the heat input required. Some of these strategies are examined in this section.

Surface cover

Heat losses from the pond surface are most heavily influenced by wind velocity and the temperature difference between the pond and the surrounding air. Any method that can be employed to reduce these values can substantially reduce heating requirements.

For outdoor pools, a floating cover is an excellent example. The use of a 127 mm floating foam cover (on the pool surface) would reduce the peak heat loss for the example pool to the values shown in Table 6.4.

This peak load is only approximately 11 per cent of the originally calculated heat loss. This is, in large measure, a result of the elimination of evaporation loss that is provided by a floating-type cover. Unfortunately, a floating cover is generally not considered practical for commercial aquaculture applications.

Pond enclosure

A pond enclosure is another (though much more expensive) option for reducing heat loss. The advantages provided by an enclosure depend to a large extent upon the construction techniques employed (covering material, degree of enclosure, presence or absence of ventilation). The variety of construction methods and materials available is too numerous to cover here. The basic advantages of an enclosure are:

- reduced air velocity
- reduced temperature difference between the pond and surrounding air
- reduced vapour-pressure difference between the pond water and air (increased relative humidity).

These effects reduce the losses associated with evaporation, convection and radiation.

Assuming an enclosure is placed over our example pond, reducing air velocity to the 3 to 10 m/min range, increasing humidity to 90 per cent and air temperature to 9 °C (about halfway between outside air and pond water temperature), pond heat loss would be reduced to the values shown in Table 6.5. This value amounts to 54 per cent of the original example.

It is often erroneously believed that the use of a pond enclosure will allow the air within it to become saturated, and thus eliminate evaporative loss from the pond surface. This is not the case in most applications. For greenhouse-type enclosures (the most common), the inside surface temperature of the roof and walls is well below the dew point of the air during the winter. This results in substantial condensation on these surfaces. As a result, moisture is continuously removed from the air, allowing more to be absorbed (through evaporation) from the pond surface.

For conventionally constructed buildings, ventilation air is normally supplied to reduce humidity in the space to a point that will protect the structure from moisture damage. Under these conditions, of course, evaporation continues to occur, and an additional heating load is imposed by the requirement to heat the ventilation air. This topic is covered in detail in Chapter 4 of the *1995 ASHRAE Handbook of Applications*.

Table 6.5 Summary of example of heat loss using pond enclosure

Heat loss method	Loss (kJ/h)	Amount (per cent)
Evaporation	37 100	48
Convection	15 300	20
Radiation	19 200	25
Conduction	5 500	7
TOTAL	77 100	100

Thermal mass

One final method for reducing peak heating requirements for pond or pool heating lies in the use of the large thermal mass supplied by the water itself. Water is an excellent heat storage medium. Assuming the example pond is 1.5 m deep and 50 m^2 in area, the total volume contained would be 75 m^3. At 1,000 l/m^3, this results in 75,000 litres or 75,000 kg of water. Because 1 kg of water yields 4,200 J for each degree it is cooled, this means that our example pond containing 75,000 kg of water could provide 315,000 kJ of offset heating requirements if it were allowed to cool 1 °C. This stored heating capacity can be used to reduce the peak heating requirement on the heating system. Using the originally calculated peak heating requirement of 143,000 kJ/h, an example of thermal storage use follows. Assume that the peak heating requirement occurs over an eight-hour period after which, because of air temperature increase and solar gain, the heating load is reduced. Further, assume that the heating system is designed to supply only 80 per cent of the peak requirement. What will happen to the pond temperature?

First, calculate the total heat required for the eight-hour period:

$$8 \text{ h} \times 143,000 \text{ kJ/h} = 1,144,000 \text{ kJ}$$

Second, calculate the heat that the system can supply based on its 80 per cent capacity:

$$8 \text{ h} \times (0.80 \times 143,000 \text{ kJ/h}) = 915,000 \text{ kJ}$$

Then, calculate the difference to be supplied by allowing the pond water to cool:

$$1,144,000 \text{ kJ} - 915,000 \text{ kJ} = 229,000 \text{ kJ}$$

Finally, calculate the drop in pond temperature caused by supplying the heat required:

$$229,000 \text{ kJ}/(75,000 \text{ kg} \times 4.20 \text{ kJ/kg °C}) = 0.73 \text{ °C}$$

As a result, the pond will have cooled by 0.73 °C. The heating system would then bring the pond back up to the temperature during the day, when higher temperatures and solar gain would reduce heating requirements.

An alternate way of looking at this is in terms of the selection of the ambient temperature to be used in the calculation of the pond losses. Use of the mean-duty temperature instead of the minimum-duty temperature would allow the design to incorporate the effect of the pond thermal mass on the heat loss. An air temperature that is higher than the mean-duty value could be used in very clear climates, where solar heat gain can be assumed to contribute to pond heating during the day.

The degree to which thermal storage can be incorporated into the heating system design is a complex issue of environmental factors, pond characteristics, and the species being raised. Some species, such as prawns, are particularly sensitive to temperature fluctuations (Johnson, 1978).

6.2.4 Flow requirements

The rate of flow required to meet the peak heating demand of a particular pond is a function of the temperature difference between the pond water and the resource temperature. The following equation can be used to determine the flow (Q) requirement and is written:

$$Q = q_{tot}/[15,040 \times (T_r - T_w)]$$

where
  Q = resource flow requirement (l/s)
  q_{tot} = total calculated pond heat loss (kJ/h)
  $= q_{EV} + q_{CV} + q_{RD} + q_{CD}$
  T_w = pond temperature (°C)
  T_r = resource temperature (°C)

Assuming that our example pond is to be heated with a resource temperature of 40 °C:

$$Q = 143,000 \text{ kJ/h}/[15,040 \times (40 \text{ °C} - 25 \text{ °C})]$$

$$Q = 0.634 \text{ l/s}$$

Again, the point is made that this is the peak requirement. The required flow at any other time would be a value <0.636 l/s. This approach is valid for aquaculture projects and resource temperatures up to levels that would prove harmful if supplied directly to the pond. Above this temperature (which varies according to species), the heating water would have to be mixed with cooler water to reduce its temperature. Two methods are possible for mixing. First, if a sufficient supply of cold water is available, the hot water could be mixed with the cold water before introduction in the pond. A second approach, which would apply in the absence of cold water, is to recirculate pond water for mixing purposes. The recirculation could be combined with an aeration scheme to increase its beneficial effect. In both cases, the quantity of cold or recirculated water could be determined by the following formula:

$$Q_c = \frac{Q_h(T_h - T_m)}{(T_m - T_c)}$$

where
  Q_c = required cold flow rate (l/s)
  Q_h = hot water flow rate (l/s)
  T_h = temperature of hot water (°C)
  T_c = temperature of cold water (°C)
  T_m = temperature of desired mixed water (°C)

The above methods are intended as an introduction to the subject of heat losses from ponds. The equations provided are simplifications of very complex relationships and should be employed only for initial calculations. In addition, losses that can occur from various aeration schemes and other activities have not been addressed. It is strongly recommended that a competent engineer be enlisted for final design purposes.

REFERENCES

ASHRAE (American Society of Heating, Refrigerating and Air-Conditioning Engineers). 1985. *ASHRAE Handbook of Fundamentals*, Atlanta, Ga., ASHRAE.

ASHRAE. 1995. *ASHRAE Handbook of Applications*, Atlanta, Ga.

AVAULT, J. W. 1996. *Fundamentals of Aquaculture*, Baton Rouge, La., AVA Publishing.

BEHRENDS, L. L. 1978. *Waste Heat Utilization for Agriculture and Aquaculture*, Knoxville, Tenn., Tennessee Valley Authority.

US DEPARTMENT OF DEFENSE. 1978. *Engineering Weather Data*. AFM 88-29, TM 5-785, NAUFAC.

JOHNSON, W. C. 1978. *Culture of Freshwater Prawns Using Geothermal Waste Water*. Klamath Falls, Ore.

LAUER, B. E. undated. Heat transfer calculations, reprinted from *Oil and Gas Journal*.

STOEVER, H. J. 1941. *Applied Heat Transmission*. New York, McGraw-Hill.

WOLF, H. 1983. *Heat Transfer*. New York, Harper and Row.

WHEATON, F. W. 1985. *Aquacultural Engineering*, Malaba, Fla., Krieger Publishing Co.

SELF-ASSESSMENT QUESTIONS

1. There are four types of heat loss from a pond: (a) evaporative, (b) convective, (c) radiant, and (d) conductive. What is the governing factor in each of these heat-loss processes?

2. (a) Which type of heat loss can be considered of least importance in a project design for a non-covered body of water, or pond?
(b) Which type of heat loss can be neglected in a covered pond?

3. The type of aeration system used in ponds will also influence heat loss. Which types of heat loss in particular will be affected?

4. The type of covering used on a pond will have no influence on one particular type of heat loss. Which type, and why?

5. The section on 'thermal mass' (section 6.2.3) explains how the body of water in a pond is an excellent heat storage medium. Taking the example given in this section, calculate the maximum acceptable heat loss from the pond during the day (sixteen hours) for the heating system to be able to bring the temperature back to its original value.

6. What steps should be taken in the case of a geothermal fluid with a high salt content?

ANSWERS

1. In (a), the governing factor is vapour pressure. In (b), (c) and (d), the governing factor is temperature.

2. (a) conductive loss; (b) evaporative loss.

3. Evaporative and convective losses.

4. Conductive heat loss. The pond covering will have no influence on conductive heat loss as the latter depends only on the temperature difference between the water and the walls of the pond. Note that in the equation given in 'conductive loss' (section 6.2.2), the temperature of the walls of the pond is estimated as the temperature of the outside air, that is, −10 °C.

5. During the day we have to recover the heat lost during the night (that is, 229,000 kJ) plus the average heat loss during the sixteen-hour day. This total cannot exceed (which probably can easily be obtained as day-time temperatures are higher than night-time temperatures):

$$0.80 \times 143,000 = 114,400 \text{ kJ/hr}$$

Thus, the maximum heat loss during the sixteen-hour day that can be tolerated is:

$$114,400 - 229,000/16 = 114,400 - 14,310 = 100,090 \text{ kJ/hr}$$

or about 77 per cent of the maximum loss determined earlier:

$$110,090/143,000 = 0.77$$

6. As some salts are potentially harmful for certain aquatic species, the geothermal fluid cannot be used directly in the breeding ponds. The heat of the geothermal fluids should be used in some kind of heat exchanger.

CHAPTER 7

Industrial Applications

Paul J. Lienau

Oregon Institute of Technology, Klamath Falls, Oregon, USA

AIMS

1. To demonstrate that geothermal energy can be used in a wide variety of applications, ranging from agriculture and the processing of agricultural products, to the production of consumer goods.

2. To give examples of industrial plants that have been in successful operation for several years; to show how the heat content of low- to medium-temperature geothermal fluids (20 to 150 °C) can be exploited in different stages of these industrial processes, including drying, process heating, evaporation, distillation, washing, salt and chemical extraction.

3. To demonstrate that, in many cases, the arrangements for the industrial processes that utilize heat in the above temperature range can be modified and adapted to the utilization of geothermal fluids, and also to show that these modifications are technically efficient and commercially convenient.

OBJECTIVES

When you have completed this chapter you should be able to:

1. Describe a wide variety of industries that could utilize geothermal heat in all or part of their cycle.

2. Give examples of industrial plants that have been specifically designed or adapted to utilize geothermal heat in their operating cycle.

3. Compare the industrial cycles using geothermal energy with similar cycles using conventional fluids.

7.1 INTRODUCTION

Geothermal energy may be used in a number of ways in the industrial field. Potential applications could include drying, process heating, evaporation, distillation, washing, desalination, and chemical extraction.

The most important energy considerations for an industrial complex are the cost, quality, and reliability. Geothermal energy may be attractive to an industry provided: (a) the cost of energy/kg of product is lower than that presently used, (b) the quality of geothermal energy is as good or better than the present supply, and (c) the geothermal energy will reliably be available for the life of the plant. Reliability and availability can only be proven by long-term use or testing.

In some situations where available geothermal fluid temperatures are lower than those required by the industrial application, the temperatures can be raised by means of integrating thermal systems (boilers, upgrading systems, heat pumps and so on). In designing geothermal energy recovery and utilization systems, alternative possibilities could be considered for various applications. The usual approach for utilization of geothermal fluid by proposed industries is to match the industry to the available fluids. An alternative approach is to match the available fluids to proposed industries. This alternate approach means that it is necessary to develop ways to economically upgrade the quality of existing geothermal fluids, or the fluids derived from them. Figure 7.1 shows application temperature ranges for some industrial and agricultural uses.

7.2 EXAMPLES OF INDUSTRIAL APPLICATIONS OF GEOTHERMAL ENERGY

While there are many potential industrial uses of geothermal energy, the number of worldwide applications is relatively small. A fairly wide range of uses is, however, represented, including heap leaching of precious metals, vegetable dehydration, grain and lumber drying, pulp and paper processing, diatomaceous Earth processing, chemical recovery, and wastewater treatment. Industrial applications largely require the use of steam, or superheated water, while agricultural users may use lower-temperature geothermal fluids. The largest industrial applications are a pulp, paper and wood processing plant in New Zealand, a diatomaceous Earth plant in Iceland, and vegetable dehydration plants in the United States. These systems provide the best current examples of industrial geothermal energy use.

7.2.1 Pulp, paper and wood processing

The integrated newsprint, pulp and timber mills of the Tasman Pulp and Paper Company Ltd, located in Kawerau, New Zealand, are the largest industrial development to utilize geothermal energy. The plant site was selected because of the availability of geothermal energy. Geothermal exploration at Kawerau started in 1952 with the main purpose of locating and developing the geothermal resource for use in a projected pulp and paper mill (Wilson, 1974). The mill produces approximately 181,000 tonnes of Kraft pulp and 363,000 tonnes of newsprint each year (Carter and Hotson, 1992; Hotson, 1995).

In 1995, the Tasman Pulp and Paper Company was using a maximum flow of 320 tonnes/h and an average flow rate of 265 tonnes/h from five wells to supply steam at two pressures, 1,379 and 690 kPa. In 1991 injection commenced in two wells. It is interesting to note that, for forty-two years of production, the reservoir pressure decline has been less than the measurement error. The geothermal steam, which is generated by separate flash plants in the bore field, is used:

- To generate clean steam in shell and tube boilers for use in the paper-making equipment. Clean steam is necessary as the small percentage of non-condensable gases in the geothermal steam can cause intolerable temperature fluctuations in paper-making equipment. These heat exchangers are the most important users of geothermal steam at Tasman.

- For an 8 MW$_e$ turbo-alternator installed in 1960, designed to exhaust to atmosphere, and for two binary electric plants of 6 MW$_e$ using separated water. In 1968, a single-effect evaporator was installed to use exhaust steam to provide additional black-liquor evaporation capacity.

- To supply high-pressure steam to the Fletcher Challenge Forests sawmill for drying timber in kilns. In line with increasing demand for kiln-dried lumber, Fletcher Challenge Forests propose installing a further five geothermally heated kilns and converting the non-geothermal kilns to geothermal.

- To provide the small quantities of steam that are also used for greenhouse heating.

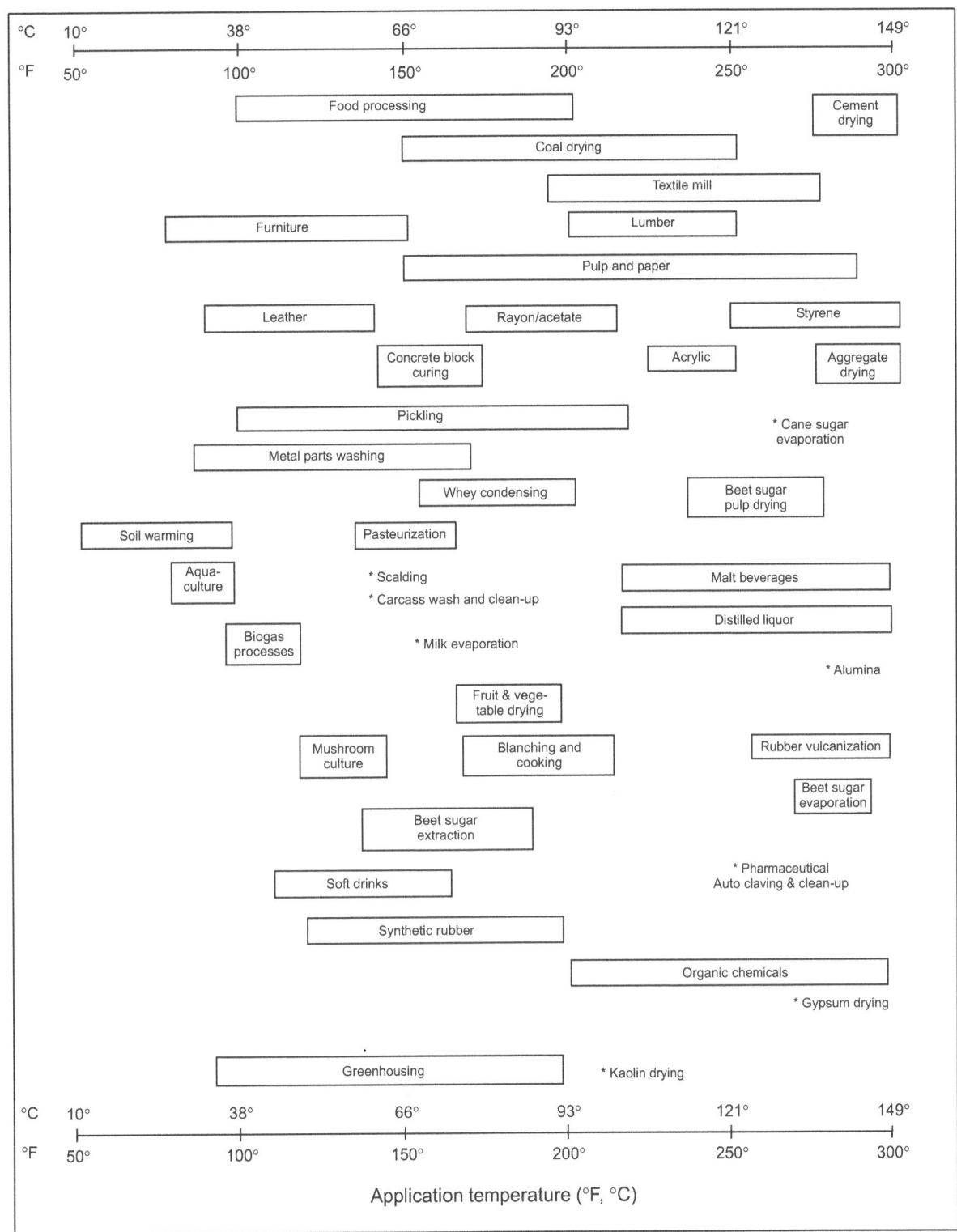

Figure 7.1 Application temperature range for some industrial processes and agricultural applications

Geothermal energy supplies approximately 30 per cent of the total process-steam requirement and up to 5 per cent of the electricity demand at Tasman. The Kawerau geothermal steam field has provided a consistent energy supply to the Tasman Pulp and Paper mill over a long period. It is predicted that producing steam at twice the current rate would be well within the sustainable capacity of the reservoir. However, it is necessary to continually improve steam-field efficiency and cost-effectiveness to remain competitive with alternative fuels such as natural gas.

7.2.2 Diatomite plant

The production of diatomaceous Earth at Namafjall, Iceland, is an important development for geothermal energy because it serves as an example of how cheap geothermal energy can make a process economic when, with conventional energy resources, the process could not be justified. The diatomaceous Earth is dredged from the bottom of Lake Myvatn by a suction dredger, and the diatomaceous slurry is transmitted by pumping through a 3 km pipeline to the plant site. Up to 45 tonnes/h of steam at 183 °C/1,014 kPa may be transmitted from boreholes 600 m away. The capacity of this plant was 25,500 tonnes/yr of diatomite filter aid in 1995 (Ragnarsson, 1996).

Steam is used both to keep the reservoirs containing settled diatomaceous Earth ice-free and in the dryer, which is a rotary steam tube type. Approximately 27 tonnes/h of steam are used for the dryer. Approximately 5 tonnes of dry diatomite are produced per hour. The moisture content of the diatomite is reduced from 88 to 89 per cent to 2 to 6 per cent in the process (Sigurdsson, 1992).

7.2.3 Vegetable dehydration

Geothermal Food Processors, a subsidiary of Gilroy Foods, located at Brady Hot Springs near Fernley, Nevada, is mainly involved in onion drying. It produces different grades of dried onion, from powdered form up to various-sized granules. The final product has a moisture content of 3.5 to 5 per cent. Geothermal fluid is used for heating requirements at the plant. The plant operates six months per year, from May to November during the harvest season. It has been operating since 1978 and there have been no major equipment failures.

Geothermal fluid is pumped from the well at a rate of 47 l/s at 154 °C/1,310 kPa, and at this condition the vapour pressure is 441 kPa. The system is pressurized to almost three times the vapour pressure to make sure that the geothermal fluid is always in its liquid state. Operating the plant at elevated pressure prevents serious formation of scale inside the hot water coils and the pipeline. The discharge temperature is 42 °C and the discharge has a pressure of 276 kPa.

The moisture content of the onions is initially 50 per cent; after going through three stages and a desiccator the final product has a moisture content of approximately 5 per cent. The product is dried in a 58 m long Proctor and Schwartz continuous conveyor food dehydrator. The drying is accomplished by passing geothermally heated air through a perforated stainless steel belt. The geothermal heat is transferred into the drying air by ten steel-tube hot-water heating coils.

Integrated Ingredients, a division of Burns Philp Food Inc., dedicated a second new onion- and garlic-processing plant in 1994. The plant, a single-line continuous conveyor dehydrator, is located in the San Emido Desert just south of Gerlach and about 161 km north of Reno, Nevada. A total of 6.35 million kg of dry product are produced annually: 60 per cent onion and 40 per cent garlic. Up to 57 l/s of the 130 °C geothermal fluids are delivered to the plant, providing 13.2 MW$_t$ (Lund and Lienau, 1994).

A 1 MW$_t$ pilot crop-drying facility using 160 °C geothermal waters has been built near the Palinpinon I geothermal field in Southern Negros in the Philippines. Both cabinet dryers and drying trenches are used to dehydrate coconut meat, fruits, root crops, spices and aquatic products. The plant, covering 650 m^2, is designed to handle about 6.4 tonnes/day of dry copra (coconut meat) (Chua and Abito, 1994).

The advantages of using a geothermal heating system in these dryers include (a) elimination of fire hazards, (b) no contamination or discoloration of the product because there are no products of combustion in the air stream, and (c) elimination of conventional fuels.

7.2.4 Other industrial uses

The oldest known use of geothermal energy for industrial applications occurred in Italy. In circa 1500 B.C., the Etruscans used geothermal energy in the Tuscany region not only for therapeutic purposes, but also for the exploitation of the salt products deposited near the edges of the 'lagoni' (fumaroles). Traces of boric salts have been found in the glaze of Etruscan plates and crockery, a fact testifying to how these people, many centuries before the birth of Christ, had already developed a high degree of artistry and technology in the grinding and chemical treatment of the borates, and also in the proportioning of these products with the other substances that composed their fine pottery.

In 1812, the first attempts were made to extract boric acid from boiling mineral springs scattered over a large area between Volterra and the mining centre of Massa Marittima. This boric acid was produced by evaporation of boric solutions in iron cauldrons, with crystallization in wooden barrels. Brick domes were built over the natural outlets of steam, forcing the steam through an orifice to feed the evaporation boilers. Francesco Larderel was founder of the boric acid industry, and in 1846 the area was named Larderello in his honour. With an increase in production, growth in trade, and refinement of the process, a wide range of boron and ammonium compounds was produced in the early 1900s. This process continued until the Second World War; after the war the plant was put into operation again and continues to this day, using imported ores, to produce boric acid with approximately 27 tonnes of steam per hour.

In New Zealand, at Broadlands, a co-operative of twelve farms is drying alfalfa (lucerne) using 184 °C steam in a large forced-air heat exchanger. The drier is a fixed-bed, double-pass drier, discharging into a hammer mill and pellet press for the final product. The plant produces 0.9 tonnes of compressed pellets from 4.5 tonnes of fresh alfalfa. It now produces 2,700 tonnes/yr of 'De-Hi', a dried product from the fibrous part of the plant, and 180 tonnes/yr of 'LPC', a high-protein concentrate produced from the extracted juices (Pirrit and Dunstall, 1995).

In Japan, Yuzawa Geothermal Drying Co. Ltd on the island of Honshu uses geothermal energy for drying timber. The drying facility consists of a vacuum dryer, a bark boiler and a forced-air unit. The plant utilizes approximately 43,000 kg/h of 98 °C hot water.

In China, low-temperature (48 to 79 °C) geothermal water is used mainly for washing in wool mills and for dyeing cloth in Tianjin, Beijing and Fengshun in Guandong Province, and Xiangyne in Liaoning Province. The Jiannan gas field in Hubei Province has for many years produced chemicals from geothermal brines. Besides a yearly production of 9,100 tonnes of table salt, the wells yield 0.45 tonnes of iodine, 17 tonnes of bromine, 36 tonnes of boron, 5.3 tonnes of aluminium carbonate, and 435 tonnes of 6 per cent ammonia water and other trace elements for use in industry.

In the United States, heap leaching in a gold mining operation in Nevada is a recent new use of geothermal fluids. Tube-and-shell heat exchangers are used to heat cyanide solutions in a heap-leaching operation. Geothermal fluids are also used as make-up water. Table 7.1 lists most of the

Table 7.1 Summary of geothermal applications

Country	No. of projects	Applications
Australia	2	Paper production, washing
China	49	Drying, dyeing, chemical, paper production, and process heating (Chin, 1976)
Georgia	1	Paper production
Guatemala	2	Drying, process heating
Iceland	27	Drying, food processing, textiles/clothing/hides, timber/paper, chemicals, and minerals
Italy	3	Drying, chemical, and mineral
Japan	1	Drying (Horii, 1985)
Macedonia	1	Drying
Mexico	2	Drying and industrial laundry
New Zealand	4	Drying, pulp and paper
Phillippines	1	Drying
Russia	3	Drying, wool washing, and paper production
Slovenia	1	Drying
Turkey	1	Chemicals
USA	7	Drying, heap leaching, mushroom growing (Rutten, 1986), and waste-water treatment
Total	105	

known geothermal industrial applications throughout the world.

7.3 SELECTED INDUSTRIAL APPLICATIONS

Pulp and paper mills, lumber drying, crop and vegetable drying, food processing, heap leaching, wastewater treatment and other industries have been studied extensively with regard to the use of geothermal energy. Examples of applications of these industries are presented below, along with some indication of how geothermal energy might be used in other processes. Greater detail can be found in referenced final reports.

7.3.1 Pulp and paper mill (Hornburg and Lindal, 1978)

The process flow diagram for a typical bleached pulp and paper mill is shown in Figure 7.2. The pulp process utilized is the Kraft, or sulfate, method.

In this typical plant all motor drives for pumps and other driven equipment are powered by steam. Steam for the process is normally generated in liquor-recovery boilers, bark-fed power boilers and oil- or gas-fired boilers.

The wood to be pulped is first debarked in the barker. The bark is used as fuel to produce process steam. Once debarked, the wood is chipped to specified chip size, which aids in packing chips in the digester. The ratio of chips to

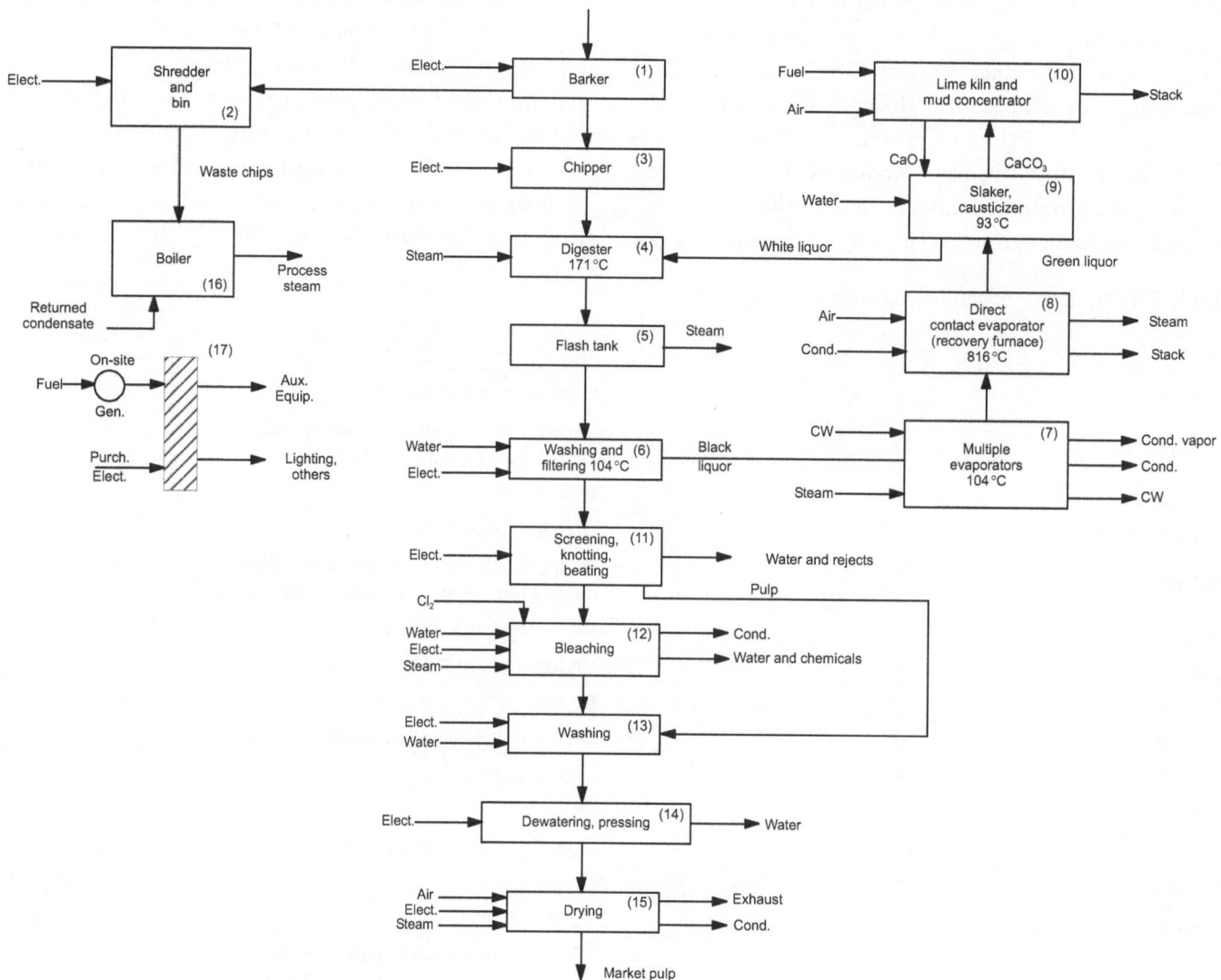

Figure 7.2 Pulp mills (Kraft process) process flow

liquor must be maintained between 2.5 and 3.5 kg liquor/kg of wood.

The cooking liquor contains essentially sodium sulfide and caustic soda. The liquor, as it is received from the recovery system, is too concentrated for proper digesting results and therefore has to be diluted. The dilution is accomplished using the weak black liquor to keep water additions to a minimum.

The digester charge is then heated either by the addition of live steam to the bottom of the digester or indirectly with steam. The time required for cooking the wood varies, depending on the end use of the pulp. The maximum cooking temperature is between 168 and 175 °C (steam pressure is 655 and 793 kPa, respectively).

At the completion of the cook, the pressure within the digester is allowed to decrease to approximately 552 kPa. Opening a quick-opening valve at the bottom of the digester then expels the pulp. The pulp then flows to the flash tank, which is fitted with a special vapour outlet. Heat is sometimes recovered from this vapour.

The pulp is then screened to remove small pieces of uncooked wood. Following screening, the pulp is washed to remove the cooking liquors. It is economically important to remove as much of the liquor as possible. The pulp washing is carried out in rotary vacuum washers. This process is so efficient that between 98 and 99 per cent of the cooking chemicals are washed from the pulp. Hot water is used for washing. The pulp leaving the washer is of relatively high consistency.

The weak black liquor washed from the pulp is first concentrated in multiple-effect evaporators and then further concentrated in direct-contact evaporators. New chemical makeup is added and the strong liquor burned to remove dissolved organic material. The smelt is then dissolved and causticized to form white cooking liquor.

The pulp is bleached in one to five or more stages. The basic steps in the bleaching process are:

1. Mixing of the chemicals in the proper ratios with the pulp.
2. Raising the pulp temperature to the required level.
3. Maintaining the mix at this temperature for a specified period.
4. Washing residual chemicals from the pulp.

Chlorine dioxide is almost always used as the bleaching chemical. The procedure is to treat the pulp with chlorine dioxide, followed by neutralization with calcium hypochlorite. This process represents the optimum for most Kraft pulp bleaching.

Before actual paper manufacture on a paper machine, the pulp stock must be prepared. Beaters and refiners are normally used to accomplish this task. The purpose of beating and refining is to change the physical form of the fibres in the pulp. The process is related to grinding. It is carried out in a number of different ways depending on the fibres desired. The overall objective is to maximize bonding strength.

Paper is made by depositing a dilute water suspension of pulp on a fine screen, which permits the water to drain through but retains the fibre layer. This layer is then removed from the screen, pressed, and dried.

Most of the process heat requirements are in the range of 121 to 177 °C, and the heating is accomplished by way of steam in tube-and-shell heat exchangers. In a conventional system the energy needs are met by generating steam at 3,100 kPa (370 °C in a black liquor recovery boiler, a bark boiler, and a conventional fossil-fuel fired boiler). Most of this steam is passed through a back-pressure/extraction turbine to generate electricity and pass-out steam at 172 kPa that is utilized in the process.

Geothermal fluids could partly accomplish water heating and heating of air for paper drying as shown in Figure 7.2. Two wash-water heaters are adopted, one using geothermal fluid at 100 °C and the other using steam at 172 kPa to heat the water to the final temperature of 99 °C. An air dryer is also used to pre-heat the air in the drying section. This section would also be designed to use steam at 172 kPa in lieu of 896 kPa as is usually the case. Other changes include use of 517 kPa steam in lieu of 896 kPa steam for black liquor heating and miscellaneous high-pressure requirements. Table 7.2 compares the process steam requirements

Table 7.2 Comparison of pulp and paper process steam requirements

Process	Conventional system (steam, kPa)	Geothermal system (steam, kPa)
Wash-water heating	172	172 and hot water
Evaporators	172	172
Miscellaneous, L.P.[a]	172	172
Black-liquor heating	896	72
Digester	896	896
Dryer	896	172 and hot water
Miscellaneous, H.P.[b]	896	517

[a] Low-pressure steam.
[b] High-pressure steam.

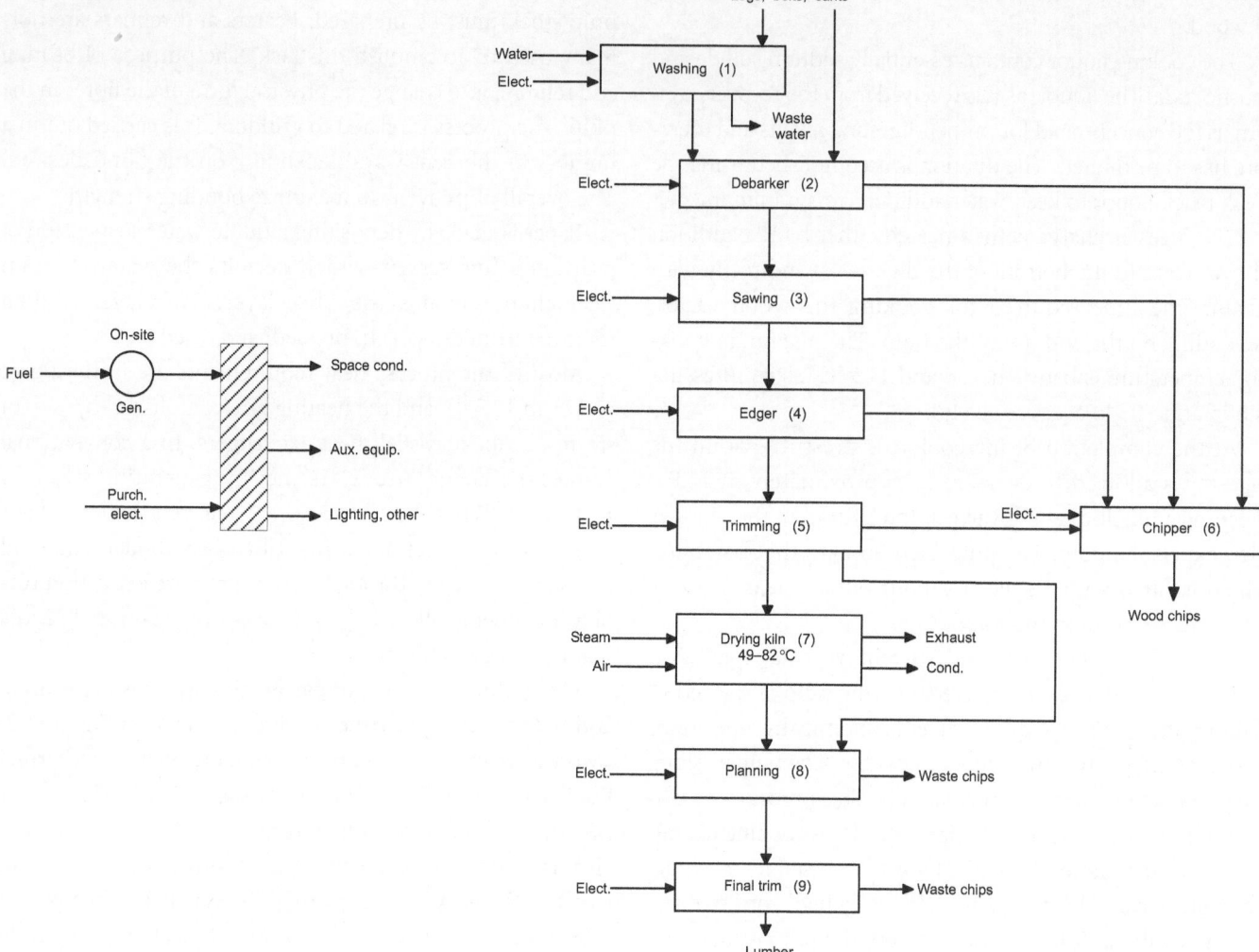

Figure 7.3 Lumber drying process flow

of a conventional system with one using a geothermal upgraded system.

The geothermal energy system could be designed to supply the energy needed as shown in Figure 7.3. In this system, the bark boiler and fuel-oil boiler have been eliminated, and the heat previously supplied by these units is now furnished by a geothermal upgrading system using geothermal fluid at 121 °C. The recovery boiler must be retained because it is needed to recover process chemicals as well as to generate high-pressure steam.

A typical pulp and paper mill could have approximately 30 per cent of its energy supplied by 121 °C geothermal fluid. Extending this to 200 °C geothermal fluid and considering that the electrical requirements could also be generated from geothermal, it is possible that 100 per cent of

the energy for a pulp and paper mill could be of geothermal origin.

The recovery boiler will generate approximately 50 per cent of the electricity required by the plant. Thus, 50 per cent of the electricity must be purchased, generated from additional geothermal fluid, or generated from steam produced from bark.

7.3.2 Drying lumber (VTN-CSL, 1977)

A process flow diagram for a typical lumber mill is shown in Figure 7.3. In small lumber mills where drying kilns are heated by steam from conventional oil-fired boilers, substitution of geothermal energy for the heating energy source can achieve substantial energy cost savings. In larger, well integrated mills, all energy from operations can be provided by burning sawdust and other wood waste products. If a market develops for the waste products or where the energy can be more economically applied elsewhere, the geothermal source may also become economical in integrated plants. Drying lumber in batch kilns is standard practice for most upper-grade lumber in the western United States. The two basic purposes of drying are to set the sap and to prevent warping.

The sap sets at 57 to 60 °C. Establishing uniform moisture content throughout the thickness prevents warping. Lumber left to dry under ambient conditions loses its moisture from exposed surfaces at a faster rate than internally. This differential drying rate sets up stresses that cause the warping. Moisture occurs in wood in cell cavities and in the cell walls. The majority of the moisture is first lost from the cavities. This loss is not accompanied by changes in the size of the cell or in warpage. When water is lost from the cell walls, however, shrinkage of the wall fibres takes place, setting up the stresses that cause warping.

In the kiln-drying process, the evaporation rate must be carefully controlled to prevent these stresses. The allowable drying rates vary from species to species, and decrease with thicker-cut sizes. Kiln drying is usually carried out as a batch process. The kiln is a box-shaped room with loading doors at one end. It has insulated walls and ceiling, and has fans to recirculate the air at high velocity through the lumber. The sawed lumber is spaced and stacked to assist free air movement, and is loaded by large fork-lifts or other specialized lumber handling trucks into the kiln. When it is fully loaded, the doors are closed and the heating cycle is started. Make-up air, pre-heated to a temperature consistent with the drying schedule, enters the kiln, where it recirculates through the stacked lumber and picks up moisture. Exhaust fans draw the moist air from the kiln and discharge it to the atmosphere. The exhaust is primarily air and water. The rates of flow and temperature are adjusted so that the temperature and the humidity in the kiln will retard the drying rate sufficiently to prevent warping. During the drying cycle, the lumber loses a large portion of its weight from evaporation of water, 50 to 60 per cent for many species.

Figure 7.4 shows a typical lumber drying kiln. The vents are over the fan shaft between the fans. The vent on the high-pressure side of the fan becomes a fresh-air inlet when the direction of circulation is reversed.

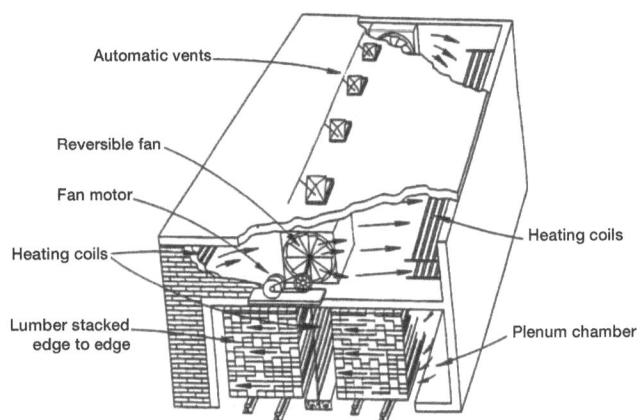

Figure 7.4 Long-shaft, double-track, compartment kiln with alternately opposing internal fans

Drying schedules are specific for each species of lumber and for size. The larger the size, the more tightly the moisture is held in the wood fibre, and the slower the schedule. Drying schedules range from less than 24 hours to several weeks per batch. Table 7.3 shows typical drying schedules for Ponderosa pine.

Green wood contains high quantities of moisture. Ponderosa pine, for example, runs approximately 60 per cent moisture. Because of the physical and chemical binding to the wood chemicals, it takes from 1.5 to 3 times the energy to evaporate moisture from wood as it does from pure water. Energy consumed in kiln-drying wood varies considerably for different species. Drying energy, therefore, varies widely with the species and sizes processed, as shown in Table 7.4.

Geothermal energy could be adapted to kiln drying by passing air over finned heat-exchanger tubes carrying hot water. The finned tube heat exchanger could be placed inside existing kilns (several arrangements are shown in Figure 7.5) so that the air-recirculating route would include a pass over the heat exchangers. The water temperature must be at least 11 to 22 °C above the ambient operating temperature in the kiln, which means that a geothermal supply temperature of 93 to 116 °C would be required. Where geothermal fluid of insufficient temperature is available (<88 °C for most uses), energy supplies could be supplemented by conventional heating systems during the final high-temperature portions of the drying schedules. Table 7.5 gives the minimum geothermal fluid temperatures for two sizes and several species of lumber.

The discharge fluid for these applications would have temperatures ranging from 71 to 82 °C and would be

Table 7.3 Typical kiln-drying schedules

Ponderosa pine	Dry bulb (°C)	Wet bulb (°C)	Time (%)	E.M.C.[a]
4/4 all heart common sort (fast on well sorted stock)	71 No conditioning	54	~ 21 h	5.8
4/4 all heart RW (conservative) common	66 66 66	54 52 54	Up to setting To 12 h 12 h till dry (24 to 28 h)	8.0 6.9 5.8
4/4 half-and-half common (mostly 8 inch)	71 No conditioning	60	40 to 50 h	8.0
Shop and select 12/4	46	42	First day	14.1
	49	43	Second day	12.1
	52	46	Third day	12.1
	54	49	Fourth day	12.1
	60	54	Fifth to tenth	11.9
	63	54	Tenth to 12th	9.5
	66	57	12th to 15th	9.5
	68	60	15th to 18th	9.4
	71 Cool	60	18th to 22nd	7.9
	82	77	About 24 h of equalizing and conditioning	11.1

[a] E.M.C. = Equilibrium Moisture Content.
Source: Kiln-drying Western Softwoods, Moore Dry Kiln Company, Oregon, Knight, 1970.

Table 7.4 Energy consumed in kiln-drying wood

Lumber	Energy use (MJ/kg)	MJ/dry (board feet)
Douglas fir	4.6–5.7	1.63–2.45
Southern yellow pine	3.7–5.0	4.8–6.7
Red oak	5.7 +	8.2 +

Source: Moore Dry Kiln Company, Oregon.

available for other applications in the mill, for heating of office buildings, for log ponds, or other cascaded uses.

7.3.3 Crop drying (Lienau, 1978)

This section describes the use of geothermal energy in crop drying of alfalfa and in grain processing.

Alfalfa processing

There are two approaches to the drying of alfalfa, from which two basic products are made, pellets and cubes. Cube production only generally requires field (Sun) drying to approximately 17 to 19 per cent moisture. When used as feed, the cubes do not require the addition of roughage. The pellets require significant quantities of heat for drying at a plant. The main advantage of this approach over Sun curing is that more vitamin A and xanthophyll (a yellow pigment present in the normal chlorophyll mixture of green plants) is retained. The latter is important in chicken and egg colouring. The xanthophyll is retained well by high heat and rapid drying.

Pellets can be processed using either high temperatures or low temperatures in combination with field wilting. The first approach, using conventional fuels, is the rotary-flame furnace, which is common in the United States, requiring temperatures up to approximately 980 °C. The second

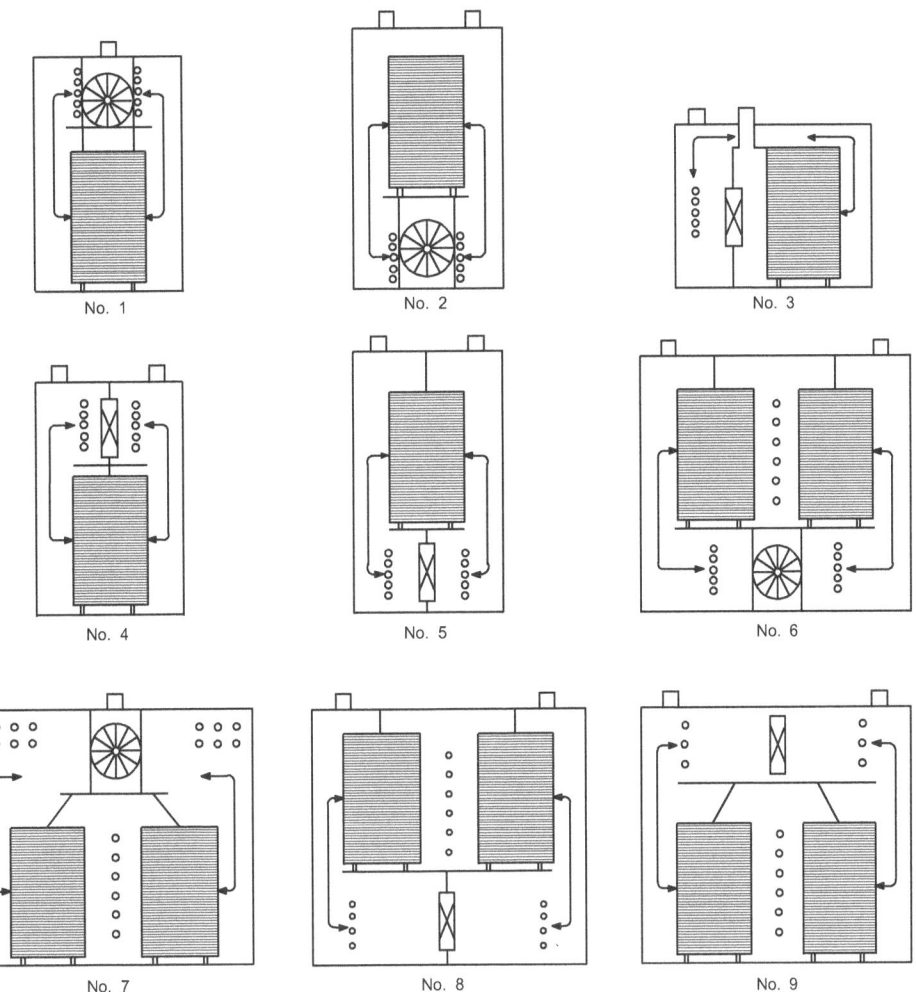

Figure 7.5 Location of fans and heat exchangers in kilns

Table 7.5 Minimum geothermal fluid temperatures for kiln drying at kiln inlet

Species	Minimum geothermal fluid temperature (°C)	
	Lumber size	
	4/4	8/4
Ponderosa pine	79	91
Sugar pine	79	79
Englemen spruce	79	—
Sitka spruce	91	91
Douglas fir	91	91
Incense cedar	85	—

Source: Moore Dry Kiln Company, Oregon.

involves field wilting to reduce the moisture content, with the remainder of the moisture to be removed in the drying plant. This process requires temperatures of about 82 to 120 °C. Figure 7.6 shows a process flow diagram of such an alfalfa drying plant.

The process starts with cutting and chopping the alfalfa in the field at approximately 70 per cent initial moisture. The chopped material is then allowed to Sun wilt for twenty-four to forty-eight hours to 15 to 25 per cent moisture content. This can easily be accomplished in areas of the western United States because of available Sun and low rainfall during the season. In the Middle West, it can only be wilted to approximately 60 per cent moisture. This short field-wilting time also prevents damage to the next crop, as the cut material is removed before the new shoots sprout and are

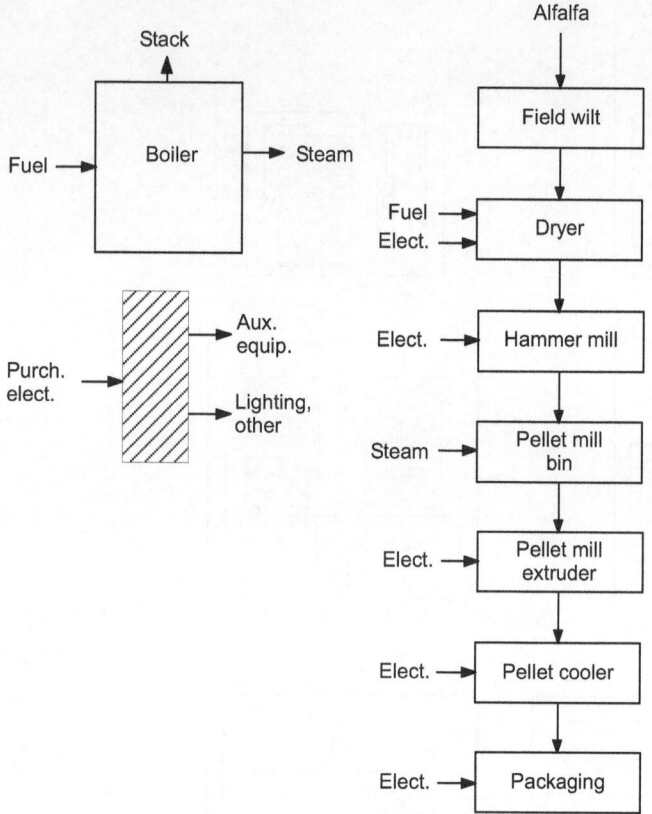

Figure 7.6 Alfalfa drying and pelletizing process

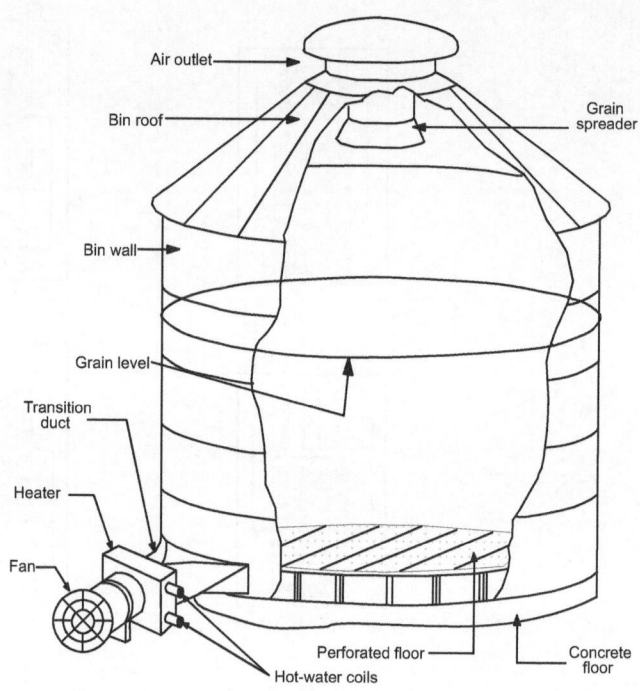

Figure 7.7 Perforated false-floor system for bin drying of grain

using at least 104 °C geothermal fluid. One well could provide the required flow for a plant producing 22,700 to 27,200 tonnes of alfalfa pellets/year (at 8 to 15 per cent moisture).

Grain drying

Significant amounts of energy are consumed annually for grain drying and barley malting. These processes can be easily adapted to geothermal energy in the temperature range of 38 to 82 °C. Most farm crops must be dried to, and maintained at, moisture content of 12 to 13 per cent wet basis, depending on the specific crop, storage temperature and length of storage. Mold growth and spoilage are functions of elapsed storage time, temperature, and moisture content above the critical value. Grain to be sold through commercial markets is priced according to specified moisture content, with discounts for moisture levels above a specified value.

The grain dryer is typically a deep-bed dryer, as shown in Figure 7.7. Most crop-drying equipment consists of (a) a fan to move the air through the product, (b) a controlled heater to increase the ambient air temperature to the desired level and (c) a container to distribute the drying air uniformly through the product. The exhaust air is vented to the atmosphere. Where the climate and other factors are favourable, unheated air is used for drying, and the heater is omitted.

crushed by equipment. The field-wilted material is then trucked to the plant, and stockpiled for no more than approximately two days. The chopped material is then belt-fed to a triple-pass rotary drum dryer. This dryer may use either natural gas or fuel oil. The alfalfa is dried at a temperature below 120 °C. Any temperature over 200 °C will over-dry the product. The actual drying temperature depends upon the ambient conditions and moisture content of the alfalfa. Dryer-temperatures can go as low as 80 °C. The material is moved through the dryer by a suction fan. The retention time is approximately fifteen to twenty minutes.

From the dryer, the alfalfa is fed to the hammer mill and the pellet meal bin. The latter is the surge point in the system. Here, the material is conditioned with steam and then fed to the pellet mill pressure extruder. The steam helps in providing a uniform product, and makes it easier to extrude through the holes in the circular steel plates. The material is then cooled and the fines removed in a scalper. Finally, the product is weighed on batch scales, packaged and stored.

A low-temperature geothermal energy conversion would require a 93 °C air-drying temperature from a triple-pass dryer

Several operating methods for drying grain in storage bins are in use. They may be classified as full-bin drying, layer drying and batch drying. The deep-bed dryer can be installed in any structure that will hold grain. Most grain storage structures can be designed or adapted for drying by providing a means of distributing the drying air uniformly through the grain. This is most commonly done by either a perforated false floor or duct systems placed on the floor of the bin.

Full-bin drying is generally done with unheated air or air heated 6 to 12 °C above ambient. A humidistat is frequently used to sense the humidity of the drying air and turn off the heater if the weather conditions are such that heated air would cause over-drying.

The depth of grain (distance of air travel) is limited only by the cost of the fan, motor, air distribution system and power required. The maximum practical depth appears to be 6 m for corn and beans, and 4 m for wheat. Grain stirring devices are used with full-bin systems. These devices typically consist of one or more open, 5 cm diameter, standard pitch augers suspended from the bin roof and side wall and extending to near the bin floor.

Conversion of the deep-bed dryer to geothermal energy is accomplished by simply installing a hot-water coil in the inlet duct using geothermal fluid in the 38 to 49 °C temperature range.

Of all grains, rice is probably the most difficult to process without quality loss. Rice containing more than 13.5 per cent moisture cannot be stored safely for long periods. When rice is harvested at a moisture content of 20 to 26 per cent, drying must be started promptly to prevent it from souring. Deep-bed or columnar dryers can be used; a columnar dryer (Figure 7.8) is considered here.

Grain is transferred from the storage bins to the top of the column dryer by bucket conveyors. The column must be completely filled before the drying operations start. The grain flows from top to bottom by gravity, and the amount of flow is controlled by the speed of the screw conveyor, located at the bottom of the column, as shown in Figure 7.8.

The two important variables in the drying operation are the air-mass flow rate and the temperature at the inlet to the dryer. Hot air is blown from the bottom, and a static pressure is maintained between columns. Air temperature is controlled by regulating the burner output from several thermocouples installed inside the column to monitor the air and kernel temperature.

Rice is loaded in the dryer at approximately 21 to 22 per cent moisture content, and the drying cycle is normally completed after three to four passes. The final moisture content should be below 15 per cent before it can safely be stored in the warehouse. After each pass, partially dried rice is stored in tempering bins for at least twelve hours before another pass

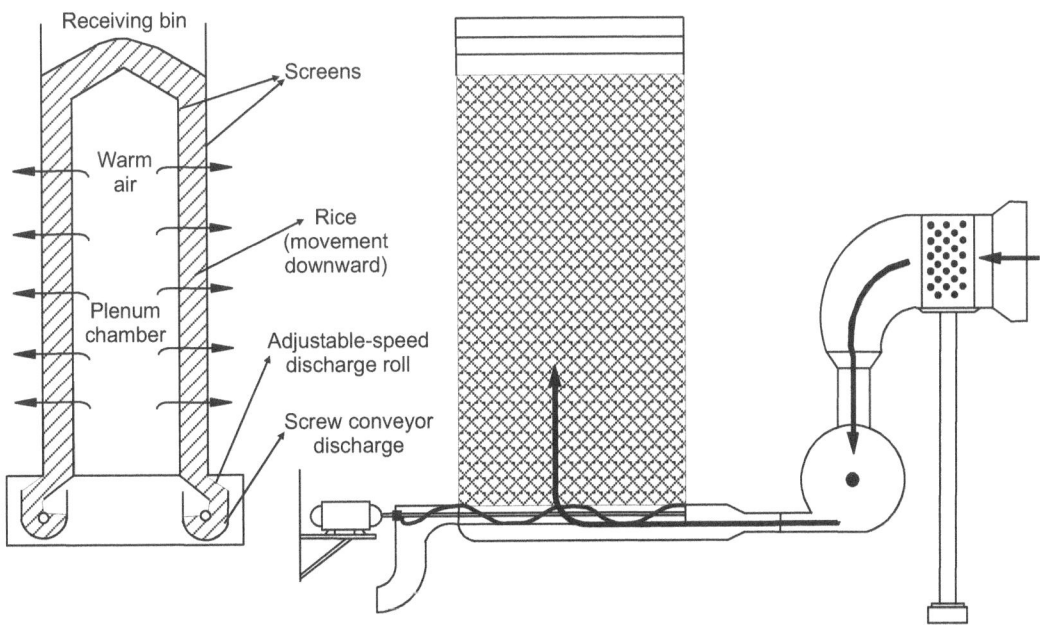

Figure 7.8 Columnar grain dryer
Source: Guillen, 1986.

takes place. The rice is tempered to equalize internal moisture content, thus minimizing thermal stresses and avoiding breakage of kernels. Kernel temperature is normally maintained at 38 °C when the moisture content is approximately 21 per cent, and at lower moisture content, <17 per cent, temperature is limited to 35 °C. At a constant grain temperature of 38 °C, air is heated to 82 to 93 °C during cold weather and approximately 60 to 82 °C during the warm season.

Converting the columnar dryer to geothermal fluids involves the installation of a hot-water coil upstream of the blower fan to obtain uniform temperature inside the plenum chamber. The air flow pattern is shown in Figure 7.8, and there is no air recirculating because of the presence of dust on the downstream side.

Air flow could be maintained at a constant rate; then the only variable would be the flow rate of the grain.

7.3.4 Vegetable and fruit dehydration (Lienau, 1978)

Vegetable and fruit dehydration involves the use of a tunnel dryer, or a continuous conveyor dryer using fairly low-temperature hot air from 38 to 104 °C.

A tunnel dryer is an enclosed, insulated housing in which the products to be dried are placed upon tiers of trays or stacked in piles in the case of large objects, as shown in Figure 7.9. Heat transfer may be direct from gases to products by circulation of large volumes of gas, or indirect by use of heated shelves or radiator coils.

Because of the high labour requirements usually associated with loading or unloading the compartments, they are rarely used except where:

- A long heating cycle is necessary because of the size of the solid objects or permissible heating temperature requires a long hold-up for internal diffusion of heat or moisture.
- The quantity of material to be processed does not justify investment in more expensive, continuous equipment. This would be the situation for a pilot plant.

The process flow diagram for a conveyor dryer, which will be considered, is shown in Figure 7.10. Table 7.6 lists the many food products that may be processed commercially on conveyor dryers.

The energy requirements for the operation of a conveyor dryer will vary because of differences in outside temperature, dryer loading, and requirements for the final moisture content of the product. A single-line conveyor dryer handling 4,500 kg of raw product/h (680 to 800 kg finished) will require approximately 6 MW$_t$, or, for an average season of 150 days, 80×10^{12} J/season, using approximately 35,000 kJ/kg of dry product.

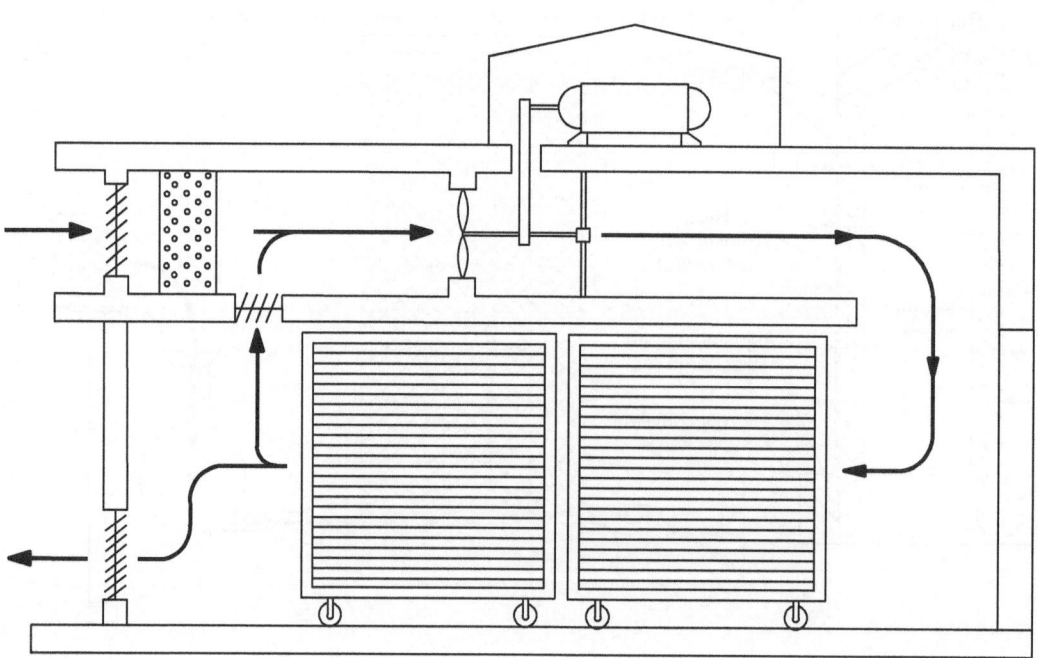

Figure 7.9 Tunnel dryer, air flow pattern
Source: Guillen, 1986.

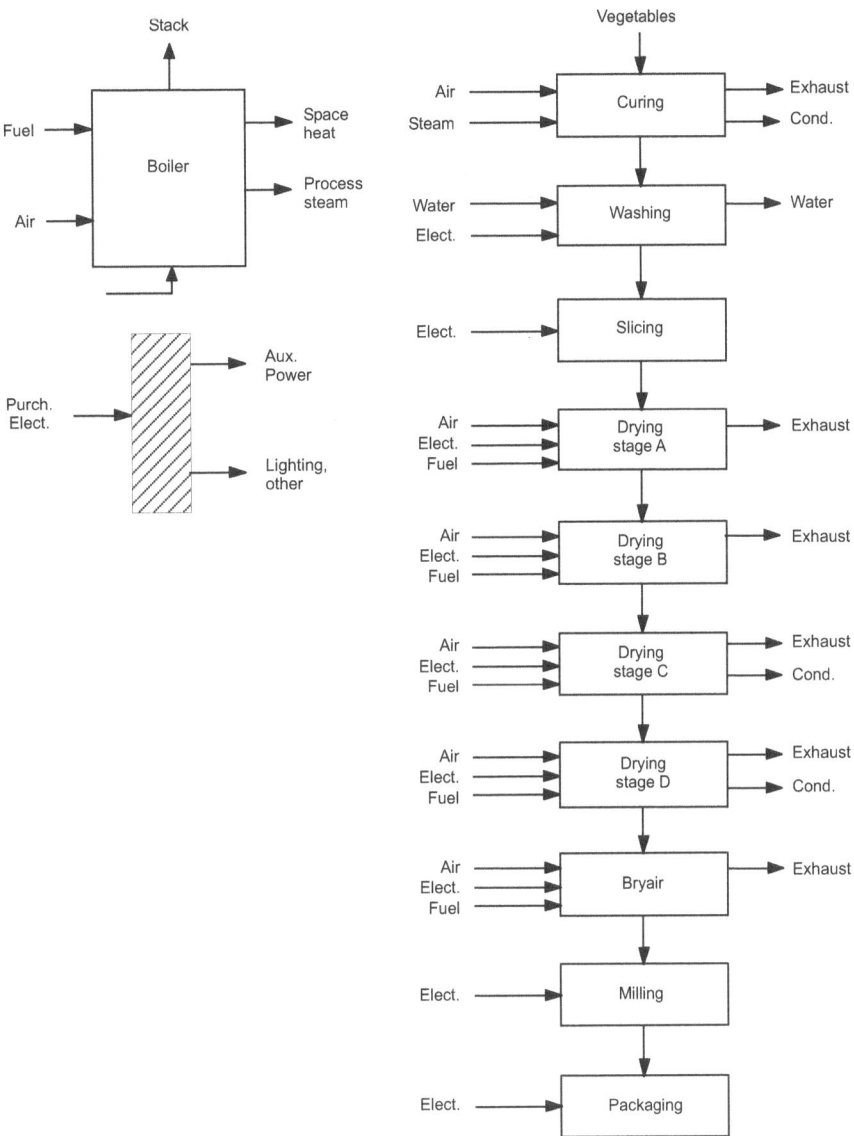

Figure 7.10 Vegetable dehydration process flow

The energy (Figure 7.10) is usually provided by natural gas; air is passed directly through the gas flame in Stages A and B, and over steam coils in Stages C and D. The steam coils are necessary to eliminate turning of the product in the last two stages.

In addition to the heating requirements, electrical energy is needed for the draft and recirculating fans, and small amounts for controls and driving the bed motors. Total electric power required for motors is from 370 to 450 kW, or approximately 1.0×10^4 kWh/day, or 2.0×10^6 kWh/season. This amounts to 2,400 kJ/kg of finished product and increases to approximately 14,000 kJ/kg when all electrical requirements are considered.

In general, four stages (A through D) are preferred; however, if the ambient air humidity is below approximately 10 per cent, Stage D can be eliminated. The temperature and number of compartments in each stage may also vary.

In summary, the total heat requirement is 6.2 to 7.6 MW$_t$ for a single-line conveyor dryer 64 m long $\times$ 3.8 m wide with an average input of 4,500 kg/h wet product, producing 680 to 800 kg/h dry product. Table 7.7 illustrates the energy requirement for each stage, using natural gas as a fuel, assuming ambient temperature at 4 °C. For ambient temperature of 18 °C, 6.2 MW$_t$ would be required.

In the example in Table 7.7, geothermal fluid is used to supply the required energy. Adopting an 11 °C minimum

Table 7.6 Product drying in a conveyor dryer

Vegetables	Fruits	Nuts	Prepared foods	Prepared feeds
Beans	Apples	Almonds	Beef jerky	Animal feeds
Onion	Raisins	Coconut	Bouillon	Pet food
Garlic		Brazil	Cereals	Cattle feed
Peppers		Peanuts	Macaroni	Fish food
Soy beans		Pecans	Snacks	Hay
Beets		Walnuts	Soup mixes	
Carrots		Macadamia		
Potato (sliced, diced, chips, french fries)				
Spinach				
Parsley				
Celery				
Okra				

Table 7.7 Conveyor dryer energy requirements

Stage	Air temperature (°C)	Heat supply	Approximate HE opening size (m²)	Estimated airflow (l/s)	Estimated[a] (MW$_t$)
A1	99	gas burners	3.1	13 700	1.5
A2	99	gas burners	3.9	13 700	1.7
A3	88	gas burners	3.6	19 300	1.1
A4	82	gas burners	4.2	19 300	1.4
B1	71	gas burners	3.9	8 000	1.1
B2	63	steam coils	3.1	9 000	0.3
C	54	steam coils	4.2	9 400	0.1
D	49	steam coils	8.1	5 000	0.2
Bryair	150	gas burners	2.3	3 000	0.3
TOTAL					7.7

[a] Assuming ambient at 4 °C, total = 6.2 MW$_t$ at 18 °C ambient.

approach temperature between the geothermal fluid and process air, a well with 110 °C fluid is required. The first-stage air temperature can be as low as 82 °C, but temperatures >93 °C are desirable.

Figure 7.11 shows a scheme using 110 °C geothermal fluid. The line has to be split between Compartments A-1 and A-2, because both require 99 °C air. A total flow of 57 l/s is required. The Bryair desiccator in Stage D requires 150 °C on the reactor side; with the result that only half of the 300 kW$_t$ energy requirements can be met by geothermal energy. Geothermal fluid will be used for pre-heating to 80 °C, with natural gas or propane used to boost the air to 150 °C. The wastewater from the Bryair pre-heater has a temperature of 90 °C, and thus could be used for cascaded uses. The wastewater could be returned to the reservoir by means of an injection well.

In compartments A-1, A-2, A-3 and A-4, four finned air–water heat exchangers in parallel would be required to

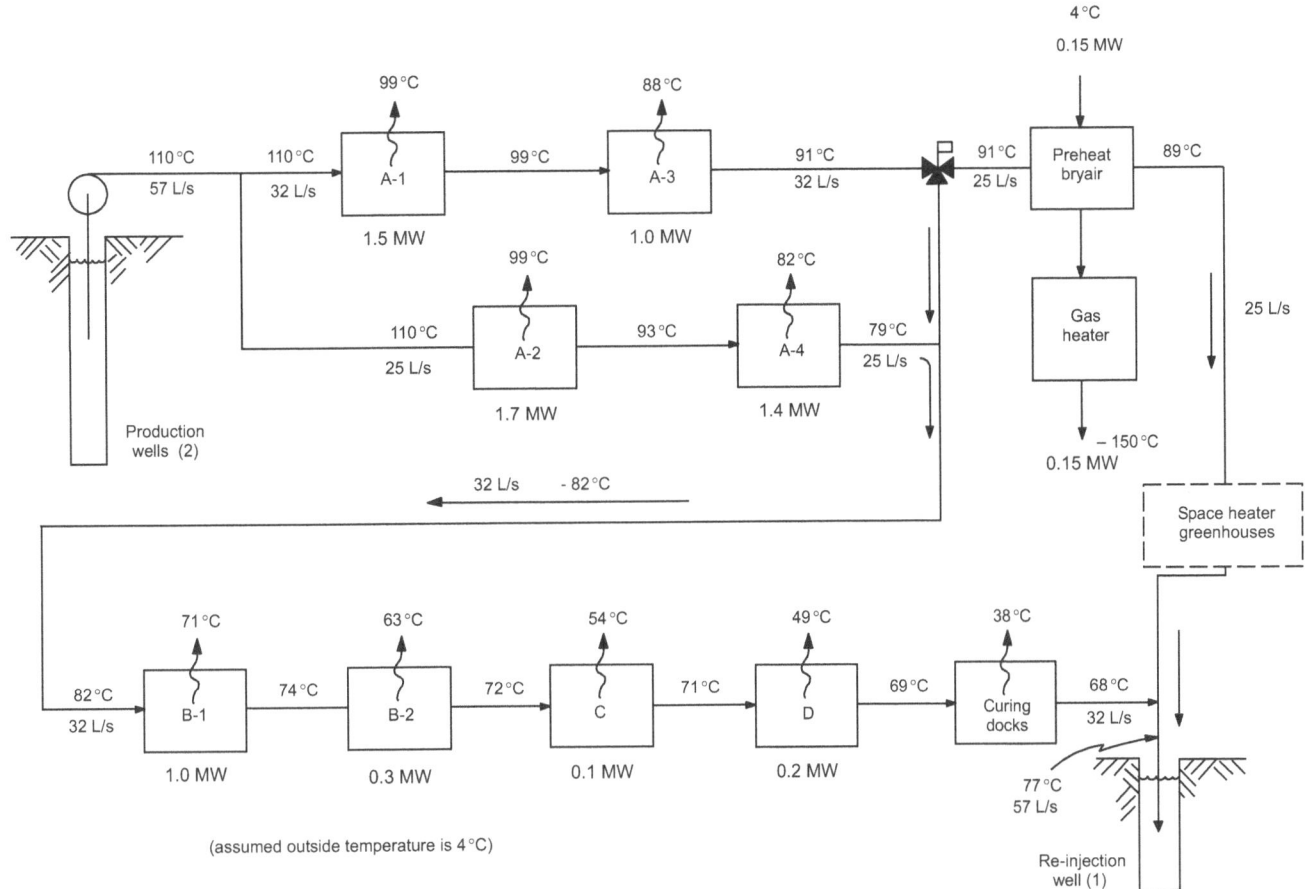

Figure 7.11 Multi-stage conveyor dryer using 110 °C geothermal fluid and 4 °C ambient air

satisfy the energy requirement and water velocity flows. The remaining stages would require from one to two heat exchangers in each compartment, depending upon the energy requirements.

If lower temperature geothermal fluids were encountered (below 93 °C), then not all of the energy could be supplied to stage A by geothermal fluid. Geothermal fluid would then be used as a pre-heater, with natural gas providing the energy for the final temperature rise.

7.3.5 Potato processing (Lienau, 1978)

Potato processing could result in a number of different types of product, including potato chips, frozen french fries and other frozen potato products, dehydrated mashed potatoes, potato granules, potato flakes, dehydrated diced potatoes, potato starch, potato flour, canned potatoes and miscellaneous other products from potatoes.

Since 1970, frozen potato products have constituted from 45 to 48 per cent of all the potatoes used for processing, or nearly one-quarter of the food use of potatoes in the United States (Talburt and Smith, 1975).

Figure 7.12 illustrates a frozen french fry processing line. Many of the processing methods used by potato processors can utilize energy supplied by 150 °C or lower temperature geothermal fluids. Typically, however, a few of the operations, notably the frying operation, will require higher temperatures than can be provided by a majority of the geothermal resources.

Potatoes for processing are conveyed to a battery of scrubbers and then moved into a pre-heater, which warms the potatoes and softens the skin, making it easier to remove. The potatoes are then chemically peeled by a 15 per cent lye solution maintained at a temperature of 60 to 80 °C.

Upon leaving the chemical peeler, the potatoes are conveyed to a battery of scrubbers, where the peeling is

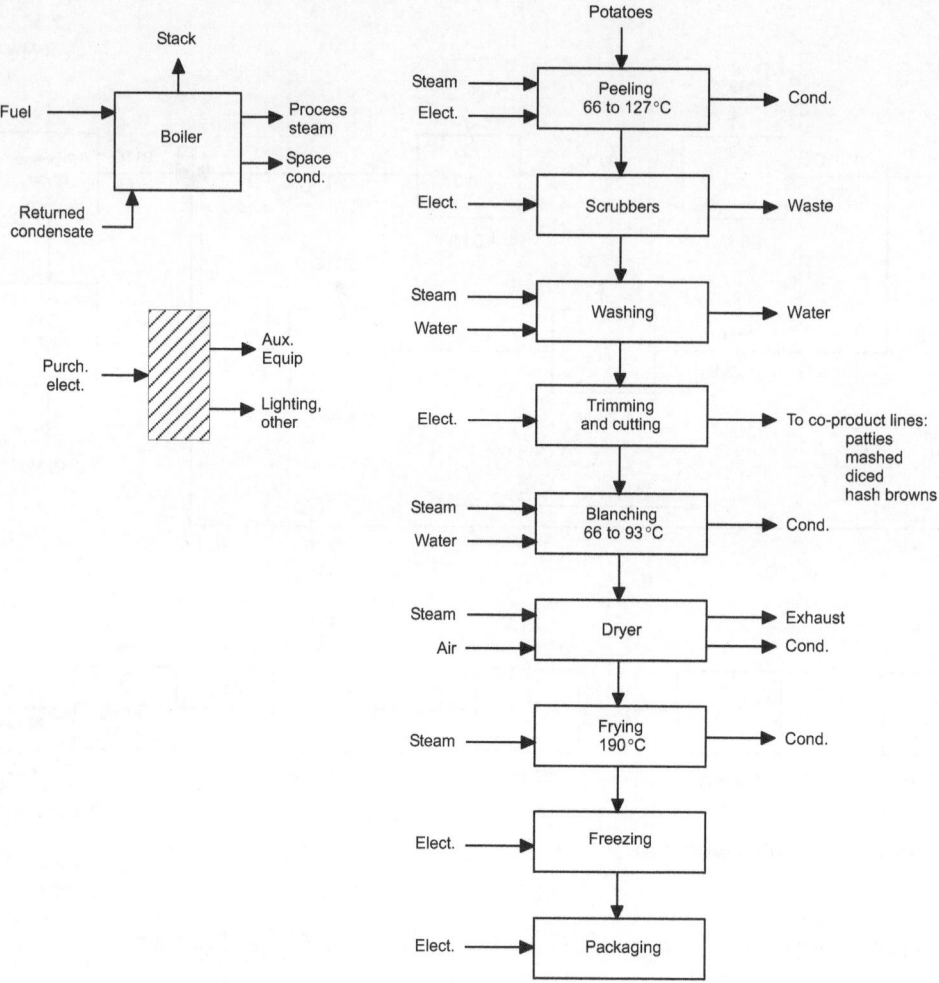

Figure 7.12 Frozen french fry process flow

removed. After the scrubbers, the peeled potatoes are subjected to another washing process and then conveyed to the trim tables by pumping. The peeling removed by the scrubbers is pumped to a holding tank and sold as cattle feed following neutralization of the lye residue.

The potatoes are trimmed for defects, and the product is conveyed to cutter areas. Shakers sort the product. Small lengths are separated and then processed into 'hash browns' or 'tater tots'. The properly trimmed and sized product is then carried by gravity to the blanching system.

After blanching, the potatoes are de-watered and fed through a sugar drag, which adds a slight amount of dextrose to the surface of the potato, imparting a golden colour when the potatoes are fried. They then pass through a dryer that removes the surface moisture before a two-stage frying process. The first stage cooks the product more completely, while the second stage gives it the golden colour. The oil in

the fryers is heated to 190 °C by heat exchangers receiving high-pressure steam at 1,900 kPa.

Freezing of the products is by continuous freezing systems powered by compressors. Freezing temperatures are maintained at a constant −34 °C.

For systems that use geothermal energy, the energy would probably be supplied to the process by way of intermediate heat exchangers. To avoid any possible contamination of the product by the geothermal fluid, or the need for treatment of the fluid, the geothermal fluid passing through these exchangers will transfer energy to a secondary fluid, usually water, which delivers the energy to the process. The secondary fluid, circulating in a closed system, then returns to the intermediate heat exchanger to be reheated.

Processes that could be supplied by a 150 °C geothermal resource are distinguished in Table 7.8 by their function and temperature requirements. The peeling process involves three distinct steps, calling for input temperatures of 120, 93 and 66 °C. The hot blanch process uses an input temperature of 93 °C and the warm blanch process requires 66 °C. Heating of hot water used for various functions also calls for 66 °C, as does the plant space-heating system.

Table 7.8 Potato processing temperature requirements

Function	Temperature in (°C)	Temperature out (°C)
Peeling	127	93
Peeling	93	66
Peeling	66	38
Hot blanch	93	38
Warm blanch	66	38
Water heating	66	10
Plant heat	66	38

Figure 7.13 suggests one possible routing of the geothermal fluid through the intermediate heat exchangers for maximum extraction of energy. Energy requirements for the high-temperature (93 °C or more) processes are satisfied by dropping the geothermal fluid temperature from 150 to 88 °C. The lower temperature processes are then supplied partially by this cascaded geothermal fluid and partially by fresh geothermal fluid.

The intermediate heat exchangers could be either of the shell-and-tube design, or the compact and versatile plate-type heat exchanger. The secondary fluid circulating to the processing tanks could either be used directly, or the fluid could pass through heat exchangers located at the processing tanks to heat the fluid in the tanks.

The energy needed for freezing at −34 °C probably could not be supplied by geothermal sources, because of the advanced state-of-the-art required to obtain such low temperatures.

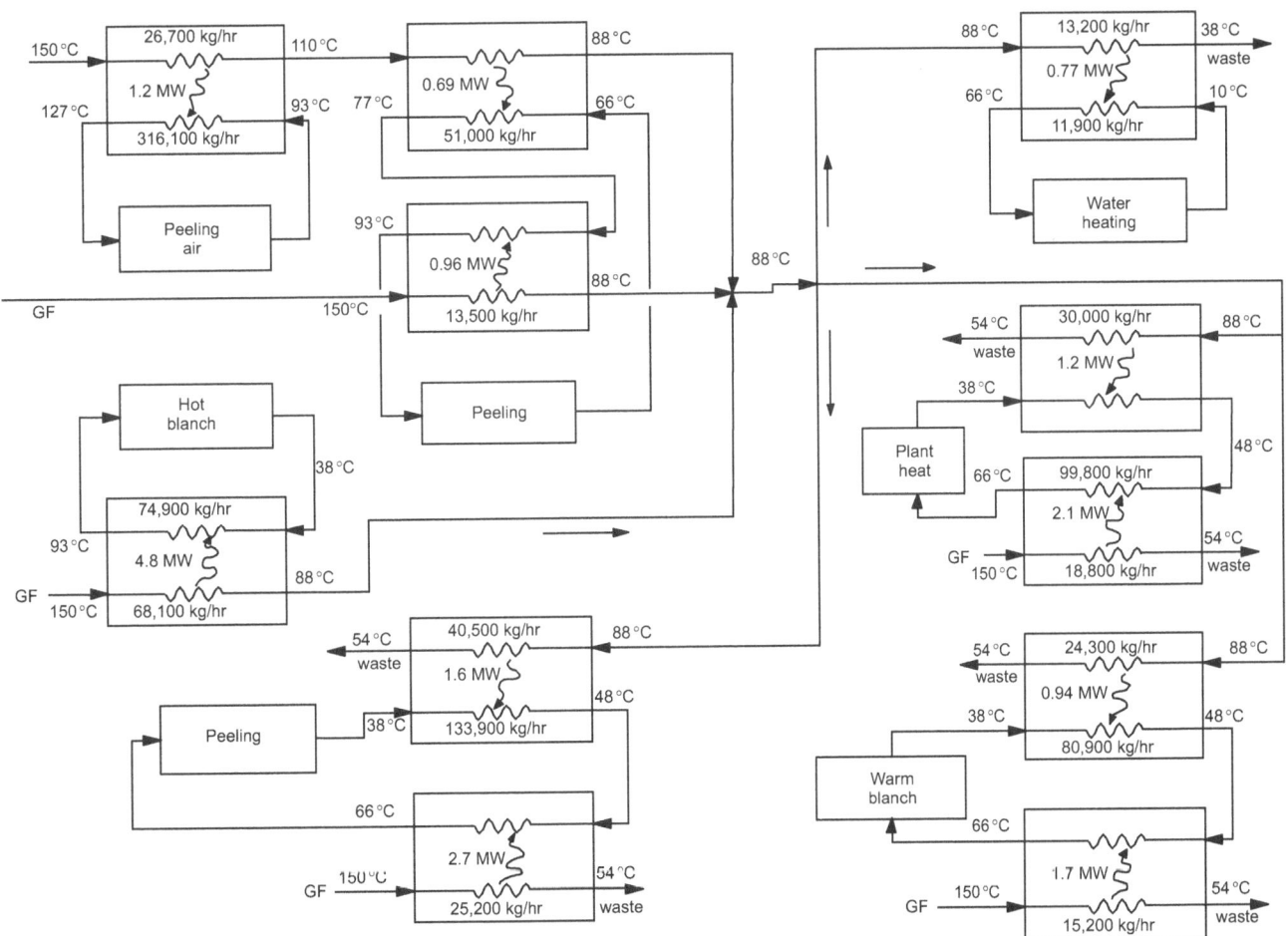

Figure 7.13 Potato processing flow diagram for geothermal conversion

Saturated frying employs heat exchangers and steam at 1,900 kPa. Typically, the fryers consume about 45 per cent of the process energy of the plant, and because the return temperature is above 150 °C, an assumed geothermal fluid supply temperature, over 50 per cent of the process energy requirements could be supplied by geothermal energy.

7.3.6 Heap leaching (Trexler et al., 1987, 1990)

Heap leaching for gold and silver recovery is a fairly simple process which eliminates many complicated steps needed in conventional milling. A 'typical' precious-metal heap-leaching operation consists of placing crushed ore on an impervious pad. A dilute sodium cyanide solution is delivered to the heap, usually by sprinkling or drip irrigation. The solution trickles through the material, dissolving the gold and silver in the rock. The pregnant (gold-bearing) solution drains from the heap and is collected in a large plastic-lined pond (Figure 7.14).

The pregnant solution is then pumped through tanks containing activated charcoal at the process plant, which absorbs the gold and silver. The now barren cyanide solution is pumped to a holding basin, where lime and cyanide are added to repeat the leaching process. Gold-bearing charcoal is chemically treated to release the gold, and is reactivated by heating for future use. The resultant gold-bearing strip solution, more concentrated than the original pregnant cyanide solution, is treated at the process plant to produce a 'doré', or bar of impure gold. The doré is then sold or shipped to a smelter for refining. Figure 7.15 is a process flow diagram for the operation.

One of the problems associated with heap leaching is low gold recovery. Commonly untreated ore will yield about 70 per cent or less of the contained gold. Crushing the ore will increase recovery, but it also increases production costs. At some mines, the ore must be agglomerated, or roasted, to increase recovery. Crushing, grinding, vat leaching, agglomeration, wasting, chemical pre-treatment, and wetting can usually increase gold recovery, depending on the ore. Gold recoveries of over 95 per cent are possible with cyanide leaching. The value of the additional gold recovered must be compared with the increased processing costs to determine the most cost-effective method.

Geothermal energy offers another means of increasing gold recovery, as heating of cyanide leach solutions with geothermal energy provides for year-round operation and increases precious metal recovery. The addition of heat to the cyanide dissolution process is known to accelerate the chemical reaction. Trexler et al. (1987) determined that gold and silver recovery could be enhanced by 5 to 17 per cent in an experiment that simulated the use of geothermal heating of cyanide solutions.

Perhaps the most important aspect of using geothermal energy is that geothermally enhanced heap-leaching operations can provide year-round production, independent of the prevailing weather conditions. Figure 7.16 illustrates a cyanide heap leach 'production window' that may be expected in central Nevada. This curve is provided for illustration purposes only and has not been substantiated by actual production data.

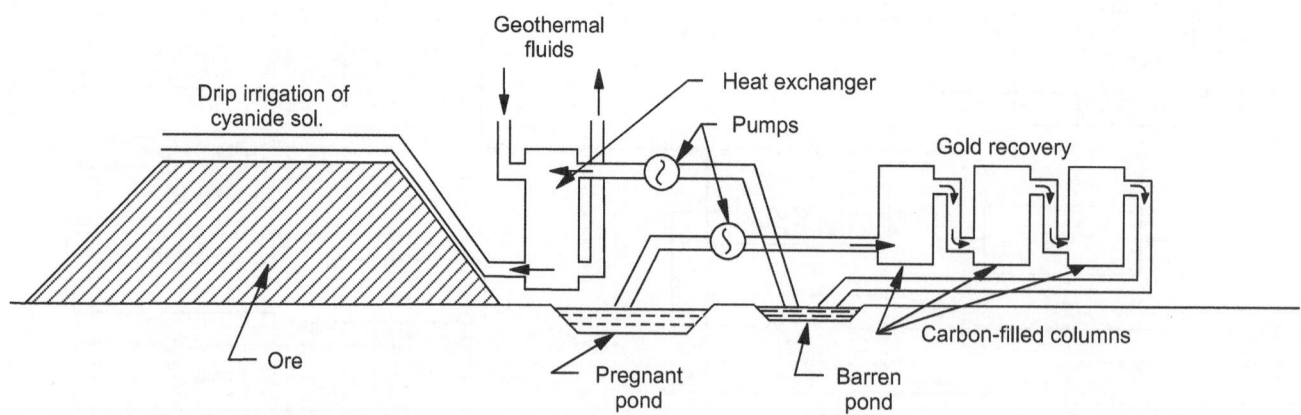

Figure 7.14 Idealized thermally enhanced heap leach
Source: Trexler et al., 1990.

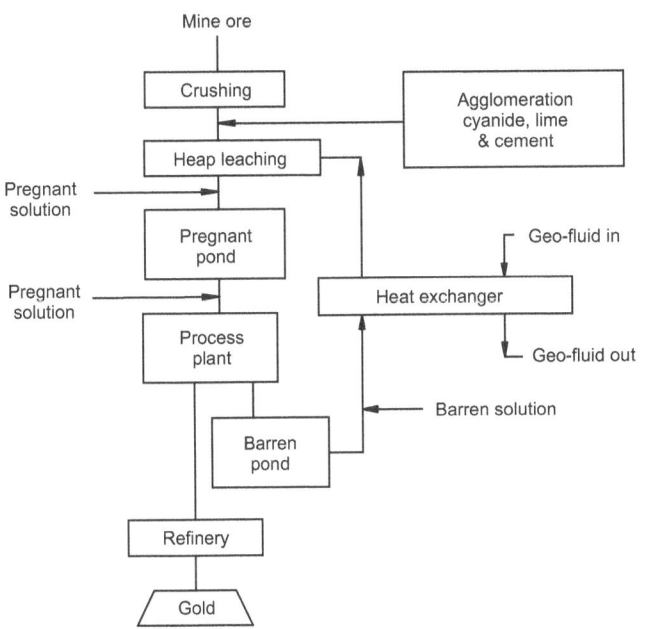

Figure 7.15 Heap leach process flow

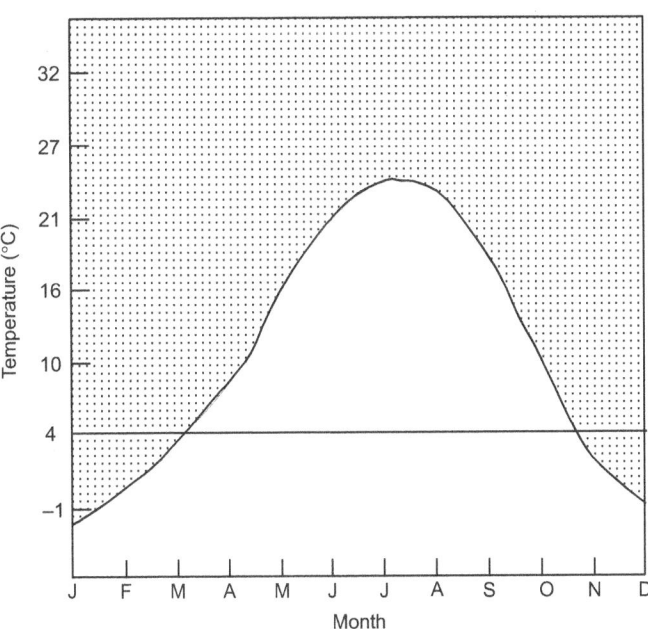

Figure 7.16 Soil temperature at a depth of 10 cm at Central Nevada Field Laboratory near Austin, Nev., (elevation 1810 m)
Source: Trexler et al., 1987.

If the production window opens at a minimum temperature of 4 °C, leaching operations may begin in mid-March and continue through late October. This has been the historical practice at Nevada mines. Since enhanced recovery of gold from heated cyanide solutions has already been established, maximum production would be restricted to June, July and August. Using geothermal fluids would substantially increase the size of the production window (shadowed area, Figure 7.16) and would provide for enhanced extraction rates on a year-round basis. The benefits include increased revenue to the mine operator, year-round employment for the labour force, and increased royalty payments for mineral leases to both federal and state governments.

Mines that incorporate geothermal fluids directly in heap leaching operations need to consider the chemical as well as the physical nature of the resource. Two aspects that must be addressed during elevated temperature leaching are the compatibility of geothermal fluids with leach solution chemistry, and the susceptibility of the heap to mineral deposit formation from geothermal fluids with high total dissolved solids (TDS).

Cyanide reacts chemically with gold and oxygen to form a soluble gold cyanate ($NaAu(CN)_2$). Silver and platinum group metals are also dissolved by cyanide in similar reactions. Non-precious metals, such as iron, copper, manganese, calcium and zinc, along with the non-metals carbon, sulfur, arsenic and antimony, also react with cyanide. Undesirable elements and chemical compounds, other than precious metals, that react with cyanide are called 'cyanocides'. Since cyanocides consume cyanide, high concentrations may interfere with the economic recovery of precious metals. To determine the compatibility of geothermal fluid chemistry with cyanide solutions, a series of consumption tests were conducted by the Division of Earth Sciences, University of Nevada, Las Vegas (UNLV), on a variety of geothermal waters from Nevada. Three major types of geothermal fluid are present in Nevada: NaCl, $NaSO_4$ and $Na/CaCo_3$.

Experimental leach columns were used by the Division of Earth Sciences, UNLV, to analyse compatibility of geothermal fluid chemistry with cyanide solutions and to determine the effects of geothermal fluid chemistry on ore permeability. Preliminary results from this work indicate that:

- Geothermal fluids do not cause plugging of the leach columns by precipitation of minerals.
- The percentage of recovery of gold is not significantly affected by concentration of the geothermal fluids in the process stream.
- Geothermal fluids with high TDS do not contain significant concentrations of cyanocides.

7.3.7 Waste-water treatment plant (Racine et al., 1981)

Potential uses of geothermal energy in the processing of domestic and industrial waste water by a treatment plant include:

- sludge digester heating
- sludge disinfection
- sludge drying
- grease melting.

Figure 7.17 is a process flow diagram for a waste-water treatment plant.

Waste water enters the treatment plant by way of sewer lines. The waste water undergoes preliminary treatment incorporating bar screens that collect debris. These are mechanically removed, and deposited into collection bins for sanitary disposal. Grit removal is accomplished by pre-aeration,

a process by which air, under pressure, is bubbled through the raw waste water to encourage floatable material and settleable material to separate more readily.

Following preliminary treatment, the waste water flows to primary treatment where organic materials are allowed to separate. This is accomplished by reducing the velocity of the waste water in the primary clarifiers, so that these substances will separate from the water carrying them. The solid material, both settled sludge and skimming, are removed for further treatment. The liquid portion, or primary effluent, then flows to the aeration system to begin secondary treatment.

Secondary treatment processes are biological processes in which living aerobic (free oxygen demanding) microorganisms feed on the suspended organic material not removed during primary treatment. The activated sludge process is accomplished in the aerators by introducing a culture of micro-organisms (activated sludge) to the primary

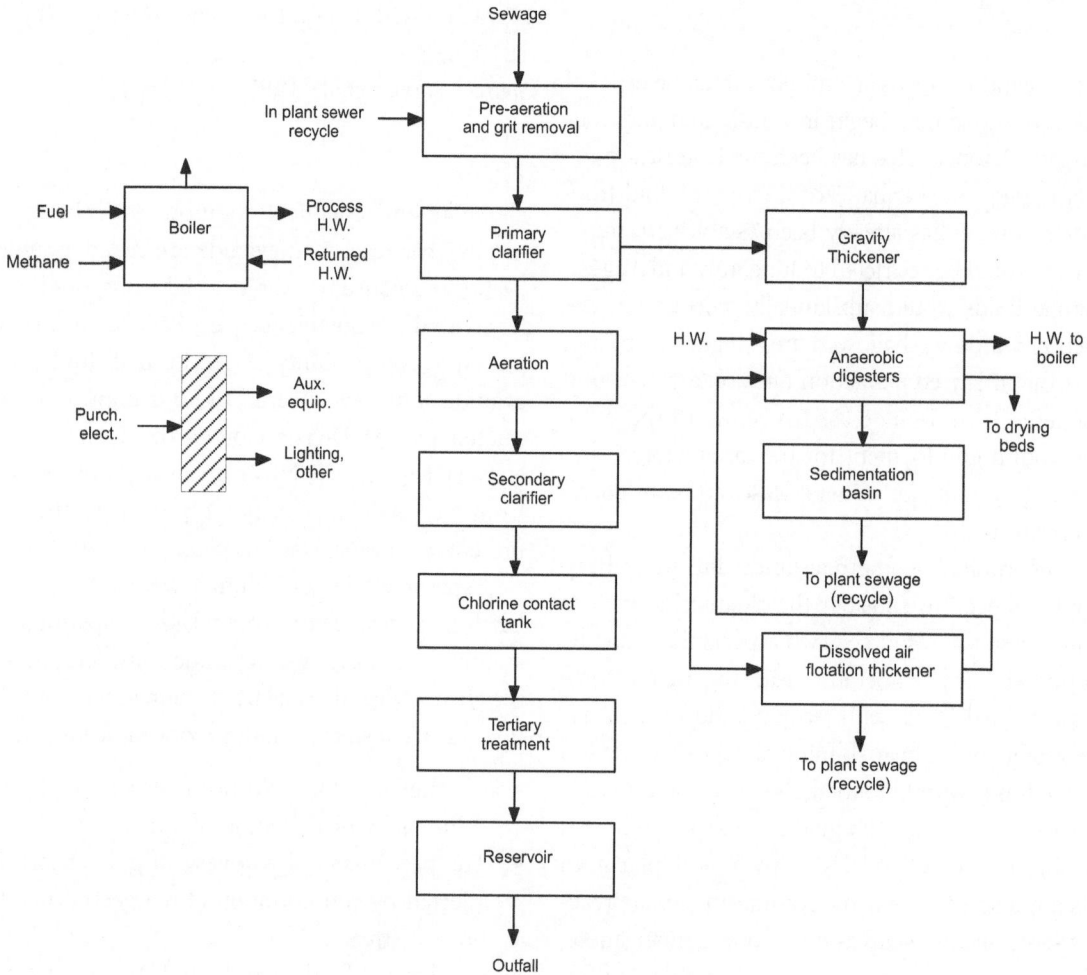

Figure 7.17 Waste-water treatment process flow

effluent, along with large quantities of air for respiration of the microbes and for turbulent mixing of the primary effluent and activated sludge.

After aeration, the mixture of primary effluent and activated sludge flows to a secondary clarifier. At this point, settleable materials are again allowed to settle and the activated sludge is pumped back to the aeration system. Gradually, an excessive amount of solids accumulates and has to be removed. This waste-activated sludge is treated with the solid material removed during primary treatment.

The secondary effluent then flows to the chlorine contact chamber and is disinfected by chlorination. In this process, liquid chlorine is evaporated into its gaseous state, the gas is injected at a controlled rate into a water supply, and this chlorine-saturated water is allowed to mix with the secondary effluent. Sufficient detention time for thorough chlorine contact is then allowed, and finally the effluent is discharged to an outfall.

A portion of this final effluent is treated for a third time at the tertiary plant, where chemical additives are introduced to help remove any suspended material remaining in the effluent. After chemical treatment in a reactor clarifier, the effluent passes through a rapid sand filter for polishing and then into a storage reservoir.

The sludges and other solids collected throughout the treatment process are pumped from their various collection points to the thickeners, where they are concentrated through settling. This thickened sludge is then pumped to the digesters. Digestion is a biological process that uses living anaerobic (absence of free oxygen) micro-organisms to feed on the organics. Processes aided by heating and mixing break down the organic materials into a digested sludge and methane gas. The methane gas is collected and can be used to fuel various in-plant engines that drive pumps and compressors, while the well-digested sludge is dried atmospherically on sand-bottom drying beds and mechanically with one belt press.

There are several uses for low-temperature geothermal fluids within a typical waste-water treatment facility. Table 7.9 presents a summary of potential heat uses which include sludge digester heating, sludge disinfection, sludge drying and grease melting. Low-temperature geothermal fluids are most suitable for sludge digester heating and sludge drying, which will be considered.

In the anaerobic digesters the contents are heated and mixed to enhance the digestion process. The sludge temperature is maintained between 32 and 38 °C, within the

Table 7.9 Waste-water treatment plant process temperatures

Process	Temperature range (°C)
Sludge digester heating	29 to 38 (mesophilic) 49 to 57 (thermophilic)
Sludge disinfection	
Pasteurization	70
Composting	55
Sludge drying	52 to 54
Grease melting	96

mesophilic range, by circulating sludge from the digester to a heat exchanger where the sludge picks up heat and is returned to the digester. Methane-fuelled or natural-gas boilers are usually used to heat water to approximately 68 °C. This water is passed through a spiral plate-type heat exchanger where its heat is transferred to sludge circulating on the other side of the exchanger. Geothermal fluid temperatures as low as 49 °C could technically be sufficient to provide heat to sludge ranging in temperature from 32 to 38 °C.

Mechanical de-watering with belt presses and drying beds usually accomplish sludge drying. The use of heat for drying may increase the sludge-handling capacity of a plant. In addition, if the sludge can be dried sufficiently, it may have commercial value as a fuel or fuel supplement. The dryer type that appears most compatible is the conveyor type using hot water coils to heat drying air. The minimum practical drying air temperature for sludge drying appears to be approximately 77 °C, which would require geothermal fluid temperatures on the order of 88 °C or above. Using the 77 °C air, approximately 25,800 J will be required to evaporate 0.45 kg of water from belt press paste (80 per cent moisture) to a dried product (10 per cent moisture).

REFERENCES

Carter, A. C.; Hotson, G. W. 1992. Industrial use of geothermal energy at the Tasman Pulp & Paper Co., Ltd Mill, Kawerau, New Zealand. *Geothermics*, Vol. 21, No. 5/6, pp. 689–700.

Chin. 1976. *Geothermal Energy in Taiwan, Republic of China.* Taipei, Taiwan, Mining Research & Service Organization, ITRI.

CHUA, S. E.; ABITO, G. F. 1994. Status of non-electric use of geothermal energy in the Southern Negros geothermal field in the Philippines. *Geo-Heat Center Quarterly Bulletin*, (Klamath Falls, Ore.), Vol. 15, No. 4, pp. 24–9.

GUILLEN, H. 1986. *Feasibility Study on the Establishment of Geothermal Food Dehydration Centres in the Philippines.* Klamath Falls, Ore., Geo-Heat Center, Oregon Institute of Technology.

HORII, S. 1985. *Direct Heat Update of Japan.* International Symposium on Geothermal Energy, International Volume, Geothermal Resources Council, Davis, California.

HORNBURG, C. D.; LINDAL, B. 1978. *Preliminary Research on Geothermal Energy Industrial Complexes.* Fort Lauderdale, Fla., DOE Report IDO/1627-4, DSS Engineers, Inc.

HOTSON, G. W. 1995. *Utilization of Geothermal Energy in a Pulp and Paper Mill.* Proceedings of the World Geothermal Congress '95, Florence, Italy, International Geothermal Association.

KNIGHT, E. 1970. *Kiln-Drying Western Softwoods.* Moore Dry Kiln Company of Oregon, Portland, Ore.

LIENAU, P. J. 1978. *Agribusiness Geothermal Energy Utilization Potential of Klamath and Western Snake River Basins, Oregon.* Klamath Falls, Ore., DOE Report, IDO/1621-1, Geo-Heat Center, Oregon Institute of Technology.

LUND, J. W.; LIENAU, P. J. 1994. Onion and garlic dehydration in the San Emidio Desert, Nevada. *Geo-Heat Center Quarterly Bulletin*, (Klamath Falls, Ore.), Vol. 15, No. 4, pp. 19–21.

PIRRIT, N.; DUNSTALL, M. 1995. *Drying of Fibrous Crops Using Geothermal Steam and Hot Water at the Taupo Lucerne Company.* Proceedings World Geothermal Congress '95, Florence, Italy, International Geothermal Association. pp. 2339–44.

RACINE, W. C. et al. 1981. *Feasibility of Geothermal Heat Use in the San Bernardino Municipal Wastewater Treatment Plant.* San Bernadino, Calif., DOE Report, Municipal Water Department.

RAGNARSSON, A. 1996. Geothermal energy in Iceland. *Geo-Heat Center Quarterly Bulletin*, (Klamath Falls, Ore.), Vol. 17, No. 4, pp. 1–6.

RUTTEN, P. 1986. *Summary of Process-Mushroom Production.* Vale, Ore., Oregon Trail Mushroom Company.

SIGURDSSON, F. 1992. Kisilidjan hf-A unique diatomite plant. *Geothermics*, Vol. 21, no. 5/6, pp. 701–7.

TALBURT, W. F.; SMITH, O. 1975. *Potato Processing*, 3rd ed. Westport, Conn., AVE Publishing.

TREXLER, D. T.; FLYNN, T.; HENDRIX, J. L. 1987. Enhancement of precious metal recovery by geothermal heat. *GRC Transactions*, (Reno, Nev.), Vol. 11, pp. 15–22.

TREXLER, D. T.; FLYNN, T.; HENDRIX, J. L. 1990. Preliminary results of column leach experiments at two gold mines using geothermal fluids. *1990 International Symposium on Geothermal Energy*, Hawaii, GRC Transactions, Vol. 14, pp. 351–8.

VTN-CSL. 1977. *Economic Study of Low Temperature Geothermal Energy in Lassen and Modoc Counties, California.* Sacramento, Calif., State of California, Division of Oil and Gas, and Energy Resources Conservation and Development Commission.

WILSON, R. D. 1974. Use of geothermal energy at Tasman Pulp and Paper Company Limited, New Zealand. In: *Multi-Purpose Use of Geothermal Energy*, Klamath Falls Ore., Oregon Institute of Technology.

SELF-ASSESSMENT QUESTIONS

1. The temperature and pressure of a geothermal fluid are both parameters that limit our choice of utilization of this resource: true or false?

2. The heat content of geothermal fluids with a temperature of >25 °C can be exploited in a variety of industrial processes, in agriculture, domestic uses and so on. Geothermal waters with even lower temperatures can, however, be exploited for their heat content. Can you name at least one application?

3. In a pulp and paper mill similar to that described in the lesson, can the geothermal fluid produced by a natural thermal spring provide on its own the quantity of heat required for this process?

4. In drying lumber processes, thermal energy is utilized to control the evaporation rate of the moisture contained in the wood, thus preventing the stresses that cause warping. What temperature should a geothermal fluid have for this type of utilization, and what are the minimum temperatures that could be required (give approximate values)? Could a cascade system be adopted for other possible applications?

5. Considering the importance of rice as a staple food worldwide, rice processing and related energy aspects are of considerable interest. How can geothermal energy be utilized in this process?

6. Geothermal heat can be utilized in vegetable and fruit dehydration. The process involves the use of a tunnel dryer or a continuous conveyor dryer. Which of these two dryers is least used and why?

7. In the chapter you were shown a diagram of a plant for producing frozen french fried potatoes (Figure 7.12). You were told that, for systems using geothermal energy, the energy would probably be supplied to the process via intermediate heat exchangers. This is often the case also in systems for treating or processing other food products. Why?

8. Waste-water treatment is an interesting potential application of geothermal heat. The different cycles of this process (sludge digester heating, sludge disinfection, sludge drying, grease melting) take place in one of the following temperature intervals. Which one?
(a) 30 to 95 °C; (b) 70 to 95 °C; (c) >95 °C

ANSWERS

1. True in most cases, but sometimes it may be convenient to upgrade the quality of the available geothermal resources by integrating with other thermal and/or mechanical energies. In this way we can modify or improve the characteristics of a geothermal fluid so as to make it suitable for a form of utilization of a higher grade than would be possible with its original characteristics. Examples of upgrading systems include (a) flashing the geothermal fluid, then heating via fossil fuels, (b) heating via fossil fuels, then flashing, and (c) mechanical compression.

2. There are several, but we will mention three of the most common:
(a) The heat content of water with temperatures below 25 °C could be extracted by means of a heat pump and utilized in processes that require higher temperature.
(b) Geothermal water at 20 °C can, in certain circumstances, be utilized in aquaculture.
(c) In cold climates, water at 15 °C or even less can be used to de-ice roads and streets.

3. No. Most of the process heat requirements are in the range 120 to 180 °C. The hot water from a natural spring, for obvious reasons, will never exceed 100 °C. The water from a thermal spring can therefore only be used in the pre-heating stages or for space-heating.

4. The geothermal fluids used for kiln drying should, ideally, have temperatures between 90 and 115 °C. The minimum temperatures required for this process vary between 80 and 90 °C, depending on the type of wood being treated. As the discharge fluid for these applications would have temperatures from 70 to 80 °C, cascade systems are possible (for example, for space heating).

5. When harvested, rice has a moisture content of 20 to 26 per cent. Grains containing more than 13.5 per cent moisture cannot be safely stored for long periods; so drying is an important phase in rice processing. The heat content of geothermal fluids can be used in the drying cycle, which requires temperatures between 40 and 90 °C. Geothermal plants for drying rice already exist in a number of countries.

6. The tunnel dryers are rarely used because of the high labour requirements, associated with loading or unloading the compartments. Tunnel dryers are used as pilot plants or in underdeveloped areas.

7. The thermal energy is supplied to the process by a secondary fluid, usually water, through heat exchangers, so as to avoid possible contamination of the product by the geothermal fluid or having to treat the fluid before using it.

8. The correct temperature interval is (a).

CHAPTER 8

Environmental Impacts and Mitigation

Kevin Brown

Geothermal Institute, University of Auckland, Auckland, New Zealand

Jenny Webster-Brown

Environmental Science, University of Auckland, Auckland, New Zealand

AIM

To raise awareness of the potential environmental impacts of geothermal energy development.

OBJECTIVES

When this chapter has been read, the reader should have an understanding of:

1. The principal environmental effects associated with geothermal development and exploitation.

2. The link between the nature of the geothermal field and type of power generation system, and the type of environmental impacts.

3. Possible ways of mitigating environmental effects.

4. The fundamental aspects of environmental legislation and its implementation.

8.1 INTRODUCTION

The environmental aspects of geothermal development are receiving increasing attention with the global shift in attitudes towards the world's natural resources. Not only is there a greater awareness of the effect of geothermal development on the surrounding ecosystems, social system and landscape, but there is also growing appreciation of the need for efficient and wise use of *all* natural resources. Most countries have now embodied their environmental concerns in legislation, and a geothermal development may now face significant costs, very significant time delays, and indeed can be completely halted if the environmental concerns of the country are ignored.

The impact of a geothermal development varies greatly depending on the type of development and the characteristics of the geothermal fluid. A small direct-use application of low-temperature, relatively unmineralized geothermal water may have a very low impact, whereas a 200 MW$_e$ power station utilizing a liquid geothermal resource can have a much more significant and diverse impact. In this chapter, we will attempt to consider all of the potential environmental impacts of a large geothermal development, despite the fact that only some may be relevant for any given field.

8.2 PHYSICAL IMPACTS

Exploration, development and utilization of a geothermal area can have a significant physical impact on the environment surrounding the resource. The impacts during initial exploration and field development will tend to be short-term: construction of access tracks for geochemical and geophysical measurements, drill pads, land clearance for pipeline routes and possibly a power station, cooling-water extraction for drilling, as well as increased noise and possibly dust levels. It is during the exploitation phase that the longer term, more serious impacts tend to occur. Natural geothermal features may decrease or increase in activity, the local climate may be affected by cooling-tower emissions, large volumes of cooling water may contribute to thermal pollution of local waterways, and some areas of land may be subject to subsidence.

8.2.1 Land and the landscape

The visual effect on the landscape depends on the type of countryside, the scale and the phase of the development. In general, the area required for geothermal development is a function of the power output of the development, the type of countryside and the properties of the reservoir. By their nature, geothermal systems are often located in volcanic environments where the terrain is steep and access difficult. Such an environment may also have severe erosion problems, particularly in high-rainfall areas.

Road construction in these steep environments normally involves extensive intrusion into the landscape, and can often cause slumping or landslides, with consequent loss of vegetation cover. The lack of vegetation then allows greatly accelerated erosion, with the possibility of further slumping or landslides, and increased suspended sediments in the surrounding watershed.

The size and impact of power-plant sites depend on what facilities are incorporated into the site. Often, as well as the turbine house and electrical switchyard, there is a separator station, cooling towers and perhaps an air-pollution abatement plant. Such installations require a large area of flat land, and in steep terrain the necessary cut-and-fill operations can have a severe impact on the landscape.

Pipeline corridors are typically 3 to 5 m in width, and depending on the pipe size, may need access roads for construction and maintenance. Transmission lines require a corridor free from overlying vegetation, and access roads are required for construction of large steel pylons.

Mitigation: Ongoing re-vegetation programmes, proper drainage for roads, and the use of settlement ponds during the construction phase can help prevent erosion and sedimentation in watersheds, and will lead to lower maintenance requirements at later stages. With modern drilling technology, a single drilling pad can house a number of deviated wells, thereby accessing a large volume of reservoir while leaving a small footprint on the land.

The visual impact of permanent features such as pipelines can be minimized by painting and avoiding changes of form and line, such as the use of horizontal rather than vertical expansion loops. Careful use of the natural vegetation and planted trees can serve to hide pipelines and small buildings, and it may be possible to use the natural form of the land to similarly reduce the visual impact of power stations, pipeline corridors and drill pads by siting them behind ridges and in hidden valleys.

8.2.2 Noise

Noise, or 'unwanted sound', is one of the most ubiquitous disturbances to the environment, and any development should seek to minimize this impact. Many geothermal developments are in remote areas where the natural level of noise is low and any additional noise is very noticeable. Residents in such areas will probably regard any noise as an intrusion into their otherwise quiet environment. Animal behaviour is also affected by noise, with reports of changes in size, weight, reproductive activity and behaviour.

Sound is measured in decibels (dB). This is a logarithmic unit based on the pressure of the sound waves. The scale runs from 0 (the limit of hearing) to about 130 (intolerable). The human ear responds differently to different frequencies (the so-called 'A' curve), and often noise is reported as dBA, which has been corrected for the human response.

Noise occurs in the exploration drilling, construction and production phases of development. Air drilling is the noisiest (120 dBA) because of the 'blow pipe' where the gases exit. Mud drilling is somewhat quieter at around 80 dBA. Vertical well discharges are very noisy (up to 120 dBA), but are often required to purge the wells and remove drilling debris. After drilling, there is normally a period of well testing. This can be suitably muffled by the use of silencers, but even here the noise is still significant. The well is then put on bleed, where the noise is around 85 dBA. The construction phase will bring the normal noise associated with heavy machinery (up to 90 dBA). During the production phase, there is the higher-pitched noise of steam travelling through pipelines and the occasional vent discharge. These are normally acceptable. At the power plant, the main noise pollution comes from the cooling-tower fans, the steam ejectors and the low-frequency turbine 'hum'.

Mitigation: Suitable muffling can reduce the noise of air drilling to around 85 dBA, and the use of silencers during well testing can reduce this noise to 70 to 110 dBA. New designs of silencers are reducing this even further. Likewise, suitable mufflers can reduce noise during well bleeding (to 65 dBA) and from heavy machinery.

Noise is attenuated by distance travelled in air. As an approximate rule of thumb, there is 6 dB attenuation every time the distance is doubled, but lower frequencies are attenuated less than higher frequencies. Consequently, careful siting of noisy operations at a distance from population centres can be beneficial. Earth and vegetation mounds or causeways can also be used to dampen noise.

geothermal systems. Because of their unique nature, these are often tourist attractions or are used by the local residents. Geothermal development that draws from the same reservoir has the potential to affect these features. These visible signs of geothermal activity are part of a country's heritage, and their response to development must be taken into account.

Currently, it is difficult to predict exactly the effect of geothermal development on any particular surface geothermal feature. However, there is now considerable experience with exploited geothermal fields, and some general comments can be made. In water-dominated geothermal systems, features that rely on the flow of deep geothermal water (generally recognized by the presence of silica sinter and a significant chloride content) have shown a tendency to decline with exploitation of the deep reservoir. On the other hand, natural features in water-dominated systems that rely on the flow of steam, such as steaming ground, acid sulfate springs and fumaroles, tend to increase in activity and may migrate during exploitation. In vapour-dominated geothermal systems, where the mobile fluid is mainly steam, a reduction in the reservoir pressure usually leads directly to a reduction in the surface discharges.

Mitigation: One possible solution is to have a system for ranking the geothermal features in order of priority of preservation. Development may or may not then proceed depending on the balance of the worth of the energy versus the heritage of the natural features. Certain hydrothermal features, such as playing geysers, could be considered to be extremely 'valuable'. Other considerations could include the preservation of a representative example of each of the hydrothermal features. The factors to be considered will vary from country to country.

Before any development takes place, the natural features associated with a geothermal field should be catalogued with as much information and for as long as possible to provide baseline data for later comparison. During the exploration phase, the natural features are normally analysed anyway for temperature, fluid flow, heat flow, chemistry and other parameters, as an exploration tool to gain information about the deep reservoir. These data should form the nucleus of an ongoing monitoring programme for the features, since geothermal features can change naturally in response to local weather patterns, local seismicity and subsidence. In particular, the activity of geysers and fumaroles is susceptible to change.

8.2.3 Natural geothermal features

Natural surface features such as hot springs, mud pools, geysers, fumaroles and steaming ground are associated with most

8.2.4 Heat-tolerant vegetation

In many geothermal fields, areas of steaming ground, springs and other features have developed special thermal vegetation habitats. In such cases, only very tolerant species can survive, and a number of very unique flora may evolve that can survive the high temperatures and acidic soil. However, the roots of most plants cannot survive temperatures much above 50 °C, and at temperatures between about 50 and 70 °C only mosses and lichens can survive. Above this temperature, vegetation is absent. Consequently, changes in the thermal areas, such as increased steam flow as a consequence of exploitation, may change the distribution of these thermally adapted plants, with the possibility of rendering some of the species very vulnerable to extinction.

Mitigation: Before field development, the distribution and nature of thermal-tolerant vegetation on the field should be catalogued, with as much information and for as long as possible, to provide a data baseline for later comparison. These data should form the nucleus of a continuous monitoring programme for the features, so that their response to altered steam-flow patterns in the ground can be assessed. Conservation and replanting programmes may need to be implemented to ensure the survival of specific vulnerable species.

8.2.5 Hydrothermal eruptions

Although relatively rare, hydrothermal eruptions constitute a potential hazard in active geothermal fields, and need to be included in the environmental impact assessment. The causes and mechanisms of hydrothermal eruptions have been reviewed by Bromley and Mongillo (1994). Eruptions occur when the steam pressure in the near-surface aquifers becomes greater than the overlying lithostatic pressure, and the overburden is then ejected to form a crater. The resulting vent can vary from 5 to 500 m in diameter, and up to 500 m in depth, although most eruptions will be relatively shallow.

The following points should therefore be taken into account in assessing the likelihood of a hydrothermal eruption hazard:

- evidence of previous natural hydrothermal eruptions
- increasing steam flow to the surface from reservoir pressure drawdown or an expanding steam zone
- vigorous dispersed or superheated steam emission and uncompacted low-density shallow formations
- near-surface aquifer temperatures close to boiling point for depth

- near-surface aquicludes (clay, sediments and so on) confining or deflecting rising steam and gas
- removal of overburden
- shallow gas pockets, kicks or blowouts during drilling.

Drilling activities can also pose a risk of causing eruptions if the casing string is set too shallow, or if the casing develops a leak.

Mitigation: Investigations of areas considered prone to hydrothermal eruption risk should concentrate on determining the depth and porosity of any shallow boiling aquifers. Temperatures of steam-heated features should be monitored carefully, as should temperatures and pressures in shallow groundwater holes to detect any increase in the heat flow to the surface.

8.2.6 Subsidence

Withdrawal of fluid from any type of underground reservoir will normally result in a reduction of pressure in the formation pore space, which in some circumstances can lead to subsidence. Subsidence has been observed in groundwater reservoirs and petroleum reservoirs as well as in geothermal reservoirs. With the vertical movement, there is usually some associated horizontal movement radially towards the point of maximum subsidence. The combination of the two ground movements can have serious consequences for the stability of pipelines, drains and well casings at a geothermal field.

It appears that subsidence in water-dominated fields is greater than in vapour-dominated fields. Probably the largest recorded subsidence has occurred at Wairakei in New Zealand (Allis, 1990), where the centre of the subsidence bowl is sinking at a rate of 450 mm/year (Figure 8.1). If the field is close to a populated area, subsidence can lead to instability in dwellings and other buildings. In more remote areas, where there may be no habitation, the local surface watershed systems may be affected.

Mitigation: There is little that can be done to mitigate ground subsidence. However, it can be monitored and the severity assessed. Before exploitation begins, a baseline levelling survey with installation of levelling stations needs to be undertaken. There should be a number of separate surveys to cover as long a time as possible before exploitation, so that the local tectonic changes in level, if any, can be subtracted from those as a result of exploitation. Ground surface level should continue to be monitored throughout development and exploitation of the field.

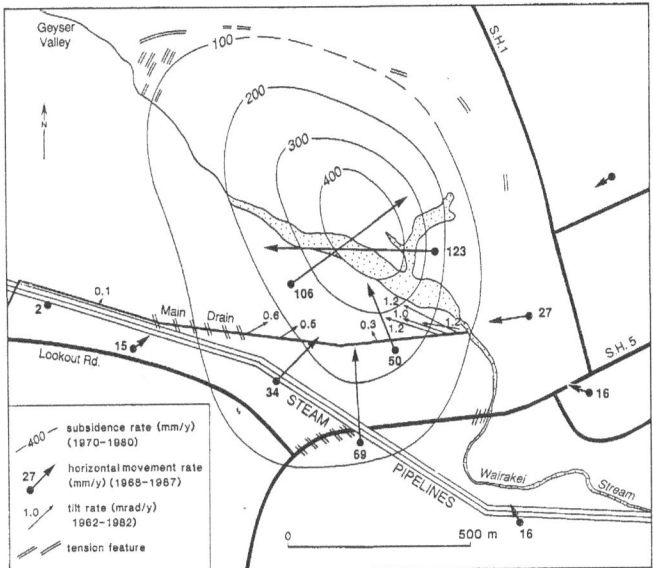

Figure 8.1 Subsidence at the Wairakei geothermal field in New Zealand
Source: After Allis, 1990.

8.2.7 Induced seismicity

By their nature, geothermal fields usually occur in regions of high seismic activity. In such a case, there is a natural occurrence of earthquakes that are not related specifically to the exploitation of the geothermal field. Micro-earthquakes are seismic events that are of very low magnitude and can only be detected by instrumentation. Micro-earthquakes have long been recognized to occur in convecting hydrothermal systems. Injection of fluids into deep formations, on the other hand, has been recognized as a cause of induced seismicity. Re-injection under large pressures at the Wairakei geothermal field had to be halted when earthquakes were felt in the local area. This re-injection was at high well-head pressures, and subsequent re-injection at saturated water-vapour pressures has produced no observable earthquake activity.

8.2.8 Thermal effects of waste discharge

Geothermal power plants utilize relatively low source temperatures to provide the primary energy for conversion to power production, and therefore the waste heat per MW of electricity generated is much larger than with other types of power generation. Thus there is a larger proportion of waste heat in geothermal systems, and this needs to be dissipated in an environmentally acceptable way. In vapour-dominated

systems, waste heat can be discharged to the atmosphere in the form of cooling tower plumes, or to surface waterways in the form of cooling-water outflows. In water-dominated systems, there can be additional waste heat discharged in waste bore water, although few geothermal developments still discharge bore water into local waterways.

Cooling towers are the principal means by which waste heat from steam is transferred to the environment, and their main impact is on the local climate. Localized slight heating of the atmosphere and an increased incidence of humidity, perhaps leading to fog, are the main consequences that have been observed. In the case of once-through use of cooling water, condenser heat will be discharged to local surface water. The use of once-through cooling-water systems with a discharge-to-the-surface environment, versus dispersal to the atmosphere, is an important environmental trade-off (Collie, 1978). Discharge to the atmosphere is most likely to affect the local climate, whereas discharge to surface waterways will more likely affect the local biological communities. In general, biological communities are considered to be more vulnerable to change, so that the trend is towards atmospheric discharge of the steam heat component.

Mitigation: There is increasing realization of the need to protect the environment from the heat input, and there should be plans to cascade the use of the heat, so that the water flowing from the development into the environment is at ambient temperature. Waste heat in the water component of water-dominated geothermal systems is being used increasingly for power generation through binary-cycle generation plants. In this way, it is further reduced in temperature and the thermal impact will be reduced.

Most geothermal developments now dispose of waste geothermal water by deep re-injection, where the environmental impact of the heat is negligible. Re-injection of waste geothermal waters, however, is not without its problems because of re-injection returns, scaling in wells and so on. There may be a swing in the future towards chemically treating the effluents so that they can be discharged to the surface.

For cooling tower discharge, plume modelling is normally now a requirement for permission prior to exploitation. A monitoring programme to define baseline weather conditions, and detect local climate change, is usually required.

8.2.9 Water usage

Water is required for drilling, for soil compaction during construction, for re-injection well testing, and for cooling water during the production phase. The impact of extracting this water from the local watershed will vary greatly depending on the locality. In arid areas, for example, the problem of a suitable water supply can be acute. In this case, geothermal water produced from the first wells may have to be used for a water supply for subsequent drilling.

Similarly, water will need to be discharged, particularly during the development phases. Besides the chemical content of the wastewater, suspended sediment levels may be high. In the design of sumps for drilling and other uses, due care must be taken to ensure that flooding and breaches are allowed for.

Mitigation: Obtaining consent for water extraction and discharge will be one of the first consent considerations. The use of holding and settling ponds (of adequate capacity) can improve water quality. Wastewater can be re-injected into the ground, rather than discharged to surface water systems. Local meteoric aquifers need to be protected by casing out these intervals during drilling, and by lining holding ponds and sumps. Regular monitoring of both chemical and physical properties of groundwater aquifers and vulnerable surface waters (see also Section 8.4) will normally be required.

8.2.10 Solid wastes

Geothermal development can produce significant amounts of solid waste, and suitable disposal methods need to be found. Because of the toxic contaminants (for example, arsenic and mercury) contained in geothermal water, these solid wastes are not environmentally benign, and if left in stockpiles on the ground surface can lead to generation of contaminated surface run-off, and possible contamination of groundwaters.

Among the principal solid wastes are cooling tower sludges. Whether there is an H_2S abatement process operating or not, these are normally predominantly sulfur with the possibility of mercury contamination. Other solid wastes may include silica scale and sinters from pipes or drains, and these may be contaminated with arsenic. If there is some treatment of the brine, then flocculated oxides may be produced.

Mitigation: These wastes can be disposed of in specially designated hazardous-waste disposal sites, if available. More commonly, however, these wastes will need to be disposed of in the local area. It may be advantageous to dispose of slurry in a deep re-injection well that is cased to protect the meteoric aquifers. If this is not possible, then total containment is normally required with no significant emissions to the air, surface water or groundwaters after burial.

8.3 IMPACTS ON AIR QUALITY

Geothermal power is often considered to be a 'clean' alternative to the use of fossil fuels, which release more carbon dioxide and sulfur gases into the air during power generation. Geothermal power generation can nevertheless affect air quality through the discharge of gaseous contaminants from wells (during drilling and testing), steam condensate pots, silencers, and most importantly, the gas exhausters of the power station (Figure 8.2). At vapour-dominated fields, and fields in which all waste fluids are re-injected, gas and steam are likely to be the only significant routine discharges.

8.3.1 Composition of gas discharges

The most important gaseous contaminants from an environmental perspective are:

- carbon dioxide (CO_2: usually the major component)
- hydrogen sulfide (H_2S)
- ammonia (NH_3)
- mercury (Hg)
- boric acid (H_3BO_3).

The discharge may also contain hydrocarbons such as methane or ethane and radon, but these constituents are less likely to have a significant adverse effect on the local environment.

Contaminant concentrations in steam will depend on reservoir geochemistry and the power generation conditions.

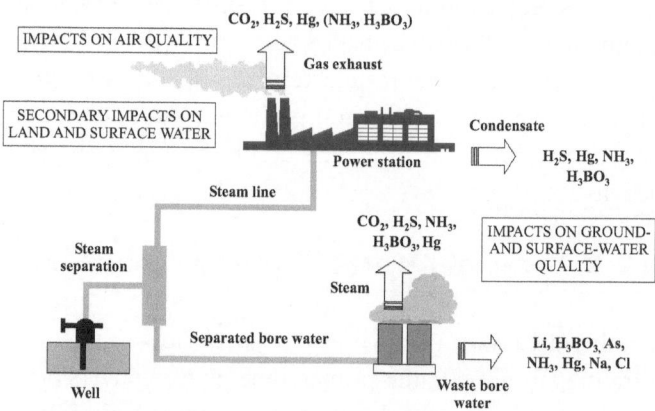

Figure 8.2 The principal contaminants potentially occurring in gas, steam and fluid discharges, shown for a model, developed geothermal field

Table 8.1 Contaminant concentrations in some representative steam and waste-bore-water discharges from developed geothermal fields. Concentration units are mg/kg, for all except mercury (µg/kg)

	CO_2	H_2S	NH_3	H_3BO_3	Hg	As	Li
In steam							
Larderello	47 500	540	150	220	—	—	—
Geysers	3 260	222	194	91	5	0.02	—
Wairakei	1 660	72	4.8	1.3[1]	1–4	—	—
Ohaaki	30 000	350	34	7.6[1]	4–8	0.02	—
Cerro Prieto	14 000	1 500	110	<0.1	7–9	0.006	—
In bore water							
Salton Sea	—	16	386	2 231	6.0	12	215
Cerro Prieto	—	0.16	127	109	0.05	2.3	—
Wairakei	—	1.7	0.20	150[1]	0.12	4.7	14
Ohaaki	—	1.0	2.1	197[1]	0.05	8.1	11.7
Los Azufres[2]	—	—	—	2 030	4.1	19	19.7
Hveragerdi	—	7.3	0.1	3.4	—	0.0	0.3

Source: 1: Glover, 1988; 2: Birkle and Merkle, 2000. Other data summarized from Ellis and Mahon (1977) and Ellis, 1978.

There can be considerable variation between fields, as shown in Table 8.1. During initial steam separation, if required, contaminants in the reservoir fluid will be partially or completely transferred into the steam phase, depending on their volatility. Boric acid and ammonia, for example, have relatively low volatility and will transfer mainly into the fluid phase. Hydrogen sulfide and mercury are more volatile, while carbon dioxide transfers almost exclusively into the steam phase at separation.

Concentrations in the gas exhaust will depend on how much of the gaseous contaminant initially present in the steam has been condensed in the condensers, intercoolers or cooling tower. Since carbon dioxide, mercury and hydrogen sulfide are relatively insoluble in water, they become important constituents of the 'non-condensable' gas fraction, which is discharged to the atmosphere. Boric acid and ammonia are more soluble, and will typically have low concentrations in the gas exhaust.

The rate at which contaminants are removed from the atmosphere by raindrops or water aerosols (for example, fog) is also influenced by gas solubility. Ammonia gas, for example, is rapidly leached from the atmosphere, because it is relatively soluble in water. Mercury vapour, on the other hand, is likely to remain in the atmosphere for relatively long periods of time and may be dispersed over a wide area. The presence of contaminants in rainwater can lead to soil or vegetation contamination, and may affect the chemistry of local surface waters, such as rivers and lakes. These effects are often referred to as 'secondary' effects of gas and steam discharge.

8.3.2 Toxic and environmental effects

The discharge of gaseous contaminants can affect the environment and human health. The latter is easier to quantify, and forms the basis for air quality criteria. In this case 'human health' includes both public and occupational health, where 'public' refers to people living outside the confines of the geothermal development, and 'occupational' refers to people working within the 'factory fence'. The principal difference is the length of time of exposure and the concentration of the contaminant.

Carbon dioxide

When discharged in the gas exhaust of a power station, and properly dispersed, carbon dioxide is unlikely to have any direct effects on human health. However, it should be noted that carbon dioxide is a greenhouse gas (as is methane), and discharge into the atmosphere may contribute to global warming. There are occupational risks associated with high carbon dioxide environments, as the gas is heavier than air, and can accumulate in pits and low depressions. It is highly toxic at high concentrations, even in the presence of oxygen, as it alters the pH of the blood. It is odourless, has a slightly acid taste, and a 5 per cent concentration in air (50,000 mg/kg) can produce shortness of breath, dizziness, mental confusion, headache and possible loss of consciousness. At 10 per cent concentrations, the patient normally loses consciousness and will die unless removed. With little or no warning from taste or odour, workers have entered a tank or pit with high concentrations of carbon dioxide and have died within a very short time.

Hydrogen sulfide

Hydrogen sulfide gas has a characteristic 'rotten eggs' odour that is detectable at very low concentrations (above about 0.3 mg/kg). Discharge can therefore lead to odour complaints from people living or working downwind. As the concentration increases, the odour becomes sweeter and finally disappears altogether at about 150 mg/kg. Thus smell is not a reliable indicator of concentration, and the effects of higher concentrations on human health are very serious. Like carbon dioxide, hydrogen sulfide is a heavy gas and will accumulate in low-lying areas. Consequently, people working within the geothermal field will normally be at greater risk. The toxic effects range from headaches, leg pains, and irritation of the respiratory tract and eyes at concentrations in the 10 to 500 mg/kg range, through to loss of consciousness for several minutes (500 to 700 mg/kg), profound coma with difficult or laboured breathing (>700 mg/kg), to death by asphyxia (>1,500 mg/kg). There is no accumulation in the body, and the gas is excreted through the urine, intestines and expired air.

The effects of hydrogen sulfide on the environment are likely to be limited mainly to the secondary effects of gas dissolved in rainwater. Dissolved sulfide will oxidize in air to form various sulfur species, some of which have been identified as components of 'acid rain'. However, no direct link between sulfide emission and acidification of local rainwater has been established.

Mercury

Discharged mercury vapour is likely to remain in the atmosphere long enough to be distributed over a wide area. No carcinogenic properties have yet been assigned to mercury, but it is known to bio-accumulate in the food chain. When inhaled, a large proportion of the mercury is retained in the body (approximately 80 per cent, compared to the 10 per cent retained when ingested) and is deposited in the kidneys, where the half-life of elimination can vary from a few days to a few weeks. Although effects on the kidneys sometimes appear first, it is the central nervous system that is the critical organ for vapour exposure. Longer-term exposure produces a fine tremor, and ultimately insanity, which may not be reversible following withdrawal of the exposure.

Mitigation: Before a geothermal field is developed, potential contaminant concentrations in both gas and steam discharges can be predicted from reservoir chemistry and a knowledge of the operating procedures to be used during power generation. There are computer models available that calculate equilibrium concentrations for all components of a geothermal fluid during steam separation or during condensation. Computer models can also be used to predict plume dispersion, and contaminant concentrations at various levels in the atmosphere, as a function of distance from the discharge point. Environmental and public health impacts can be minimized by ensuring that these concentrations will not exceed international or national guidelines for air quality (for example, WHO, 1987).

If it appears that air-quality guidelines will be exceeded, proposed operating conditions or power station design might need to be changed. For example, a system for removing contaminants from the gas exhaust may need to be included. Such measures will also need to be taken if an existing power station discharge is exceeding air-quality guidelines.

Occupational health and safety standards can be used to protect the health and well-being of people working on the geothermal field (see Section 8.6). For carbon dioxide and hydrogen sulfide exposure, for example, typical exposure standards are 5,000 to 15,000 mg/kg, and 2 to 10 mg/kg, respectively, which reflects the relative toxicity of the gases. Personnel required to enter risk areas should carry individual monitors for these gases.

8.4 IMPACTS ON WATER QUALITY

The discharge of waste bore water or condensate from a geothermal field (Figure 8.2) into a local river or stream will alter water chemistry, affecting both aquatic ecosystems and terrestrial communities using the water as a resource. Although the same contaminants are contained in the fluids of natural geothermal features, in such hot springs and pools contaminants are attenuated to a greater degree by precipitation in sinters or fixation in soils and sediments.

Even if waste fluids are re-injected on the field, contaminants may still reach surface waters via the groundwater system. Contamination of groundwater can occur during re-injection of waste bore water or condensate, or seepage from holding ponds for drilling fluids and well discharge. Surface water can also be contaminated by chemical spills or leaks during power-plant operation. These are unpredictable, usually avoidable problems, but the environmental effects of drilling fluids, fuels, lubricants, biocides, anti-scalants and other specialist chemicals stored on site must be considered during an environmental impact assessment.

8.4.1 Composition of fluid discharges

As with steam and gas discharge, contaminant concentrations in the fluid discharge are determined by the geochemistry of the reservoir, and the operating conditions used for power generation. The most important contaminants present in waste bore water and condensate include:

- lithium (Li)
- boric acid
- arsenic (As)
- mercury
- hydrogen sulfide
- ammonia.

Waste brine may also be highly saline and have measurable concentrations of the other trace elements typically associated with geothermal fluids, such as antimony, thallium, silver and selenium.

The steam condensate will typically have higher concentrations of the more volatile contaminants (for instance, dissolved gases), while the waste bore water will contain higher concentrations of the non-volatile, or only slightly volatile, contaminants such as lithium, arsenic and boron. The concentrations of these contaminants in the waste bore waters of selected developed geothermal fields are shown in Table 8.1. Different chemical forms of the same contaminant can have very different impacts on the environment, and it is important to know which form is present. For arsenic, for example, the arsenite ion (H_3AsO_3) is generally more toxic than the arsenate ion ($H_2AsO_4^-$). Likewise, toxic hydrogen sulfide is

readily converted to non-toxic bisulfide ion (HS^-) or oxidized sulfur species, under favourable conditions.

After discharge, contaminants may be removed from surface or groundwater by adsorption onto sediment and soil surfaces or by precipitation. Arsenic, boron and mercury, for example, are adsorbed onto clay and oxide surfaces, and onto organic matter. This may not be a permanent arrangement. Adsorbed contaminants can often be re-released from the sediment when chemical conditions change, or as a consequence of bacterial activity.

8.4.2 Toxicity and environmental effects

A deterioration in surface-water quality can have wide-ranging effects on the environment, including damage to:

- aquatic life
- stock, if the water is used for stock watering
- crops, if the water is used for irrigation water
- humans, if the water is used for drinking water or if contaminated aquatic animals or plants, stock or crops are consumed.

Lithium, boric acid and high salinity

If the surface water is used for irrigation, crops will be adversely affected by high lithium, boron, sodium and chloride concentrations. The effects depend on crop sensitivity and soil type, but range from foliage damage to alteration of the structure and properties of the soil. These contaminants are unlikely to seriously affect human or stock health, although high boron concentrations can cause weight loss in stock and gastro-intestinal problems in people. Highly saline water is generally unpalatable, and can adversely affect freshwater aquatic life.

Arsenic

The relative toxicity of the various forms of arsenic is organic arsenic < arsenate < arsenite. Very high concentrations of arsenite can lead to chronic or even acute poisoning, in humans, stock and aquatic life. Aquatic plants can accumulate high concentrations of inorganic arsenic and may therefore be toxic if ingested. Terrestrial crop plants do not usually accumulate arsenic from irrigation water to a level dangerous to consumers, but foliage damage can occur.

Arsenic is also documented as a human carcinogen. A relatively high incidence of skin and possibly other cancers has been noted in populations using drinking water with elevated concentrations of arsenic.

Mercury

The fundamental problem with mercury is its tendency to accumulate through the aquatic and terrestrial food chain, a process known as 'bio-accumulation'. Mercury is accumulated in aquatic plants and animals predominantly as methylmercury, which is the most toxic form and which affects the central nervous system. This places humans, and other species near the top of the food chain, at the greatest risk. Human ingestion of mercury occurs mainly through food, rather than direct uptake from drinking water, and the toxic effects have been outlined in Section 8.3.2. Animals appear to be more sensitive than plants to both inorganic mercury and methylmercury.

Hydrogen sulfide

It is unlikely that dissolved hydrogen sulfide in drinking water would ever reach sufficiently high concentrations to adversely affect the health of humans or stock. However, as a dissolved gas, it is very toxic to fish as it affects the uptake of dissolved oxygen through the gills.

Ammonia

Dissolved ammonia in drinking water does not directly affect human health, although it can cause taste and odour problems, nor is it a problem for stock or crops. As a dissolved gas, however, ammonia is acutely toxic to freshwater organisms, particularly fish. The toxicity of total ammoniacal nitrogen decreases with increasing acidity, as ammonia is converted to the non-toxic ammonium ion.

Mitigation: Groundwater contamination can be avoided by casing re-injection wells through the groundwater aquifer, and by lining holding ponds to prevent unacceptable seepage rates. Contaminant concentrations in fluid discharges to surface waters can be predicted from the geochemistry of the reservoir, and knowledge of the operating procedures to be used during power generation. Computer models can be used to calculate contaminant concentrations in a geothermal fluid, and it is a simple matter to calculate the receiving water contaminant concentrations after dilution has occurred. Geochemical speciation computer models can also be used to predict the concentrations of the more toxic species of these contaminants (for example, arsenite ion).

Environmental and human health impacts can be minimized by ensuring that receiving water concentrations will not exceed national or international guidelines for water quality. Most countries have developed or adopted guidelines to protect their own aquatic ecosystems, and to protect the water for specific purposes such as drinking, irrigation or stock watering. Such guidelines are frequently based on criteria developed by the US Environmental Protection Agency (for instance, USEPA, 1986), or the World Health Organization (for instance, WHO, 1993). The WHO drinking water guidelines (WHO, 1993), for example, have formed the basis for the drinking water guidelines and standards of many countries, and are often adopted with few, if any, alterations.

Water-quality guideline values apply to the *receiving* environment, although a suitable mixing zone downstream from the discharge is generally allowed before compliance is required. A geothermal discharge containing 4 mg/kg arsenic would, for example, need to be diluted at least 400 times by an *arsenic-free* river water in order to meet the most recent WHO (1993) drinking water guideline value of 0.01 mg/kg of arsenic. However, when the receiving environment contains significant levels of a contaminant prior to discharge, this must also be taken into account.

8.5 SOCIAL IMPACTS

A geothermal project development in an area will have consequences for the surrounding human communities. A development in close proximity to such a community will have considerable social implications as well as more direct physical and chemical impacts. Often, the first cause for conflict is the use of the land for geothermal development as opposed to other uses such as agriculture, recreation and housing. Compatibility between different land uses has already been demonstrated at various developed geothermal fields around the world, but each individual development will have its own incompatibility problems. The ownership of the land and legal consent to develop a geothermal resource should obviously be one of the first issues considered, negotiated and settled in any field development. The participation of the local population in this process is vital. This highlights the most important aspect of the social impact, which is that the *local* people (if any) should be involved from the very start of the development. The larger the development, the more the local people will need to be involved.

In the past, the rights and obligations of both developers and the indigenous population have varied considerably between countries. However, recent global agreements such as the *Earth Charter* and *Agenda 21* have provided a framework for interaction between developers and the local population. These agreements contain specific environmental clauses, such as:

- 'The state shall develop laws on liability and compensation for victims of pollution and other environmental damages' (*Earth Charter, Principle 13*)

but also contain general guidelines such as:

- 'To strengthen the role of indigenous people/tribes and their communities' (*Agenda 21, Chapter 26*) (de Jesus, 1995).

As well as these global agreements, there is often national and local legislature to cover the rights of the indigenous people, and the obligations of the developer.

The impacts that a geothermal development can have on the local community can be separated into three distinct groups (de Jesus, 1995):

- *Physical and chemical changes*, giving rise to housing relocation, landscape alteration, health problems and increased traffic. Some of the resulting social problems could be lower production, inappropriate skills in a new location, destruction of social structure and changes in aesthetic values.
- *Economic changes* that may lead to a flow of capital into the local area, and an increased per capita income. Social changes arising from this could be increased employment, a reduction of resources as they are used for the development, and changes in lifestyle from a rural to an industrial economy.
- *Institutional changes*, giving rise to changes in social structure, population changes, adaptation of the local culture and changes in other human interest areas.

Mitigation: The social impacts can be minimized by following a similar approach to that of chemical or physical impact mitigation. Before development, there needs to be baseline monitoring to determine the 'background' or current social structure. The affected communities can be defined using physico-chemical or cultural boundaries. The likely effects of development have to be predicted and assessed using social criteria. The measurement of socio-economic parameters then provides a base for analysis of the social impacts. Social and environmental costs can then be quantified, allowing a cost–benefit comparison for the developer with other energy sources. Finally, measures can be taken to minimize the impacts and resolve any conflicts through negotiation, and suitable mitigation methods can be implemented or compensation can be negotiated. During the production phase of the development, further monitoring is required to ascertain the success of the social measures.

8.6 WORKPLACE IMPACTS

The environmental impacts considered so far have been those potentially occurring in the general environment, affecting fauna, flora, surrounding human populations and landscape. This section will examine the environmental risks posed to people working within the confines of the geothermal field itself, where they are exposed to hazards such as airborne contaminants, liquid and solid contaminants, noise and heat. This is the area covered by occupational safety and health legislation (OSH), which is applicable 'inside the factory fence'.

8.6.1 Exposure to airborne contaminants

Airborne contaminants include gaseous contaminants, which mix in all proportions with air, and aerosol contaminants. The units of measurement of the gaseous contaminants are usually parts per million (ppm) by volume. Aerosols are usually reported gravimetrically as mg per cubic metre of air (mg/m^3). There can also be dusts, which are discrete solid particles suspended in the air, fumes, which are airborne particulates with very small diameters (<1 μm), and mists, which are droplets of liquid suspended in air. Airborne particulates are associated with biological effects that can be classified as infectious, carcinogenic, fibrogenic, systemic, allergenic or irritative.

Although airborne contaminants are normally taken up by inhalation, some are able to penetrate and be absorbed through the skin. This is usually not the major method of absorption into the body, but certain vapours and liquids can have significant skin absorption.

The main gas hazards are carbon dioxide and hydrogen sulfide. Other gas hazards that may be present are mercury, and to a lesser extent ammonia. The toxic effect of these gases has already been described. Hydrogen sulfide and carbon dioxide are likely to occur near gas exhausters in power stations and in fumarolic geothermal areas. Three hazardous solids that may be found in the air during geothermal development are asbestos in insulation (in older developments), silica flour in drilling and waste brine lines, and mercury-rich sulfur, sometimes found in cooling towers. Asbestos occurs mainly as insulation in geothermal development, and its use has now become much less common in most countries where other insulating materials are available. Asbestos is a collective name for a number of fibrous minerals. Most of the asbestos problem is a consequence of inhaled fibres, which lodge in the lung causing a disease called 'asbestosis', which is a fibrosis of the lung. Asbestos is a confirmed human carcinogen causing bronchial carcinomas.

Inhalation of silica dust can lead to a condition called 'silicosis', which is a progressive fibrosis of the lung. The principal symptom for silicosis is shortness of breath on physical exertion. The form of silica most often encountered in geothermal development is amorphous silica, which may be found, for example, as a deposit in re-injection pipelines. As a general rule, amorphous silica particles are more benign in their interaction with the lungs than small particles of crystalline quartz.

Note that for chemicals such as asbestos, which cause cancer, there is often a large time lag between the initial exposure and the diagnosis of disease. Evidence for the connection between exposure and disease is obtained from epidemiological and animal studies. Practical considerations and the difficulty of obtaining reliable exposure limits have inhibited the classification of carcinogens. Most OSH publications list two types of carcinogen: those that have definitely been proven as carcinogenic, and those that are 'justifiably suspected' of having carcinogenic potential.

Mitigation: Exposure to airborne contaminants should clearly be minimized as much as possible and areas of high risk clearly signposted (Figure 8.3). The maximum concentrations to which workers should routinely be exposed are defined in OSH standards (see Section 8.6.5). However, these criteria only cover routine conditions; short-term exposure to high contaminant concentrations resulting from an accidental discharge can still occur. In high-risk areas continuous monitoring is required, and personnel required to enter risk areas should carry individual monitors (for example, for hydrogen sulfide).

Figure 8.3 A sign warning of the potential danger posed by hydrogen sulfide in a well-head pit at Wairakei geothermal field in New Zealand

In general, exposure to carcinogenic materials should be kept to an absolute minimum, and if technically feasible, substitution with a less hazardous substance is the preferred option.

8.6.2 Exposure to liquid contaminants

Liquid hazards include caustic soda (NaOH), which is used in cooling towers and in gas-sampling flasks, and can cause burns to the skin. Similarly, acids such as hydrochloric acid (HCl), sulfuric acid (H$_2$SO$_4$) and hydrofluoric acid (HF) are used in chemical sampling, acid treatment of waste brine and acidification to prevent scaling in re-injection lines. Acids can cause serious burns to the skin, as well as to the lungs and mucous membranes if the fumes are inhaled. The chemicals typically used as biocides and calcite anti-scalants are, however, usually rather benign.

Mitigation: Care must be taken in the handling of corrosive substances, and the appropriate protective clothing worn. Chemicals should be stored on site in such a way that any accidental spillage is contained to a small area and can be neutralized quickly. The chemical required for neutralization should be stored in the same vicinity. Safety showers should also be installed in these areas. Training programmes are essential to ensure that proper handling and storage procedures are followed.

8.6.3 Exposure to noise

The noise levels associated with an operating geothermal power station can often be extreme (as described in Section 8.2). Sources in the field include drilling rigs, vertical well discharges, and well testing and bleeding, while at the power plant the main noise pollution comes from the cooling-tower fans, the steam ejectors and the low-frequency turbine 'hum'.

Mitigation: Noise is sometimes dealt with by wearing appropriate hearing protection. However, many countries also have an OSH standard where, regardless of hearing protection, a limit is set on the average noise level in a workplace. These are set as limits for the acceptable risk for hearing impairment. A typical such noise-exposure level is 85 dBA for an averaged eight-hour day, with a peak level of 140 dBA. The exposure level is a time-weighted average. It may, for example, be 85 dBA for eight hours, or 88 dBA for four hours, or 91 dBA for two hours.

8.6.4 Exposure to heat

One of the principal physical safety hazards on a geothermal field is burn, which can be caused by any one of the number of high-temperature hazards found in geothermal developments.

Heat stress is the adverse effect of heat on the human body that occurs when the body is not able to dissipate sufficient heat to the environment. The heat can derive from an external heat source, as when working close to steam lines, or can be derived internally, where the heat stress is due to the insulation of clothing (usually special protective clothing required for some specific purpose). The result of a body's inability to lose the required amount of heat is a rise in the core temperature, which can lead to serious harm and even death.

Mitigation: Burns can be prevented by fencing to isolate the area, such as boiling-water drains or active geothermal areas, or by insulation of pipelines and by a training schedule to raise the awareness of the hazards present. There is a 'hot environment standard', which measures the heat stress of workers required to work in hot situations. This can be used to ensure overheating does not occur.

8.6.5 OSH criteria and standards

These criteria are mainly designed to provide a limit for the concentration of a contaminant in air, because it is considered that breathing the air is the most likely way of coming into contact with the contaminant. The criteria do also record whether the contaminant can be absorbed through the skin, and whether it is a likely allergenic substance.

There are a number of different methods of defining criteria for occupational safety and health. These have such names as 'workplace exposure standard' or 'threshold limit values' (TLV) or 'recommended exposure limit'. The criteria are often further refined to define the type of exposure, such as 'ceiling' (not to be exceeded during any part of the day), 'TWA' (a time-weighted average calculated for a set period such as an eight-hour day, that is, long-term exposure) or 'STEL' (short-term exposure limit to protect workers against chronic effects of exposure for short periods, typically fifteen minutes).

As well as the criteria for air concentrations, there are criteria called 'biological exposure indices' (BEI), which provide a complementary measurement of the effects of hazardous substance exposure. These are the maximum permissible quantities of a compound or its metabolism products that can be found in the human body, and are normally given as blood or urine concentrations. In general, if a worker is exposed to the workplace standard and is engaged in moderate work over long periods of time, then the BEI represents the expected biological level. Depending on the metabolism of the substance, however, the BEI may indicate a recent acute exposure, an average over the last few days, or long-term steady state or cumulative exposure. Biological monitoring has been widely used to monitor the uptake of cumulative toxins such as lead. It has also been used to monitor those substances able to be absorbed through the skin.

Some of the major organizations in the world that are responsible for setting the standards in their own countries are given below. Normally data for a workplace hazard, if available, can be found in the publications of these organizations.

- American Conference of Governmental and Industrial Hygienists (ACGIH), which sets standards in the United States and compiles frequently updated TLV lists based on scientific and toxicological data. Examples include *Threshold Limit Values and Biological Exposure Indices for 1993–1994* and *Documentation of the Threshold Limit Values and Biological Exposure Indices*, 6th Edition. (ACGIH publications can be obtained from: ACGIH, Kemper Woods Center, 1330 Kemper Meadow Dr., Cincinnati, Ohio 45240, USA) See also www.acgih.org/home.htm

- Occupational Safety and Health Administration (OSHA), again a US organization, which publishes lists of permissible exposure limits (PELs). See also www.osha-slc.gov

8.7 LEGISLATION AND EIA

Geothermal energy production generally has a well-deserved image as an environmentally friendly energy source compared with fossil fuels and nuclear energy. However, continuing justification for this reputation will rely on the conscience of the developer and on legislation to protect the environment. While this legislation varies in detail from country to country, the overall requirements, and the purpose and need for the legislation, are similar worldwide. The principal difference lies more in the administration of the legislation than its content.

In general, legislation will require:

- environmental impact assessment prior to development
- consent application/award for the development
- monitoring of discharge quality and effects during production.

As discussed in this chapter, geothermal development can have an impact upon the social, physical, chemical and biological components of the surrounding environment. Many legislative decisions concerning permission to develop a field will depend on the balance between the advantages to be gained from geothermal development and the damage that may be inflicted on the local environment. The decisions involve both local issues, such as the possible physical impact on the local habitat, and national issues, such as the preservation of unique geothermal features that might be considered as part of the entire nation's heritage. Consequently, an environmental impact assessment (EIA) is one of the first and most important documents to be compiled in the course of a development, and will underpin subsequent consent applications and monitoring-programme design.

8.7.1 Environmental impact assessment

The minimum requirements of an EIA will include:

- *Impacts on air quality*: including existing background concentrations of contaminant gases in the atmosphere, predicted gas exhaust and steam discharge chemistry, contaminant

dilution and dispersion in the atmosphere, comparison with air-quality criteria, wet and dry precipitation rates, and secondary effects on vegetation and surface waters.
- *Impacts on land*: including surveys of local land use and human population distribution, inventories of local vegetation, terrestrial ecosystems, soil types, potential impacts of gas discharge (for example, secondary effects) and surface water contamination (for example, irrigation), potential for subsidence, and nature of existing natural geothermal features. Where appropriate, existing background levels of geothermal contaminants in soils should also be measured. Particular attention should be paid to the presence of native plant or animal species protected by law or peculiar to that environment, as well as those taken for food or for commercial income, or protected for recreation (for instance, game).
- *Impacts on water quality*: including existing background concentrations of contaminants in surface and groundwaters, predicted waste-bore-water and condensate-discharge chemistry, contaminant dilution and dispersion in the surface or groundwater, comparison with water-quality criteria (covering all potential uses of the water bodies concerned), and a detailed survey of predominant aquatic wildlife species including plants, invertebrates and fish.
- *Social impacts*: including proximity of human populations, previous exposure to geothermal gases and noise, cultural or practical importance of natural geothermal features, requirements of land and water likely to be affected by the development, a description of the consultation process, identification of concerns and how these might be addressed, or appropriate compensation measures.
- *Potential workplace hazards*: including an inventory of the chemicals to be used on-site, necessary precautions in handling and storage, a survey of potential exposure to geothermal contaminants, comparison with OSH standards and criteria, identification of high-risk areas, and measures to be taken to prevent unacceptable exposure to toxic substances.

8.7.2 The consent application and award process

The actual procedure will vary between different countries, but the process has some common elements (Figure 8.4).

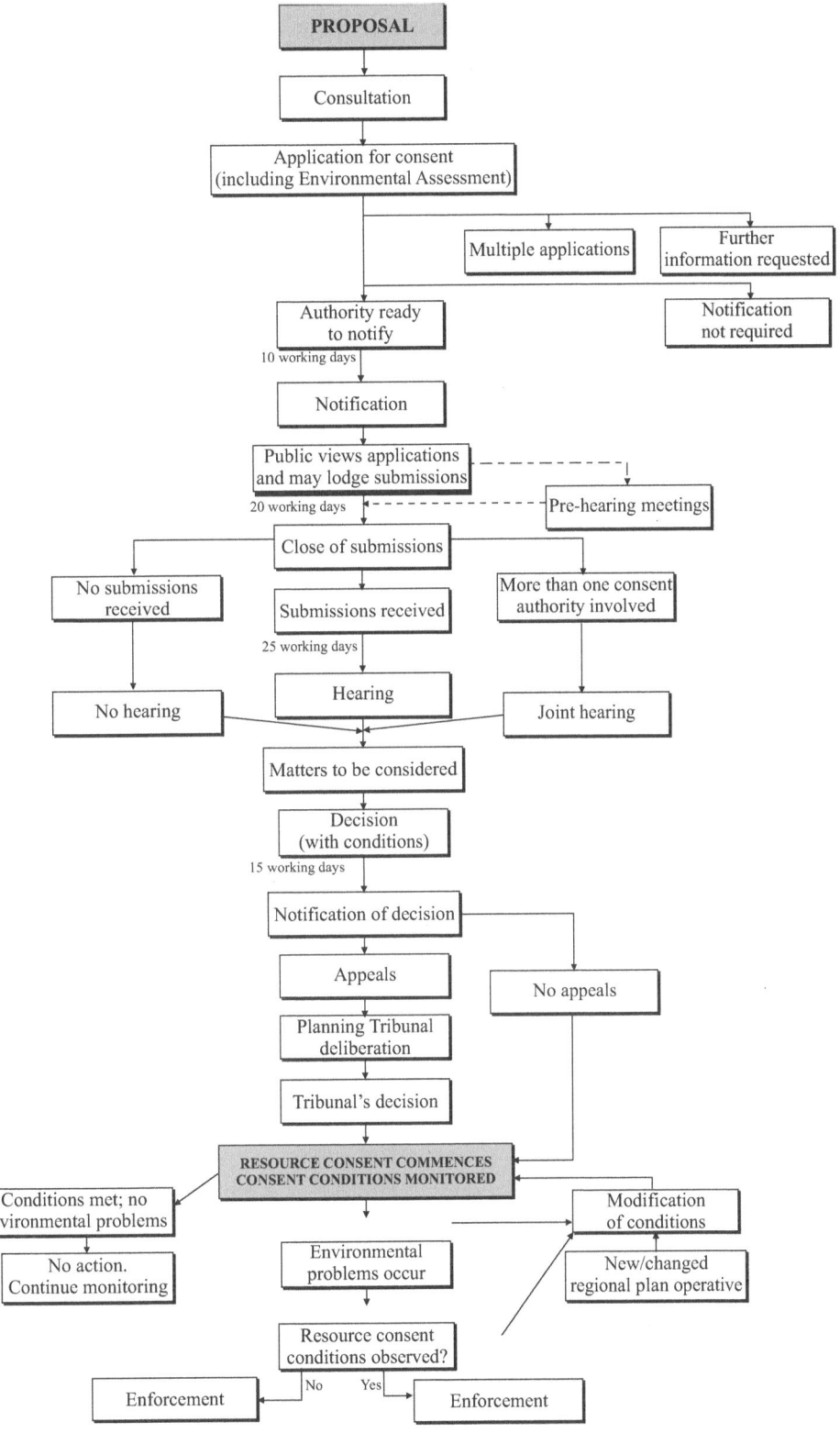

Figure 8.4 Flow diagram showing the consent application process under the Resource Management Act, New Zealand's principal environmental legislation
Source: After Milne, 1992.

The EIA will be submitted to a regulating authority that will normally then open the document to public scrutiny. At this stage, the public is invited to lodge submissions. After a period of time sufficient to allow proper public participation, submissions are closed, and a public hearing may be called. Following replies to the submissions, the regulating authority will reach a decision as to whether the development will proceed, and will place limits on various environmental parameters. This decision may be open to appeal at this stage, and may go to a higher authority for a ruling. If the development is permitted, then the regulating authority outlining the conditions under which the development is being allowed to proceed will issue 'permits' or 'consents'. This will normally define very strictly the environmental parameters to be monitored, and the values of the parameters to be met.

8.7.3 Monitoring programmes

Consent to discharge will include requirements to monitor discharge chemistry, and all potential impacts on the environment, including receiving environment (air or water) quality, land subsidence, natural features, vegetation, social impacts and a variety of other impacts. These monitoring programmes will detect the *actual* effects, as opposed to the potential effects, of geothermal field development and exploitation.

For all monitoring programmes, it is extremely important to have previously collected reliable background data for all aspects of the environment likely to be affected by geothermal power generation. If pre-development conditions have not been fully characterized it becomes very difficult to quantify impacts at a later date, and even more difficult to protect the developer from incorrect actions of environmental damage. These data become the baseline for future monitoring, and data collection should extend over the entire region that might be affected, and include sufficient surveys to cover seasonal variations.

The design of a monitoring programme, that is, the frequency and type of sampling and analysis required, will vary. Water- and air-discharge-quality monitoring programmes, for example, should be designed to ensure that permitted discharge conditions are being met, and therefore require regular monitoring at frequent intervals (for example, hourly, daily or weekly). The monitoring should also be designed such that, if consent conditions are not being met, a contingency plan can be implemented and the discharge contained in a holding pond or similar construction.

Biological and chemical monitoring of receiving environments, on the other hand, will not require monitoring as frequently as discharge chemistry. Biological impacts are unlikely to show up in the short term. A six- or twelve-month timescale is usually more appropriate, although this will depend on the information required. It is important to monitor receiving-water quality at the same time of year, and under the same physical and chemical conditions, if possible, so that results can be compared with those of the previous year or season. Physical impact monitoring may be on a six-month or twelve-month timescale, but if significant changes are being observed, more frequent monitoring may be required. Standard methods for the chemical and biological examination of waters are given in APHA (1992), and advice on the design of a monitoring programme in Ward and Loftis (1989).

8.7.4 The future

With a greater global awareness of environmental issues, the efficient use of natural resources is increasingly desirable. Technological advances have meant that geothermal development can follow this trend. Through the use of binary generation plants, for example, development of water-dominated geothermal resources is moving towards greater use of the heat content available from geothermal systems. Solutions to silica scaling problems are becoming available, which may further increase the use of the total heat content of the geothermal fluid. In addition, extraction of minerals from the geothermal fluid as saleable products is now being actively investigated at a number of geothermal developments.

As further technology comes to hand, better recovery and utilization of the heat and chemicals in geothermal fluids will not only reduce environmental impacts, but will also allow for an integrated and more efficient development of geothermal resources.

REFERENCES

ALLIS, R. G. 1990. Subsidence at Wairakei field, New Zealand. *Geothermal Resources Council Transactions*, Vol. 14, pp. 1081–87.

APHA. 1992. *Standard Methods for the Examination of Water and Wastewater*. 18th ed. Washington D.C., United States, American Public Health Association.

BIRKLE, P.; MERKEL, B. 2000. Environmental impact by spill of geothermal fluids at the geothermal field of Los Azufres, Michoacan, Mexico. *Water, Air and Soil Pollution*, Vol. 124, pp. 371–410.

BROMLEY, C. J.; MONGILLO, M. A. 1994. Hydrothermal eruptions: a hazard assessment. *Proc. 16th Geothermal Workshop 1994*, Auckland, New Zealand, pp. 45–50.

COLLIE, M. J. (ed.). 1978. Geothermal energy recent developments. *Energy Technology Review*, No. 32, Park Ridge, N.J., Noyes Data Corporation.

ELLIS, A. J. 1978. *Environmental Impact of Geothermal Development.* Report prepared for the United Nations, New York, United Nations, October.

ELLIS, A. J.; MAHON, A. 1977. *Chemistry and Geothermal Systems.* New York, Academic Press.

GLOVER, R. B. 1988. Boron distribution between liquid and vapour in geothermal fluids. *Proc. 10th New Zealand Geothermal Workshop*, Auckland, New Zealand, pp. 223–7.

DE JESUS, A. C. 1995. Socio-economic impacts of geothermal development. In: K. L. Brown, convener, *Environmental Aspects of Geothermal Development*, International Geothermal Association short course, Florence, Italy.

MILNE, D. A. 1992. *Handbook of Environmental Law.* Wellington, New Zealand, Royal Forest and Bird Society of New Zealand.

USEPA. 1986. *Quality Criteria for Water 1986.* US Environmental Protection Agency Report, EPA 440/5-86-001.

WARD, R. C.; LOFTIS, J. C. 1989. Monitoring systems for water quality. *Critical Reviews in Environmental Control*, Vol. 19, pp. 101–18.

WHO. 1993. *Guidelines for Drinking Water Quality.* 2nd ed. Geneva, World Health Organization, Geneva, Switzerland.

WHO. 1987. *Air Quality Guidelines for the European Region.* Copenhagen, Denmark, World Health Organization Office for Europe.

RECOMMENDED LITERATURE

FREESTON, D. H. 1993. *Geothermal Production Technology.* Geothermal Institute Course Notes.

HOUGHTON, B. F.; LLOYD, E. F.; KEAM, R. F. 1980. *The Preservation of Hydrothermal System Features of Scientific and Other Interest.* Report to Geological Society of New Zealand (to contact the society: see http://www.gsnz.org.nz/gsnz.htm)

SPYCHER, N. F.; REED, M. H. 1990. *Chiller: A program for computing water–rock interactions, boiling, mixing and other reaction processes in aqueous-mineral-gas systems.* Eugene, Ore., Report from the Dept. of Geosciences, University of Oregon, Oregon 97403.

SELF-ASSESSMENT QUESTIONS

1. Changes in the activity of natural geothermal features are almost always present following geothermal development. In a water-dominated geothermal system, what is the likely impact on chloride features (springs and geysers) and steam features (fumaroles, steaming ground) during sustained production?

2. What controls the concentration of gaseous contaminants present in the gas exhaust from a power station, and in the atmosphere after discharge?

3. After steam separation, what type of contaminants will occur in (a) waste bore water, and (b) condensate discharge?

4. How can the potential impacts of contaminants in gas and fluid discharges on the environment be assessed before production begins?

5. In a large geothermal development, what is the first issue that needs to be considered?

6. What areas of a geothermal field present the greatest risk to workers in terms of exposure to toxic geothermal gases? How can they act to prevent health effects?

7. What documents set the environmental limits allowed during production from a geothermal field?

ANSWERS

1. Often, natural features that contain the deep chloride water are reduced in activity following a period of production. Steam-heated features often increase in activity following production.

2. The solubility of the gas in cooling water, and in rainwater rate after discharge.

3. (a) Waste bore water will contain elevated concentrations of contaminants that have low volatility, such as arsenic, lithium and boron. (b) The condensate will have elevated concentrations of contaminants that had initially separated into the steam phase, such as hydrogen sulfide and mercury.

4. By predicting contaminant concentrations in the discharges, then after dilution in the receiving environment. These values are then compared with relevant criteria for the protection of the environment of interest.

5. The first priority is to identify potential issues relating to land use and ownership, and to open communication with the local community.

6. Pits and depressions where heavy gases such as carbon dioxide and hydrogen sulfide accumulate. Workers should wear personal monitors to detect the presence of these gases.

7. The limits of the environmental parameters that are allowed during production are set in the 'consents' or 'permits' issued by the regulating authority.

CHAPTER 9

Economics and Financing

R. Gordon Bloomquist
Washington State University Energy Program, Olympia, Wash., USA

George Knapp
Squire, Sanders & Dempsey L.L.P., Washington D.C., USA

AIMS

1. To introduce the reader to the complex array of interrelated factors that must be understood and taken into consideration in order to assess the economic viability of a proposed geothermal project.

2. To describe the institutional prerequisites for successful development and financing of a geothermal energy project.

OBJECTIVES

When this chapter has been completed, the reader should be able to:

1. Understand the importance of taking all relevant factors into consideration when determining the economics of a given project.

2. Understand the unique characteristics of geothermal projects relative to fuel supply.

3. Understand the multitude of ways in which geothermal resources can be

used, and how each presents its own unique set of economic factors.

4. Determine the design of the most economically robust system to ensure that revenue requirements can be met.

5. Understand the basic complexities of project financing, and how adequate fuel supply, proper system design and construction, operation and maintenance (O&M), and consideration of all relevant legal, institutional, and environmental factors all play a significant role in obtaining financing for a project.

6. Identify those constitutional, legislative or regulatory changes that are needed in order to facilitate development of geothermal energy projects by the private sector in a given country.

7. Understand financing options available to private sector sponsors of geothermal energy projects.

8. Analyse key project agreements from the perspective of project investors and lenders.

9.1 INTRODUCTION

The factors that must be considered when assessing the viability of a geothermal project vary from project to project, from conversion technology to conversion technology, and especially from electrical generation to direct use. There are, however, a number of factors common to all projects, although actual cost and impact on project economics will be, to a large extent, dependent upon resource characteristics and national or even local political and economic circumstances.

The economic factors that are common to all projects include provision of fuel, that is, the geothermal resource; design and construction of the conversion facility and related surface equipment, for instance, the electrical generation plant together with required transformers and transmission lines; the generation of revenue; and of course, financing. The cost of obtaining the required fuel supply, together with the capital cost of the conversion facility, will determine the amount that must be financed. Revenue generated through the sale of electricity, by-products, thermal energy, or product produced, for example, vegetables, plants or flowers from a greenhouse, minus the cost of operation and maintenance (O&M) of the fuel supply and conversion facility, must be sufficient to meet or exceed the requirements of the financing package.

Because financing is such a critical factor in the economics of any project, an entire subchapter is devoted to this subject with the express aims of describing the institutional prerequisites for successful financing and development of a geothermal energy project, summarizing the debt and equity structures for such a project, with emphasis on the broad range of structuring options and funding sources, and surveying the key issues in project agreements that should be addressed so that financing can take place. For many new projects, the largest annual operating cost is the cost of capital (Eliasson et al., 1990). In fact, the cost of capital can be as high as 75 per cent of the annual operating expense for a new geothermal district energy project, with O&M (15 per cent) and ancillary energy provision (10 per cent) making up the balance.

9.2 ECONOMIC CONSIDERATIONS

9.2.1 Provision of fuel

For most projects that require a sustained and economically attractive fuel supply, the project sponsor must only contact a supplier and negotiate a long-term supply of natural gas, oil, propane or coal. To help guarantee low and stable fuel supplies, more and more project sponsors are purchasing gas fields, or oil or coal reserves. For projects that depend upon biomass (wood), fuel can be contracted for from a wood-products mill, or the mill may even become a partner in the project, providing an even more secure supply. Long-term availability of biomass can be determined from long-term timber holdings within a geographically defined area, and/or plans for harvesting as defined by a state or federal land management authority. With municipal solid waste, fuel supply can be assured through local government action requiring that all material be controlled by one authority and delivered to a specific facility for a given time period.

In the case of geothermal resources, however, the fuel cannot be purchased on the open market, legislated into existence, bought from a local utility or transported over long distances from a remote field.

Whether the steam or hot water is to be provided by the project sponsor, that is, the steam field and conversion facility are under one ownership, or the steam is to be provided by a resource company, the geothermal fuel is only available after extensive exploration, confirmation drilling and detailed reservoir testing and engineering. Once located, it must be used near the site and must be able to meet the fuel requirements of the project for the lifetime of the project. Even before exploration can begin, however, the project sponsor may incur significant cost, and a number of extremely important legal, institutional, regulatory and environmental factors must be evaluated fully and their economic impacts considered.

Obtaining access and regulatory approval

In order to obtain rights to explore for and develop geothermal resources, access must be obtained through lease or concession from the surface and subsurface owners. In many countries, the state claims rights to all land and to all mineral and water resources. In other countries, land and subsurface rights can be held in private ownership. Unless the geothermal developer has clear title to both surface and subsurface estates, an agreement for access will have to be entered into with the titleholder of these estates. Such access will normally require a yearly lease fee and eventually royalties upon production. In areas where there is significant competitive interest, competitive bidding may be used to select the developer. Competitive bids can be in the form of cash bonuses or royalty percentages. Royalties can be assessed on energy extracted, electrical or thermal energy sales, or even product sales. Whatever the system, it will have an impact upon project economics and should be carefully considered in terms of overall economic impact. In particular, developers of direct-use projects, because of the limited rewards that can be expected, must carefully evaluate how royalties will be calculated. In a number of instances royalties, if assessed, would comprise up to 50 per cent of annual operating cost, making projects uneconomic to pursue.

The second factor that will have an impact on overall project economics is obtaining all regulatory approvals, including the completion of all environmental assessments and the securing of all required permits and licences, including, if necessary, a water right. Increasing concern for the environment in nearly all countries of the world has resulted in sharply increased cost for preparing the necessary environmental documents. A complete environmental impact statement for any proposed development is now required by federal land management agencies in the United States, and cost for preparation can exceed US$1 million. It is not uncommon to invest up to 40,000 to 60,000 person-hours in completing all necessary environmental documents and obtaining required licences and permits for a major electrical generation project.

Because so many environmental decisions are now contested, a contingency to cover the legal costs related to appeals must be included in any economic analysis; depending upon the issues and the financial and political power of those appealing a decision, the cost of obtaining necessary approvals can easily double. Because most direct-use projects are more limited in scale, and therefore in environmental impact, these costs may be only a small fraction of the cost incurred by the proposal for a major power-generation project. Even such a reduced cost can, however, be significant in relationship to the scale of the project, and the economic impact should not be underestimated. Unfortunately for the project sponsor, most of the cost related to obtaining access and environmental and regulatory approval must be incurred early in the project, and in many instances even before detailed exploration or drilling can begin, and with no clear indication that any of the costs will or can be recovered.

Exploration

Once access has been secured and all necessary regulatory approvals have been obtained, the developer may initiate a detailed exploration programme, refining whatever data was initially gathered in the reconnaissance or pre-lease phase of the development process, and sequentially employing increasingly sophisticated techniques that will lead to the drilling of one or more exploration wells. Hopefully these wells will be capable of sustaining a reservoir testing programme, and possibly also of serving as preliminary discovery and development wells. Reconnaissance, in all likelihood, will include such activities as a literature search, temperature gradient measurements in any existing wells, spring and soil sampling and geochemical analysis, geologic reconnaissance mapping, air-photo interpretation, and possibly regional geophysical studies. Costs incurred may range from a low of a few thousand dollars to US$100,000 or more, depending on the prior work in the area, geological complexity, and of course the scale of the proposed project and whether or not the intended use is electrical generation or direct application.

Once the area of principal interest has been selected, the exploration programme can be more intensely focused, with the primary objective of siting deep exploration wells. Techniques likely to be employed include detailed geologic mapping, lineament analysis, detailed geochemical analysis, including soil surveys and geochemical analysis of all springs and wells, temperature gradient and/or core drilling, and geophysical surveys, including resistivity, magnetotellurics, gravity and seismic. Costs increase with the complexity of the techniques and as the detail of the surveys becomes more focused. For large, direct-use projects, costs of US$100,000 or more can be incurred. For projects directed toward electrical generation, the cost of this phase of the work can easily exceed several hundred thousand dollars, and may exceed several million dollars.

The final phase in any geothermal exploration programme involves the siting, drilling, and testing of deep exploratory wells, and subsequently production and injection wells.

Well drilling

Well costs can vary from a low of tens of thousands of dollars for small, direct-use projects to several million dollars per well for wells required to access high-temperature resources for electricity generation. Success ratios for exploration wells can be expected to exceed 60 per cent; however, the risk of dry holes in the exploration phase remains high and can have a significant economic impact. Even in developed fields, 10 to 20 per cent of the wells drilled will be unsuccessful (Baldi, 1990). Drilling costs are typically 30 to 50 per cent of the total development cost for an electrical generation project, and variations in well yield can influence total development cost by some 25 per cent (Stefansson, 1999). Prospective developers must anticipate and prepare for the eventuality that, despite an investment ranging from a few hundred thousand dollars to several million dollars in lease fees, environmental studies, licences and permits, and exploration and drilling activities, an economically viable geothermal reservoir may not be discovered.

If, however, drilling is successful, the reservoir must then be tested to determine its magnitude, productivity and expected longevity. Only after such testing can a determination be made as to the eventual size and design of the generating facility or direct-use application.

Well field development

Well field development for an electricity generation project can last from a few months to several years, depending upon the size and complexity of the project, the speed at which procurement contracts can be let (Koenig, 1995), and the availability of drill rigs. At this stage it also becomes of increasingly critical importance to collect detailed data and to refine the information available on the reservoir. Of course, for most projects this will include both production and injection wells. Many projects experience unnecessary difficulties and delays in financing or in milestone review because of incomplete or inaccurate data collection, analysis and/or interpretation (Koenig, 1995). Such difficulties and delays can seriously affect project economics, and can have a catastrophic economic impact if delays result in contract forfeiture, or if contracts contain a penalty clause tied to milestone completion. Coincidental with well field development will be the construction of well field surface facilities.

Costs associated with both drilling and the construction of well field surface facilities will be affected by the availability of skilled local labour and by geologic and terrain factors. Labour costs can be expected to increase by 8 to 12 per cent in areas where most of the labour must be brought in, or a construction camp erected to provide housing and meals. Terrain and geologic factors can add from 2 to 5 per cent if special provisions must be made for work on unstable slopes, or where extensive cut-and-fill is required for roads, well pads, sumps and so on.

Over half of the total production cost over the lifetime of the project will in fact be expenses associated with the well field. Because of this, it is imperative that wells be properly maintained and operated to ensure production longevity. But even with proper O&M, many wells will have to be periodically worked over, and for most power generation projects, 50 per cent or more of the wells will likely have to be replaced over the course of the project, adding considerably to the initial well field cost and, of course, to the cost of generating power. For example, if 60 per cent of the wells needed to be replaced over the economic life of the plant, it would have the effect of increasing the levelized cost of electricity by 15 to 20 per cent (Parker et al., 1985).

For small to medium-sized direct-use projects requiring only one or two production and injection wells, costs will generally be much lower. Because the water chemistry of most geothermal resources that are developed for direct-use applications is of generally higher quality than that available for power production, well life can be expected to be much longer and few, if any, wells will have to be worked over or re-drilled during the economic life of the project.

9.2.2 Project design and facility construction

The power plant

Just as there are numerous geothermal resources throughout the world exhibiting differing temperatures and chemical characteristics, there are numerous power plant designs. These include direct steam, flashed steam, double-flashed steam and binary cycle, each able to best meet the specific requirements of a particular reservoir. The selection of the most economically viable power conversion technology can only be accomplished through a thorough evaluation of the differing strengths and weaknesses of various technologies relative to the characteristics of the resource and local circumstances, including environmental and regulatory requirements (for example, requirements for non-condensable gas emission abatement or fluid injection).

Terms of the power sales contract can also have a major influence on power plant design. For example, are there premiums paid for availability during certain times of the year or even times of the day, are there advantages to being able to operate in a load-following manner, or is the capacity factor of paramount importance? Another major consideration is the manner in which steam is provided; for example, are the steam field and the power plant under the same ownership, or is steam purchased from another party? If purchased, the terms of the steam purchase contract can have a profound impact on economics, and thus on design. For example, if steam is paid for as a percentage of the selling price of electricity, there is little incentive to achieve high steam use efficiency and a strong incentive to minimize capital cost. On the other hand, if steam is purchased on the basis of dollars per kilogram delivered, then achieving the highest possible fuel use efficiency becomes extremely important (Bloomquist and Sifford, 1995).

In order to achieve maximum steam use efficiency, some developers have adopted equipment procurement evaluation criteria that penalize offerings for inefficient use of steam and/or electricity at a capitalized rate of US$X thousand per kilogram of additional steam required and US$X thousand per kilowatt of parasitic load (Kleinhaus and Prideau, 1985).

Cycle selection
Direct steam: Although extremely rare in nature, where available, direct steam will result in the lowest power plant cost. The steam is directed from the well head, expanded through the turbine, and condensed, or in certain circumstances exhausted to the atmosphere. If condensed, the condensate can be used for cooling water make-up and/or injected back into the reservoir.

Flash steam: In the case of high-temperature, liquid-dominated resources, a flash-steam plant is the most economical choice. The hot water or liquid vapour mixture produced from the well head is directed into a separator where the steam is separated from the liquid. The steam is expanded through a turbine, and if condensed can be used as cooling water or injected, together with the separated brine, back into the reservoir. The brine could, however, be used in another application, such as space or industrial-process heating and/or agriculture, in a technique known as 'cascading'.

Double-flash steam: A double-flashed steam cycle differs from a single-flash cycle in that the hot brine is passed through successive separators, each at a subsequently lower pressure. The steam is directed to a dual-entry turbine, with each steam flow flowing to a different part of the turbine. The advantage is increased overall cycle efficiency and better utilization of the geothermal resource, but at an overall increase in cost. The decision as to whether or not a double-flash plant is worth the extra cost and complexity can only be made after a thorough economic evaluation based on the cost

of developing and maintaining the fuel supply, or cost of purchasing fuel from a resource company, plant costs, and the value of the electricity to be sold.

Binary: With a binary cycle, the heat from the geothermal brine is used to vaporize a secondary or working fluid which is then expanded through the turbine, condensed through an air condenser, and pumped back to the heat exchanger to be re-vaporized. Binary cycles can more economically recover power from a low-temperature (<175 °C) reservoir than can a steam cycle. In addition, binary plants may be more easily sited where environmental concerns are paramount, and where either gas emissions or cooling tower plumes need to be avoided. Recent developments in adding spray cooling to air condensers can improve summer efficiency by as much as 25+ per cent, greatly improving the economics of such operations (Sullivan, 2001, personal communication). The brine can be used in other cascaded applications and/or injected back into the reservoir.

Other design considerations: In addition to temperature, fluid chemistry is extremely important in cycle selection and power plant design. Many high-temperature resources are highly aggressive brines, with high contents of total dissolved solids (TDS), and bring a host of other problems that affect both design and economics.

A number of techniques have been adopted to recover power from problem brines. Design options include the use of a crystallizer reactor clarifier and pH modification technologies. The use of either technique can add considerably to capital costs as well as to plant O&M cost. If pH modification is used for scale control, corrosion could also become more severe. Of course, metallurgy of system components thus also becomes crucial, and can add significant cost to the plant if more exotic materials such as titanium must be specified.

The use of binary cycles in the presence of high TDS or corrosive brines is limited by the fact that tube-and-shell heat exchangers can easily be fouled, or suffer rapid deterioration from corrosion.

The availability of cooling water is also an important consideration in plant design. In a condensing direct-steam or flashed-steam power plant, the condensate is used for cooling water make-up. The plant can thus take advantage of the low wet-bulb temperatures that may be present, even though the ambient dry-bulb temperature may be quite high. A water-cooled cycle capable of approaching the wet-bulb temperature presents a significant advantage, as far as overall power generation is concerned, in comparison to a dry-cooled binary cycle that approaches the dry bulb instead of the wet bulb (Campbell, 1995). If, however, limited water is available, it may be used to improve the overall efficiency of a dry-cooled binary plant by injecting a fine spray or mist through the air condenser or onto a fibrous material (for instance, fibreglass) which can be used to enclose the sides of the air condenser (Sullivan, 2001, personal communication). This could be especially attractive where there is a premium for peak summer power. In an area lacking any source of water for cooling, the optional economic cycle may shift from a binary cycle to a flashed-steam cycle (Campbell, 1995). In fact, the terms of the power sales agreement may have a profound influence upon conversion cycle selection, cooling system design, and eventually plant operation.

Equipment selection
Steam cycle: The turbine generator set is the most expensive piece of equipment in a steam-cycle power plant. For direct-steam and single-flashed cycles, a single admission steam turbine is appropriate. In turbines up to approximately 30 MW_e, a single-flow turbine is usually selected. However, larger turbines generally incorporate double flow, that is, the steam is introduced into the middle of the turbine and flows in both directions, thus balancing thrust. Single-flow turbines generally exhaust from the top, allowing the condenser to be located to the side and at the same elevation as the turbine, thus minimizing cost. With the double-flow turbine, the steam exhausts downward, requiring the turbine to be mounted above the condenser. This arrangement increases capital cost, but that cost is more than justified by the increase in turbine efficiency. Other efficiency considerations include the number of turbine stages, blade length, and whether the plant will operate as a base load unit, will be used for load following, or must be dispatchable.

If load following is desirable for either resource or contractual considerations, incorporation of partial-arc admission into the turbine design is critical. Partial-arc admission, as the name implies, allows for steam to enter the turbine through only a portion or 'partial arc' of blades under certain operating conditions, and to enter the turbine through the full arc of blades during other conditions. Partial-arc admission allows a turbine to be operated at various output levels while maintaining a much higher level of operating efficiency than would be possible if the turbine were controlled through the use of a single throttling valve. In fact, when the

plant is operating at the minimum output allowed by the partial-arc arrangement, it will be only 5 per cent less efficient than at full output. This operational flexibility ensures the use of the minimum amount of steam possible for any given level of output. The use of partial-arc admission also allows for plants to be ramped up very quickly, that is, from minimum output to full output in only a few minutes (Bloomquist, 1990). Partial-arc admission can also provide the ability to increase output significantly from a single machine if higher-pressure steam is available. For example, at one plant at The Geysers each of the turbines produces 40 MW_e at 8 bars inlet pressure, while either machine can produce 80 MW_e at 11.6 bars. This arrangement has allowed this particular plant to maintain both capacity and availability in the high 90 per cent range (Bloomquist and Sifford, 1995).

Two major categories of condensers are used with steam cycles: the surface condenser and the direct-contact condenser. In a surface condenser, the cooling water is circulated through the inside of heat-transfer tubes, with steam condensing on the outside of the tubes. In contrast, in a direct-contact condenser the cooling water is sprayed into the condenser, where it directly contacts the steam from the turbine discharge. The primary advantage of the surface condenser is that contamination of the cooling water with constituents of the well-head steam is avoided, an important factor where hydrogen sulfide abatement is required (Campbell, 1995). The direct-contact condenser, however, is less expensive and is less prone to maintenance problems, and would thus be the most economical choice if hydrogen sulfide were not a problem.

The selection of direct-contact or surface condenser will also have an impact on pumps and pumping requirements, that is, parasitic power requirements. In a surface condenser, the condensate from the condenser is collected in a hot well, and a condensate pump is required to pump this condensate from a vacuum of about 0.098 bara up to the top of the cooling tower. The other major pumps required for surface-condenser operation are the cooling-water pumps, located at the base of the cooling tower and used to circulate cooling water through the tubes of the condenser and back to the cooling tower.

Because the condenser itself is under a vacuum, no pump is required in a direct-contact condenser to move cooling water from the cooling-tower basin into the condenser. However, a pump is required to pump cooling water and the condensate back into the cooling tower. Because of the usually high content of carbon dioxide and other contaminates in a direct-contact condenser, stainless steel pumps are normally specified to resist corrosion.

Non-condensable gases must be removed from the condenser in order to reduce back-pressure and optimize steam use efficiency. Non-condensable gas removal, however, results in a significant parasitic load, in terms of either steam used in jet ejectors, or electricity used to power compression or vacuum pumps. Steam jet ejectors have by far the lowest capital cost, but are relatively inefficient in comparison with liquid-ring vacuum pumps or mechanical compression. A commonly used arrangement employs one or two steam jet ejectors in series, followed by a liquid-ring vacuum pump, thus taking advantage of the low capital cost of the initial stage with the higher efficiency final stage.

If the non-condensable gas contains concentrations of hydrogen sulfide that require removal, a number of options are available, including liquid reduction-oxidation using an iron chelate solution such as is employed in the Dow Sulferox process, and the Wheelabrator Lo-Cat process. The Stretford process is another option that has been successfully used with geothermal power generation. Inclusion of hydrogen sulfide abatement can increase the capital cost of a steam cycle plant by 10 per cent or more, and will also result in an ongoing increased cost for O&M.

The cooling tower design can also have a major impact on capital cost, O&M and cycle efficiency. The most commonly used cooling-tower designs include cross-flow, cross-flow with high-efficiency fills, and counter-flow. The counter-flow tower yields more efficient heat transfer and greater depression of water temperature than the cross-flow design. The high-efficiency fill not only increases efficiency at a lower cost than conventional towers, but also the tower can be shorter, thus resulting in a lower parasite load for pumping cooling water to the top of the tower. On the downside, high-efficiency fills have a tendency to become clogged, and cooling-water chemistry must be carefully controlled. Biocides are generally added to minimize algae and other biological growth, and corrosion inhibitors are added to protect the system (Campbell, 1995). Although dry cooling towers (see 'Binary cycle' below) can be used with steam systems, efficiency considerations will generally discourage their use.

Binary cycle: Selection of the right working fluid is the most critical design decision in the development of a binary-cycle power plant. The selection must achieve a good match between the heating curve of the working fluid and the cooling curve of the geothermal heat source. The cooling curve of liquid brine is a relatively straight line, whereas a two-phase flow of liquid and vapour will give a curve of a different shape. Working fluids used in binary plants fall into two broad categories: light hydrocarbons and freons. The light hydrocarbons include butane, propane, isopentane, isobutane, and even hydrocarbon mixtures designed to find the most efficient match of working fluid to resource. In terms of freons, R11 and R22 have both been successfully used with low-temperature resources.

The light hydrocarbons have the disadvantage of being highly flammable, requiring installation of fire control equipment. The use of R11 has been banned because of its adverse impact on the ozone layer, and R22 will be phased out over the next several years. However, more environmentally friendly replacements are now available, and work is now being directed toward the development of other even more efficient and environmentally acceptable replacements. There is also increasing interest in ammonia as a working fluid, and a number of demonstration applications are already planned or on-line.

Because the heat content of the geothermal resource is transferred in the binary cycle to the working fluid, the heat exchanger becomes an additional critical equipment component, and can account for a significant capital-cost increase over a steam cycle plant. The heat exchanger is generally of shell and tube design, with the geothermal brine pumped through the tubes and the working fluid on the shell side. Because of the heating curve, counter-current flow is desired and achieved by laying out the heat exchangers in series, with single-pass flow on both shell and tube sides. Material selection is critical to avoid problems of both corrosion and erosion of the heat transfer tubing. For most applications, carbon steel is acceptable if oxygen can be kept out of the system, and has the lowest capital cost. The cost of the heat exchanger can escalate rapidly if stainless steel or even titanium is required. The use of a direct-contact heat exchanger would reduce capital cost and limit the problems associated with erosion and corrosion of the heat exchanger tubing. However, problems associated with contamination of the working fluid by corrosive constituents in the brine and non-condensable gases can result in serious problems downstream in the turbine and condenser. Loss of working fluid to the spent brine is less of a problem, but must still be taken into account by including recovery equipment.

Another major cost that is specific to the binary cycle is the number of pumps required and the significant parasitic load they place on the plant. Because the binary cycle operates much more efficiently if brine from a liquid-dominated reservoir can be maintained as a single-phase flow through the heat exchanger, production well pumps are used. Standard production pumps are multi-stage, vertical-turbine pumps driven by a motor at the surface. Down-hole pumps could be an attractive and economical alternative, and recent advances could soon result in commercially available down-hole pumps, but the high temperature of the geothermal brine has so far limited their applications. Improvements in vertical shaft turbine pumps now make multi-year runs between servicing possible, and have significantly improved operational economics and reduced downtime.

The second major requirement for pumps stems from the need to pump the working fluid through the heat exchangers and to the turbine inlet. The pumps are usually multi-stage, vertical canned pumps. Multiple stages are used to achieve the required turbine pressure. In addition to the additional capital cost attributable to the need for production and/or working fluid circulating pumps, pumping requirements result in a parasitic load of 10 to 15 per cent of the power that is generated, a significant reduction in the amount of power that is available for sale.

Power is generated in the binary cycle using either a radial inflow or axial flow turbine. The radial inflow turbine can achieve efficiencies as high as 90 per cent, and is usually the preferred option.

Once expanded through the turbine, the working fluid must be condensed before being returned to the heat exchanger in a continuous cycle. To date, a majority of binary cycle plants have employed air-cooled condensers. In the air-cooled condenser, the condensing working fluid is directed through the heat transfer tubes, and air is forced across the tubes to remove the heat. The air-cooled condenser can be extremely large and expensive both to build and to operate.

A water-cooled system is an alternative, and can provide significantly increased efficiency under certain operating conditions. The water-cooled system, however, does have a number of drawbacks. The most critical of these is probably the fact that the binary cycle does not in itself generate a source of cooling water, so an external source of cooling water is required. If an external source can be obtained, an evaluation must be made of capital and operating costs versus electrical output. The combined cost of the condenser, cooling towers and cooling-water pumps of the water-cooled system will be less than the cost of the air coolers of the air-cooled system. However, the cost of water, chemicals and disposal of blow-down, coupled with the parasitic load associated with cooling tower pumps and fans, can exceed the cost of operating the air-cooled condenser. Other advantages of the air-cooled option are avoidance of the cooling tower plumes and cooling tower emissions, factors that are often critical to obtaining the necessary permits and meeting regulatory mandates. The best of both systems may be the hybrid based on the use of air coolers but with injection of a water mist into the airflow of the air cooler, or the spraying of water onto fibrous metal used to enclose the walls of the cooling tower. This can significantly increase cycle efficiency and output during peak demand periods in a summer peaking area. Testing at a facility in California in 2001 resulted in a 25+ per cent increase in power output (Sullivan, 2001, personal communication).

Power plant construction

A number of factors related to power plant construction can have a significant influence on project economics, including geologic conditions, terrain, accessibility, labour force, economies of scale, and site or factory assembly of major components.

Geologic conditions and terrain, for instance slope stability and need for extensive cut and fill, can be expected to increase the cost of construction by 2 to 5 per cent. The need to build or reinforce roads to carry heavy equipment will also be affected by both geologic conditions and terrain factors.

The availability of an adequate and skilled labour force can also impact construction cost. If the site is located in a rural area with little or no skilled construction labour force, most construction personnel will have to be brought to the site, and in fact, depending upon the commuting distance, a construction camp may have to be established to provide living quarters and meals for the workers (Sifford et al., 1985).

Economics of scale will often favour the larger power plant; however, a number of factors can virtually eliminate the initial capital cost advantage of opting for the larger power plant and these factors may also provide operational features that greatly increase plant availability and the capacity factors. The most important of these are modular design and factory assembly of major components. Modular design will often allow for factory assembly of major components, virtually eliminating most weather-related delays, minimizing the need to upgrade roads to carry extremely heavy pieces of equipment, and helping to ensure more consistent and higher quality workmanship, possible because of the controlled environment where the work is taking place. Modular design may also allow for staged start-up of generation, providing for a revenue stream much earlier than with the larger, site-erected plant, and minimizing interest during construction. For example, a 110 MW$_e$ power plant, made up of two 55 MW$_e$ turbine generators, at The Geysers has a normal construction period of three years. Modular plants of approximately 25 MW$_e$ and less can often be on-line within one year of the start of construction, with subsequent modular plants coming on-line at six- to twelve-month intervals. The generation of considerable revenue during the construction period more than offsets any advantage that economy of scale may provide. In addition, the ability to bring units on-line sequentially is often a major benefit which makes it possible to better track the load growth of the utility. However, within the size range of most modular constructions, for instance, less than 25 MW$_e$, economy of scale does apply. For example, it would be more cost effective to erect five 5 MW$_e$ modules rather than ten 2.5 MW$_e$ modules.

Direct use

A discussion of project design and facility construction relative to direct-use projects is much more difficult than the previous discussion relating to power generation. A direct-use project may be supplying the needs of a greenhouse or aquaculture complex, an industrial facility, or a district energy system supplying multiple commercial, industrial and even residential customers. (Note: individual systems to heat and/or cool a single residence or greenhouse, or projects directed toward balneology, are not considered here.)

The three uses mentioned above, however, share a number of design considerations and even some equipment components, all having a bearing on the economics of the project. All are highly dependent upon resource characteristics, including temperature and flow, hydrostatic head, drawdown and fluid chemistry. The characteristics of the resource will dictate not only the type of project that can be developed, but also the scale of the project and the metallurgy of the components selected. Direct-use projects must be located near enough to the resource site to allow for economic transport of the geothermal fluids from the wells. For very large district energy systems, however, this distance may be several tens of kilometres. If the well(s) does not flow

artesian, well pumps will be required, and at the resource temperatures at which most direct-use projects operate, either line shaft or down-hole pumps may be used. Because of variations in flow requirements to meet seasonal loads, inclusion of variable speed drives should be considered in order to minimize electrical costs.

Piping from the well(s) to the application site will be dependent upon temperature, pressure and distance. Insulated pipe may or may not be required, and will depend on distance, and whether or not some temperature loss is acceptable. The pipes may be constructed above ground, but local regulations may require burial.

Another major design consideration is whether or not the heating system should be based on meeting the peak heat demand entirely with geothermal energy, or whether the system should rely on a fossil-fuel (oil, propane, natural gas or even coal) boiler for peaking and/or backup. In many instances, a strategy where the geothermal system is designed for 'base load only' operation may be the most economical. For both greenhouse applications and district energy systems, designing the geothermal system to meet 50 to 70 per cent of the peak-heating load will still allow the geothermal system to meet 90 to 95 per cent or more of the annual heating requirement in most climatic zones. This is because a system that is designed to meet peak-heating load operates only a few hours of the year under those conditions. For example, if a district energy system is to meet peak demand solely with geothermal, the number of wells will have to be doubled and the size of the distribution piping increased by approximately 30 per cent to accommodate the requirement for increased flow. Another strong argument for meeting peak demand with a non-geothermal system is the need for back-up for both greenhouse applications and district energy systems. Although back-up can be provided through the use of standby wells and back-up generators to run pumps, a fossil-fuel system may be the most secure alternative and also the most cost-effective. Whether or not to include fossil-fuel peaking for an aquaculture or industrial application will depend upon the particular requirements of the application.

In addition to careful design consideration in the selection of the most appropriate and economical heating system, similar consideration should also be given to the provision of cooling. For most greenhouse operations, cooling can be provided through a combination of shading and the use of evaporative coolers. However, if a more sophisticated cooling system is required, or there is a need for refrigeration, absorption cooling may be an option worth evaluating. New advances in double- and even triple-pass absorption equipment allow for a coefficient of performance (COP) significantly above 1 to be obtained, and even at geothermal resource temperatures as low as 80 to 100 °C, absorption cooling may be the answer to meeting the needs of both greenhouse operators and providers of district energy services.

Equipment selection

Most, if not all, systems will require the inclusion of a heat exchanger to separate the geothermal fluids from the in-building circulating loop, because of the potential for corrosion and scaling associated with most geothermal fluids. Both plate-and-frame and shell-and-tube heat exchangers have been employed successfully in such applications. Despite the higher cost, a number of factors tend to favour the plate-and-frame exchanger. Approach temperatures across the plate-and-frame exchanger are somewhat better at 3° to 6 °C versus 8° to 11 °C for shell-and-tube. Another major consideration in the selection of a plate-and-frame heat exchanger is the ability to easily add plates in order to expand the heat exchanger capacity, and the fact that the exchanger can be opened easily for cleaning. (Note: this is not true for brazed plate-and-frame exchangers.) Materials include various grades of stainless steel and titanium.

Selection of the piping material is especially important in applications that have extremely long pipe runs, such as is common to all district energy systems. If the geothermal fluid is to be circulated through the distribution-piping network, material selection and even carrier-pipe wall thickness become crucial decisions. For example, in the case where geothermal fluids are circulated, thin-walled, pre-insulated district heating pipe, so common to most district energy systems in Europe, may not be appropriate. If, however, the heat is transferred to a secondary fluid that is circulated in a closed loop – and where addition of inhibitors is practical – the thin-walled, pre-insulated pipe is probably a logical choice. Other points to consider include the choice between metallic and non-metallic pipes, and whether flexible pipes should be used. Flexible piping is only available in the smaller size ranges, but the decrease in cost associated with its installation may make providing heat to areas with relatively low heat-load density economically viable. If non-metallic piping is selected, care must be taken to ensure that it has an oxygen barrier, or that areas served with non-metallic pipes are separated by a heat exchanger from areas served with metallic pipes. If this is not done, severe corrosion problems may occur in the metallic pipe portions of the system due to oxygen infiltration.

Other system components and design considerations are very application-dependent and beyond the scope of this chapter. The reader is, however, referred to the other chapters of this book and to the *Geothermal Direct-Use Engineering and Design Guidebook* published by the Oregon Institute of Technology in Klamath Falls, Oregon (see listing at: http://geoheat.oit.edu).

Project construction

For greenhouses and aquaculture projects, construction of the geothermal portion of the project is usually a very minor part of the entire project, and consists primarily of wells, pumps, heat exchangers, peaking and/or back-up equipment, piping and controls. However, with a district energy system, the thermal energy transmission and distribution piping system will comprise 60 per cent or more of the total construction budget. District energy systems may include multiple heat exchange and peaking or back-up stations, thermal storage tanks and extensive control systems. In the majority of district energy applications, the geothermal fluid is most often used to heat a secondary fluid that is circulated to meet customer needs. In some cases, however, the geothermal fluid is circulated directly to each customer, where the heat exchange takes place. The principal cost during construction is related to pipelines, and includes excavation, backfilling, and repaving if necessary. The installation of the piping system can run from an equivalent of US$300 per metre to as high as US$9,000 per metre in highly developed urban areas. A major problem for most developers of district energy systems is that the transmission piping must be sized to meet the needs of the system at full build-out, although revenue will increase only slowly as the system expands and as the customer base increases. This dilemma is by far the most important economic consideration in determining the feasibility of introducing a geothermal district energy service into an existing community. The use of computer models for determining the economic viability of constructing a new district energy system or expanding an already existing system is now available. For one such model see HEATMAP©5GEO (http://www.energy.wsu.edu/software/heatmap/). In a new community or a new area of a community, much of the cost of constructing the distribution system can be shared with the developers of other utility services, including sewers, water and electricity.

9.2.3 Revenue generation

For power generation projects, the power sales contract establishes the legal framework for revenue generation. For direct-use projects, however, the revenue stream to support the project may well come from the sale of a product, for instance: flowers, plants or vegetables from a greenhouse project; fish or shellfish from an aquaculture project; value-added service, for example dehydration in an industrial process; or thermal energy sales for a district energy project. Considerable interest in so-called 'co-production' is increasing rapidly as a means of improving the economics of geothermal power generation by providing an additional revenue stream. Co-production involves the extraction of valuable by-products from the geothermal brine before re-injection. These by-products may include zinc, manganese, lithium and silica, all of which have a relatively high market value.

Electricity generation

Ultimately, the economic viability of a particular power generation project will depend on its ability to generate revenue, and revenue can only be generated from power sales. Such sales must be equal to or exceed that required to purchase or maintain the fuel supply, to cover debt service related to capital purchases, and to cover operation and maintenance of the facility. The output from the plant, and hence the source of revenue generated, will be highly dependent on how well the plant is maintained, how it is operated, and the ability to take maximum advantage of incentives to produce at certain times or under certain conditions. For example, a plant selling into a summer peaking service area must be able to provide maximum possible output when a premium is being paid for output.

With the increase in the number of geothermal power plants by private sector developers, O&M has assumed even greater importance. Competition means that margins for profit are slimmer, making O&M costs all the more critical. Because most power sales contracts are output-based, and because geothermal generation costs are predominantly fixed as opposed to variable, the unit cost of geothermal power decreases rapidly as the capacity factor increases: that is, maximum operation yields maximum return to the owner. For example, as the plant capacity factor increases from 50 to 90 per cent, the levelized cost of producing electricity could be expected to decrease by nearly 50 per cent (Parker et al., 1985).

A number of innovative approaches have been adopted to ensure the highest possible capacity factor, and thus maximum revenue to the plant owner. The most common of these is the use of redundant or back-up equipment, including spare wells, cooling-water pumps, non-condensable gas removal equipment, and the use of multiple-turbine generation sets. The presence of redundant equipment allows for routine or even forced maintenance to be accomplished without taking the plant off-line, or at least the entire facility off-line. The use of multiple modular turbine generators is a prime example of a strategy to achieve the maximum capacity factor. In many instances, the steam or brine can be routed from the downed unit to other operating units capable of operating at slightly over design, thus providing the possibility of covering the entire load of the unit that is out of service.

One of the most innovative uses of this philosophy is at the Santa Fe plant in The Geysers geothermal field. As a result of Santa Fe's common ownership of the plant and well field, and regulatory restrictions and contractual incentives, it was highly desirable to obtain highest allowable capacity and availability for maximum revenue generation with minimum per-unit fuel use and production cost. Maximum capacity was ensured through the use of two two-flow turbines, each capable of producing the maximum regulatory allowable 80 MW$_e$. The use of two turbine generator sets allows for more frequent refurbishing of the turbine blades to maintain performance at or near design value without significant loss of revenue during downtime and without the need to maintain two spare rotors. Higher efficiency was also ensured through the selection of a turbine design that incorporated partial-arc admission and the capability to slide to 50 per cent overpressure. The use of the partial-arc admission resulted in reduced throttling losses under normal operating conditions (two-turbine operation, each at 40 MW$_e$ net output) and improved single-turbine operation, that is, 80 MW$_e$ net output. At eight bars throttle pressure at (or near) the first valve point, the larger valve is fully open and the smaller valve is fully closed. Under these operating conditions, each machine produces 40 MW$_e$ of net output. When only a single turbine generator is operable, the small valve is operated at 11.6 bars throttle pressure, and one turbine can produce approximately 80 MW$_e$ net (McKay and Tucker, 1985). According to the design engineers, Stone and Webster (personal communication), the cost of the second turbine generator would be covered by an additional two weeks of operation per year, approximately equal to the time required for routine maintenance. Efficiency is further enhanced through the use of a large, multi-pressure condenser that guarantees a low average condenser pressure, and condenser bypass that allows full steam flow through the condenser of the operating unit, as well as the use of the entire cooling tower, so that design back-pressure can be maintained during single-turbine operation (McKay and Tucker, 1985). Through such innovative approaches, not only is maximum capacity and potential for revenue generation ensured, but efficient use of steam is also achieved.

Revenue can also be affected by plant availability, dispatchability and load-following capability. Many power purchase contracts provide incentive payments for:

- availability, that is, the ability to generate at certain levels or during certain peak demand periods
- dispatchability, that is, the ability to go off-line or curtail production when the power is unneeded
- load-following capability, that is, the ability to match power output to the need for power of the receiving utility.

Availability, much like plant capacity factor, can be achieved through the highest possible flexibility and reliability in plant operation, and as with capacity, it is often achieved through the use of redundant equipment. However, perhaps as important in terms of revenue generation is the ability of the plant to quickly come on-line after a forced outage, after being tripped off-line, or after a request of the utility to curtail production. In many areas, being tripped off-line means shutting in wells to avoid unabated hydrogen sulfide emission and a lengthy restart because major components have to be brought up to temperature slowly. The use of a turbine bypass and computerized well-field control can individually or, ideally, together minimize both these effects and help maximize on-line availability. Because the steam flow can be routed past the turbine and directly into the condenser, it is possible to remove the non-condensable gases, and any hydrogen sulfide can be removed and treated in the hydrogen sulfide abatement system. Without the turbine bypass, the wells would have to be shut in or vented to the atmosphere, and it could take up to several hours to bring the wells and plant back to full production. In its first full year of operation, Santa Fe Geothermal, one of the first plants to use the turbine bypass, achieved a capacity factor of 98.6 per cent of its 80 MWe operating permit and had an availability of 99.9 per cent (Fesmire, 1985).

Other factors that can affect revenue generation include plant dispatchability and load-following capability. Although the commonly held philosophy is that geothermal power plants, because of the ratio of fixed to variable costs, must operate in a base-load manner, utility requirements and/or reservoir concerns may require that the plant be operated in a load-following or dispatchable manner. Reservoir depletion at both The Geysers and Larderello has forced load-following, and some utility contracts provide incentives for dispatchability that more than offset any loss of revenue while the plant is operated below design capacity.

A number of plant features, including partial-arc admission, turbine bypass, and computerized well-field operation, all mentioned before, help maximize revenue generation during load-following operation. The Italians have also found that remote operation can play a significant role in meeting the demands of load-following operation while at the same time significantly reducing labour costs, costs that become increasingly important when a plant or plants are operated below design capacity. It is also important to note that availability takes on increasing importance when operated in a load-following mode, inasmuch as severe economic penalties may be imposed if the plant is not available when needed.

The direct link between revenue generation, plant availability and capacity also places greater emphasis on O&M. Plant operation costs and on-line performance are under increasing scrutiny by purchasing utilities, direct electrical service customers and those who provide financing. Indeed, because investors and financiers are typically more conservative than developers, an experienced, big-name company able to provide both O&M has strong appeal to backers, and that appeal translates directly into slightly lower financing costs, which are a major economic consideration. It is extremely important that the O&M provider or in-house staff be retained at an early stage in the development process and provide review and input into plant design, participate in plant construction and start-up, and conduct system checks. The contractor or plant staff should also be capable of and required to perform a *post-hoc* analysis of all significant events, including root-cause analyses for future planning (*Independent Power*, 1989).

The increase in partnerships for developing projects also highlights another O&M trend: affiliates of financiers and/or partners are often highly competent facility operators. A vested interest in plant performance provides a motivating influence to the O&M provider. Such motivation, in turn, provides security to financiers. Other incentives to peak performance, however, do exist. A bonus for good operation, tied with a penalty for not meeting minimum performance requirements, helps ensure optimum performance, guarantees achieving output to match contractual requirements, and generates maximum revenue and profit. But good O&M goes beyond maximizing current profits, to an efficient use of the reservoir in order to prolong life and assure supply. Smart developers also know that a good performance record will be critical to obtaining both future power sales agreements and financing for future plants at attractive rates.

Co-production

Co-production (that is, the production of silica and other marketable products from geothermal brines) is rapidly becoming not only a very viable source of additional revenue for power plant owners, but also a key technique for improving power plant economics by reducing operation and maintenance costs. The removal of silica may allow additional geothermal energy extraction in bottoming cycles, or additional uses of low-grade heat that are presently prohibited because of problems associated with scaling.

Precipitated silica has a relatively high market value (US$1 to $10 per kilogram) for such uses as waste and odour control, or as an additive in paper, paint and rubber (Borcier, 2002, personal communication; Borcier et al., 2001). Initial estimates from Salton Sea geothermal fields place the market value of extracted silica at US$84 million a year.

Silica removal also opens the door to the downstream extraction of, for example, zinc (Zn), manganese (Mn), and lithium (Li), all with relatively high market values. The first commercial facility for the recovery of zinc from geothermal brine was built in the Salton Sea geothermal area of southern California in 2000. The facility is designed to produce 30,000 metric tonnes of 99.99 per cent pure zinc annually at a value of approximately US$50 million (Clutter, 2000).

Silica removal has the additional benefit of helping to minimize re-injection problems and, in one case in California, could allow use of the spent brine as the source of cooling water needed to improve summer power-plant performance. Initial studies indicate that power plant efficiency of an air-cooled binary plant could be increased by 25+ per cent through the use of spray cooling (Sullivan, 2001, personal communication).

Direct use

Most large-scale direct-use projects tend to fall into three broad categories: provision of district energy; industrial processes, including dehydration; and agriculture, including greenhouses and aquaculture. In all except the provision of district energy, revenue is generated from the sale of a product, such as potted plants from a greenhouse, or from a value-added service rendered, for example, the drying of onions in a dehydration plant. Ultimately, in both cases, revenue generated and economic viability are totally dependent on the value and marketability of the end product. Long-term contracts for sale of these products are almost never available. Geothermal may be the most economic form of energy for any given application, and may even provide certain other benefits such as fuel-price stability or constant heat, but the economic viability of the project will seldom be driven by the cost of developing and/or operating and maintaining the geothermal source. The geothermal resource developer must therefore not only have a thorough appreciation of the costs involved in developing and operating a geothermal project in an economical manner, but fully understand what factors ultimately determine the economic viability of the products produced.

With district energy, on the other hand, revenue is generated solely from the sale of thermal energy in the form of either hot water or chilled water. Long-term sales contracts to customers are the norm, and most contracts call for both capacity (fixed) payment and variable payment components. The capacity or fixed portion of the payment is based upon the capital invested, including wells, heat exchangers, thermal storage units, back-up or peaking boilers, and the transmission and distribution network. The variable portion of the amount charged relates to O&M, including personal cost, cost for fossil fuels used in the back-up and/or peaking boilers, and re-drilling of wells. In most systems, charges are based on usage, metered either as flow or thermal demand, that is, kW/hour. Some systems, however, use a fixed orifice and charges are based upon the orifice size.

Because weather conditions will, to a large extent, determine thermal energy usage by residential and commercial customers, it is extremely important that rates are structured in such a way as to ensure that revenue is always able to cover both fixed and variable costs.

9.3 FINANCING CONSIDERATIONS

Once a geothermal project has demonstrated technical feasibility and economic viability, it faces a critical test, financeability. Whether a project is financeable depends on such matters as:

- the intent and permanence of legal, structural and economic reforms
- the availability of debt and equity capital
- the risk allocation reflected in project contracts.

The following is an analysis of some of the critical issues to be examined when considering development of a geothermal energy project. The primary focus of this analysis is those issues that most often arise in the context of a proposed geothermal power project. Some of these issues will not be applicable to a direct-use or district heating project using geothermal energy. Nevertheless, the basic institutional framework, financing option, and risk allocation principles described below provide a useful benchmark for analysis of geothermal energy projects that do not involve the production of electricity.

9.3.1 Institutional framework

Facilitating private investments in geothermal energy projects requires that governments create a legal environment that is conducive to long-term investment. Such an environment consists not only of governmental restraint in affecting the settled expectations of private investors, but also of explicit governmental authority for the contemplated investment. In this context, some of the basic legal reforms undertaken to facilitate financing of geothermal energy projects may include:

- Empowering a private party to engage in a given 'public' activity (for example, electricity generation or supply of thermal energy) together with, or on behalf of, the government.
- Empowering a public entity to select the private party that will perform the activity and empowering the public entity to enter into legally-binding relationships designed to produce a firm, reliable stream of payments to support a project financing.
- Reconciling the financing arrangements for the project with the terms of any applicable public procurement rules.
- Authorizing the private party to have a legal entitlement to payments made by the ultimate consumers of the goods/services to be provided.

Through establishment and implementation of a reform programme, governments have also taken a proactive role in creating an environment conducive to foreign investment in energy sector projects.

In most countries, privately owned geothermal energy projects are new entrants in a market that previously consisted of government-owned electricity or thermal energy suppliers. The encouragement of these new market entrants may also require fundamental constitutional, legislative and regulatory reforms.

In some countries, private ownership of power generation facilities or thermal energy systems has been prohibited by the constitution. Many countries have instituted constitutional, statutory and regulatory reforms to authorize private ownership.

Attention often focuses on the need to institute sound economic reforms in utility regulation: cost of service pricing and an end to government subsidies (which results in electricity or thermal energy being priced below the marginal cost of production) are among the most important of these reforms. From a legislative perspective, however, two types of reforms stand out as essential predicates to attracting private capital: a (1) commitment to refrain from upsetting the settled expectations of contracting parties by imposing retroactive and economically burdensome regulations and/or taxes, and (2) a willingness to enable the support of utility or direct-customer purchase commitments by authorizing central bank commitments in the form of currency convertibility assurances (where necessary) and liquidity instruments (for instance, letters of credit) to enhance the credit of the utility or direct-use purchaser.

Discussed below are seven key areas where the extent (and certainty) of policy reform is critical to the success of privately owned and financed power plants, or thermal energy supply and distribution systems.

Power procurement approach

A basic issue that needs to be clarified in most countries is the framework for purchasers to acquire new supplies of electric or thermal energy. Options include the following:

- *Sole-source negotiations*: the purchaser is free to negotiate with proponents of a proposed project, without the need for formal submission of proposals and ranking.
- *Formal competitive bidding*: the purchaser establishes criteria for ranking and selection, with bids to be submitted

and evaluated (and contracts awarded) on a specified schedule.

- *Uniform rates and contracts*: a standard rate and master contract is developed, with contracts available for projects meeting pre-qualification criteria (if any).

Variations of these approaches exist. However, project development has been most difficult in those countries that have shifted approaches with some frequency.

In addition, attention needs to focus on the types of market opportunities that are to be encouraged. The four basic types of market opportunities: electric power sales to a utility under the build–operate–transfer (BOT) or build–own–operate (BOO) models, sales of electric or thermal energy to 'end-use' customers, unit re-powering, and operations contracts, are discussed below.

Central power plants: the BOT and BOO models

Egypt, the Philippines and Pakistan are among countries that have allowed privately owned entities (with or without government equity participation) to build, own and operate major power plants to sell electricity to the government-owned electric utility. At the end of a specified term (ten to twenty years, usually), it is contemplated that project ownership will be transferred to the government. Project documentation, including technology transfer agreements, must take account of the transfer. A critical issue in power contract negotiations is the transfer provisions, which may establish certain performance criteria to be satisfied as of the time of transfer, and which may specify liquidated damages for failure to meet those criteria. Some of the variations on the BOT theme include: concession agreements, with transfer at the end of the concession term; no transfer of the facility at the end of a specified term (the BOO model); construction by the government utility and sale to a private owner/operator; and lease by the government utility of a privately constructed and owned plant.

Direct sales

Many countries allow only a limited class of privately owned facilities, those serving the industrial user that owns the facility. Other countries allow a broader category of privately owned facilities, but allow them to sell electricity or thermal energy only to utility wholesalers. In some countries, markets have been liberalized and sales can be made to any purchaser, including industrial and commercial end-users.

Purchase, re-powering and refurbishment of existing facilities

In addition to the construction of new facilities, many countries are offering to sell existing power generation and transmission and distribution system assets (or thermal energy production and distribution systems) to private investors, including foreign investors. The private purchaser is then enabled or required to re-power, refurbish, or expand the existing facilities.

Operator contracts

While it does not involve private ownership per se, a number of governments have been willing to enter into operation contracts providing the operator with a participation in the outcome of the operations (for example, in cost savings), with or without a responsibility for hard costs of operation and maintenance. Such contracts may pertain to particular facilities, or to entire electric energy or thermal energy systems. There may be accompanying obligations or opportunities to improve performance by investment and refurbishment of the facilities.

Energy regulation

In countries that allow privately owned electricity generation facilities or thermal energy systems, an important question that must be addressed is the choice of a regulatory model. While there is no standard model, it is important to establish a transparent regulatory framework, not subject to retroactive changes, that addresses certain basic matters.

Prices

In some countries, the price for sales of electric or thermal energy is a product of conventional 'cost of service' regulation. In essence, profits are controlled, and revenues are a product of a cost-plus-reasonable-profit calculation. Countries that have deregulated pricing for sales of electric or thermal energy follow a different approach: the seller can charge the price established through bilateral contracts, spot-market sales, or competitive bidding (for long-term sales), or electricity trading transactions. An alternative approach is to establish a standard price for all sellers, with some variation in pricing to reflect differences in fixed and variable costs of certain technologies.

Market entrance

Some countries also require owners of new generating units or thermal energy facilities to obtain licences or concessions. Significant issues can arise under a licence and concession regime, including such matters as grounds for revocation, level of royalty payments or licence/concession fees, and restrictions on transfer of shares.

Performance

In many countries, regulations may cover such matters as dispatch, interconnection, metering and quality of service. These regulations need to address any specific exemptions (or modifications) that need to be given to certain power generation or thermal energy technologies.

In reviewing an existing regulatory framework, or establishing a new regulatory system, it is important to determine whether the regulations meet intended policy goals. For example, if a country desires to encourage the use of renewable technologies using indigenous resources, care must be taken to assure that price and dispatch regulations do not act to give preference to technologies using non-renewable resources.

Privatization

Privatization, in the classic sense of the sale of government-owned utilities to the private sector, is not a necessary and essential prerequisite to success in encouraging new, privately owned electric power generation or thermal energy facilities. Certainly, constitutional and legislative provisions may need to be modified to allow for private ownership of generating facilities or thermal energy systems. But a successful private energy sector programme need not wait for completion of a broad privatization programme.

The critical issue is the interplay in any country between (1) the introduction of privately owned facilities and (2) the transition, through privatization, from government-owned, integrated electric or thermal energy utilities to a system of publicly and privately owned generation, transmission and distribution systems. Development and financing of new generation or thermal energy systems is difficult during a transition period, when the identity, let alone credit quality, of the wholesale purchaser is uncertain.

Environmental regulation

An important area for governmental action involves legislative and/or regulatory initiatives aimed at clarifying the environmental standards to be applied to independent power or thermal energy projects. In some countries, environmental standards are a 'moving target'.

Given the inherent complexity of environmental laws, and the many difficulties that will arise in trying to interpret

these laws in the absence of a lengthy history of implementation, it is important to determine the prospects for changes in environmental law that would be made applicable to projects in development or operations. At a minimum, attention needs to focus on the power or thermal energy sales contract implications of potential changes in environmental law.

Commercial and tax laws

In some countries, fundamental corporate and finance law reform is essential to facilitating private investment in power or thermal energy projects. Preferred forms of business and ownership of particular projects or operations may not be available in a given country. Some countries limit the scope of corporate activities generally, or those of foreign corporations, in ways that may not be acceptable. For example, some statutes may not allow corporate entities to be stockholders in a corporation. Similarly, corporate entities may not be allowed to be partners in a partnership. Certain entities, such as limited partnerships and limited liability corporations, may not exist under local law. Local nationals may be required to hold a part or a majority of the equity.

Tax issues arise in two contexts. First, care must be taken to identify all existing and potential local taxes, to ensure that potential sales rates and pro forma projections adequately account for the array of income, sales, excise, withholding and import taxes characteristic of most projects. As a related point, attention has to focus on the potential risk of change of tax law. A second block of tax issues relates to the ability to implement tax structures that maximize project profitability. In part, this relates to the question of whether tax payments made to foreign tax authorities are recognized for purposes of home country taxation of profits from project investment.

Currency issues

One potentially daunting issue faced by project developers is the risk associated with the potential unavailability of hard currency to pay for the output of the project. Unfavourable balance of trade conditions in many countries have required government officials to direct their other hard currency reserves to the repayment of long-standing foreign debt and the purchase of basic essentials (for instance, food, medical supplies and basic manufacturing equipment).

The assured availability of hard currency is critical to obtaining project financing, which is premised on creditworthy contractual arrangements designed to produce a firm, long-term revenue stream. If project revenues are payable in local currency, or if the country has inadequate access to hard currency, lenders will insist on appropriate credit supports (for example, stand-by letters of credit or parent guarantees) to mitigate perceived currency risks. Equally important, a shortage of hard currency may impair the ability of the developer to repatriate profits from the project.

An additional currency issue exists if payment is to be made in local currency. Assuming satisfactory arrangements are made for availability of hard currency, attention must also focus on potential volatility in exchange rates. This is especially important to the extent equity investors and lenders require payment in hard currency, and seek protection against local currency devaluation.

If the purchaser of electric or thermal energy prefers to pay in local currency, two related considerations arise:

- Is there a central bank commitment to making available hard currency reserves to facilitate currency conversion (or to offer inconvertibility insurance)?
- What are the consequences of changes in exchange rates?

As discussed below (in Section 9.3.3, 'Contracts and risk allocation'), contractual approaches exist by which governments can provide meaningful assurances on currency issues.

Guarantees of payment

In the international context, a key issue is the practical ability of a utility purchaser of electric or thermal energy to make full payments on a timely basis. In some countries, the issue focuses on the question of availability of payment backstops, guarantees or supplemental measures that alleviate concerns about reliability of payment.

Central bank support of the payment obligation of a utility may have to be reconciled with existing constitutional and International Monetary Fund (IMF) restrictions. For example, countries with constitutional barriers on pledging of credit for private purposes may not be able to offer central bank guarantees of utility payment obligations, or even if offered, those guarantees may cease to have legal effect if the utility is privatized. Similarly, IMF limitations on central bank borrowing or allocation of hard currency from export earnings (for purposes other than debt service) may present similar difficulties.

9.3.2 Financing approaches and sources

Geothermal energy production continues to advance in industrialized and emerging economy countries. While a variety of factors have increased the demand for new geothermal energy facilities, the relative success of a project relates to the practical availability to the project sponsors of the wide range of financing approaches, project structures and funding sources that can be used.

One question often asked is why a particular financing approach is used for a given project. As a practical matter, a project sponsor will access those capital sources that will result in the lowest possible overall cost of capital. In some instances, such as projects involving market or technology risks, debt capital may not be available at attractive terms. Thus, the choice of project structure and capital sources will reflect an ongoing examination of the attractiveness of the project to debt and equity sources.

In a similar vein, there is no constant differential in overall costs of capital between differing financing options. Certain rule-of-thumb principles, of course, exist. 'All equity' financed projects in most cases will have a higher cost of capital than 100 per cent debt-financed projects; however, 'all equity' financing may be necessary when debt capital is unavailable (or, if available, is priced – due to financial market conditions, risk uncertainties, or otherwise – at a level too high for a project sponsor). Debt and equity capital usually has a higher cost when raised at the early development stage than that of funds available when construction is ready to commence. Likewise, debt and equity capital will have a lower cost when a project has a favourable operating history.

Traditional corporate or sovereign debt

Through the early 1980s, power supply and thermal energy projects were financed through traditional debt-financing techniques. For example, investor-owned utilities financed new power projects through a combination of newly-issued corporate securities and first-mortgage debt.

Government-owned electric or thermal energy utilities financed construction and operation through two vehicles. In some countries, the government-owned utility or an authorized authority issued debt that was either guaranteed by the issuing authority or secured by the revenues of the project owner (with a related obligation to charge ratepayers at rates sufficient to pay the debt). In most countries, the debt (which in turn may have been borrowed from the

World Bank) was backstopped by a sovereign (that is, country) guarantee of repayment.

These traditional sources of financing improvements in the electric or thermal energy sector are not generally available to meet the full need for financing new projects. In the power sector, one reason for the change in sources of finance is the sheer magnitude of capital requirements for new projects. A government-owned utility may not have sufficient net worth to support a corporate credit financing. Another is that the borrowing capacity of the developing countries is substantially exhausted. A third is that the World Bank, which has financed or subsidized much of the electric sector of the developing world, has made a deliberate decision to greatly reduce its government-to-government lending, while promoting private-sector investment in electric power generation or thermal energy technologies that are environmentally sensitive.

Export credit and trade assistance

In the past decade, project owners have increasingly focused on the availability of 'tied assistance' financing in making decisions on equipment suppliers and construction contractors. The export credit agencies of Canada, France, Germany, Italy, Japan, Spain, Denmark, Sweden and the United States have been active (in varying degrees) in offering subsidized debt financing for projects that use equipment fabricated in their country or construction contractors located in their country. In addition, specialized trade agencies, such as the US Trade and Development Agency (TDA), offer financing for some of the costs of developing a project. TDA makes funds available for projects that potentially will use US goods and services (and where there is the existence of competition from non-US companies that are receiving financing assistance from their home country).

Project finance

The term 'project finance' is generally used to refer to the arrangement of debt and equity for the construction of a particular facility in a capital-intensive industry in which the lenders look chiefly to revenues generated by the facility – rather than the general assets of the developer of the facility – to cover the repayment of the debt, and rely on the assets of the facility as collateral to secure payment of the debt. Thus, in simplest terms, project financing is essentially corporate financing secured substantially by revenue-producing contracts, rather than the credit of the developer.

In addition to the financial strength of the project itself, external credit support may be provided by the participants in the project, such as the developers, the construction contractor, the purchaser of the electric or thermal energy, or others who may benefit through the sale of equipment to the project. This indirect support may take the form of completion agreements, contracts to purchase the output, and so forth. While the objective of the project developers may be to have non-recourse financing (that is, non-recourse to the general credit of the developers), that objective must be reconciled with the requirements of the lenders. Ultimately, most project financings are not truly non-recourse throughout all phases of the project.

There are various risks in financing a project. Project financing involves identifying and evaluating these risks and allocating them appropriately. Risks can be allocated to the project developer, to another participant in the project, or to the lender. All of the parties that might be willing to accept a specific risk will have a limit as to the degree of risk they will be willing to accept, and their price for doing so. In making an informed judgement on risks to be accepted, a lender attempts to understand each risk thoroughly. Specific ratios or standards that might be applied to a given category of risk may be unacceptable if they do not give proper weight to the interrelationship with other risks.

Typically, project financing is used by entities that want to achieve any or all of the following objectives:

- elimination of, or limitation on, the recourse nature of the financing of the facility
- off-balance-sheet treatment of debt financing
- leveraging debt so that existing equity need not be diluted.

Each of these objectives is briefly discussed below.

Non-recourse financing

The objective of project developers to eliminate the recourse nature of a financing is often an important one. Classic non-recourse project financing provides a structure that removes the developer from any obligation to stand behind the project debt if the revenues of the project are insufficient to cover principal and interest payments on the underlying debt. The non-recourse nature of project financing insulates other projects and assets owned by the developer from the liabilities associated with any particular project. A typical non-recourse loan includes this type of general provision:

> Notwithstanding any other provisions of the loan documents, there shall be no recourse against the borrower or any affiliate of the borrower nor any of their respective stockholders, agents, officers, directors or employees for any liability to the lender arising in connection with any breach or default under this loan agreement except to reach project security, and the lender shall look solely to the project security in enforcing rights and obligations under and in connection with the loan documents.

Occasionally, the terms 'non-recourse' and 'limited recourse' are used interchangeably. It is important to note, however, that typically project financing is with recourse to the developer, to a limited extent, since the developer remains liable for fraudulent representation and warranties made in connection with the financing.

Off-balance-sheet treatment of project debt
A second objective of some project financings is the off-balance-sheet treatment of the non-recourse project debt, which generally is available where the project promoter does not control the project entity. The balance-sheet liabilities of a project entity that is controlled more than 50 per cent by the project promoter are consolidated on a line-by-line basis with the balance-sheet liabilities of the project promoter. If this control test is not satisfied, the project promoter reflects its interest in the project entity as a line item on its balance sheet on an 'equity' or 'cost' accounting basis.

It should be noted that since the bankruptcy of Enron Corp. in November 2001, the structuring techniques utilized by the financial community and the rules applied by the accounting profession are the focus of greater scrutiny. As a result, the market's view on the range and acceptability of off-balance-sheet financing approaches is in a relative state of flux.

Non-dilution of existing equity to finance the project
The third, and perhaps most significant, objective of project finance is to enable the developer to finance a project using highly leveraged debt. Typically, the leverage percentage is between 70 and 90 per cent. Developers thereby obtain needed financing for a project without diluting existing equity.

The percentage of the equity contribution varies from transaction to transaction, and is influenced by many factors, including whether project participants have provided subordinated debt to the project.

Financing sources: multilateral and bilateral agencies

A variety of international and country-specific agencies provide assistance to individuals and entities pursuing power development or thermal energy projects. Among its many functions, an international agency or organization may provide direct loans or grants, help an enterprise to pursue adequate financing, and insure investments against commercial and non-commercial risks.

Several types of financing assistance and investment insurance programmes are available through bilateral, multilateral (for example, the Multilateral Investment Guarantee Agency and the International Finance Corporation), regional (for example, the Inter-American Development Bank), export credit and trade assistance agencies.

Financing sources: commercial banks

Commercial banks have also been lenders in private power project and thermal energy system financings. They have served as construction and take-out (for example, permanent financing) lenders, and less frequently as lenders for resource evaluation and development projects (those funds most typically are provided by the project sponsors). Commercial bank willingness to finance projects in a specific country is a product of traditional project finance analysis, coupled with evaluation of lending risks in a specific country.

Financing for energy facilities typically is provided in two tranches (with, perhaps, an additional tranche available to finance resource development). The initial tranche comprises the financing necessary to construct the facility; the second and final tranche comprises the permanent financing for the facility. The construction financing is intended to fund development of the project on a relatively short-term basis. The construction lender extends this financing with the intention of being repaid in full out of the proceeds of the permanent loan at such time as the construction of the facility is completed and commercial operation commences.

The construction lender typically will loan the project developer the full amount of funds necessary to construct the project. In contrast, the permanent lender generally will be willing to loan the developer only a portion of the funds required to repay the construction lender, and will require that the balance of such funds be provided by the developer in the form of equity. To the extent that the developer itself lacks or is unwilling to invest the requisite amount of equity, it will have to be obtained from institutional or other sources. Construction lenders require that a stipulated amount of equity be firmly committed by a creditworthy equity source or subordinated lender prior to closing the construction loan.

Projects have generally relied on two forms of funding options from creditors:

- fixed-rate loans by insurance and finance companies with pre-payment penalties
- floating-rate loans by banks at rates based on spreads over the prime rate, the London Interbank Offered Rate (LIBOR), or other established rates.

Banks generally are more flexible in their lending practices and documentation provisions than insurance and finance companies, and floating-interest rates are generally lower than fixed-interest rates. Accordingly, the trend has been toward bank borrowings, at least during the construction period when sensitivity to rates is usually less. However, as increased levels of bank reserve and capital adequacy charges drive up rates available from banks, projects have been attracted to the commercial paper market.

Subordinated debt also serves a role in the project finance marketplace. Subordinated (or high-yield) debt issues have helped the project financing market by allowing capital to be raised from funding sources with a larger appetite for credit risk of a type that used to be apportioned totally to equity owners, such as:

- inability to provide security or liens to the creditors in a project financing
- accepting specific risks in a project such as refinancing, fuel, or operating risk
- providing part of the returns of the vendors and suppliers through the subordinated claims instead of cash payment
- lowering equity funding requirements.

Notwithstanding the benefits of subordinated debt, the presence of more than one class of creditors will complicate the transaction substantially.

Funding sources: equity

Private power or thermal energy projects need a substantial amount of equity. Equity investment can come from energy companies, institutional investors such as specialized funds and insurance companies, and members of the project consortium.

Participants in project-financing transactions must be knowledgeable about the legal and business consequences of selecting a particular type of entity to own the project. The type of ownership entity selected impacts on the overall project finance structure and the drafting and negotiation of project finance documentation, as well as the regulatory permissions process. For example, permits granted to a developer and later transferred to a partnership in which the developer is a partner may no longer be valid in some jurisdictions. Also by example, a contract that does not contain a full assignment clause may preclude a later transfer to a limited partnership organized as a means for the needed equity contribution. Factors that may be significant in determining the equity structure for a project financing include:

- the permissible ratio of equity to debt
- tax considerations affecting the project participants
- concern with management of the project.

Funding sources: public capital

A number of international power or thermal energy projects have accessed the public capital markets to raise debt or equity capital. In a typical structure, the project company intends to issue securities in the capital markets in the United States, Europe or Asia. In some countries, there may be active interest in the local stock markets for flotation of shares in the project company.

9.3.3 Contracts and risk allocation

The ability to obtain commercial agreements that support available financing options is the key to project development, and for most developers, project financing will be the financing technique of first choice. The relationship of contracts to project financing is straightforward – the ability of the project to produce revenues is the basis of most project financings. Consequently, participants in project financing must pay particular attention to the contracts negotiated by the project developer. Each of the underlying contracts necessary to construct and operate a project must be sufficient to support the financing structure.

The project contracts are the means by which risk allocation is memorialized in a form acceptable to lenders and equity investors. The basic risks that arise in investment in power generation or thermal energy production projects differ during the three stages of project investment: development, construction and operation.

Project risks

During the development stage, the following risk factors are applicable:

- *Market risk*: The need for power or thermal energy (and applicable price) may change before a purchaser can be identified that will commit to a long-term contract. Similarly, the relative price advantage of a given project over alternative sources of thermal or electric energy may change over time.
- *Regulatory changes*: Anticipated regulatory approvals may be denied, or applicable energy and environmental laws may change without grandfathering provisions, rendering the project site unusable or the project concept/structure meaningless.
- *Unforeseen price increases or delay*: Often as a consequence of changes in market profile or interest rates, project economics can be altered in a manner that renders the project uneconomic. While other risks may apply, these factors most often are the reason for project failure and a related write-off of investment.

At the construction phase, the potential for additional cost exposure, or incurrence of project losses, is a product of four events (as well as the continuing potential occurrence of adverse regulatory or market changes, as described above).

- *Cost overruns*: Whether due to change in law, unexpected site conditions, or contractor or equipment supplier error, construction costs can increase in a manner that jeopardizes project economics.
- *Construction delays*: Either because of *force majeure* or contractor/supplier error, a project can fail to meet scheduled completion dates, requiring increased payment of interest during construction during periods of incurrence of operating losses due to minimal receipt of project revenues.
- *Substandard performance*: Even if construction deadlines are satisfied, the completed project may not produce electricity, or be able to supply thermal energy, at the expected unit availability, may not produce power at the expected level of unit capacity, or may face output shortfalls as a result of inadequate flow or quantity of geothermal fluids.
- *Finance cost increases*: In a period of volatile interest rates, project debt costs (if unhedged) may fluctuate to a point where interest payments exceed projected levels.

These construction period risks can leave a project owner with the unpalatable choice of either making substantial equity contributions (or loaning needed additional funds) or abandoning an uneconomic project.

At the stage of project operation, in addition to regulatory and market changes, substandard operations may jeopardize project investment. Whether as a consequence of *force majeure*, operator fault or equipment failure, the power plant or thermal energy system may operate at insufficient levels of output or substandard availability. During the operation period, the project may also face risks similar to those applicable in the prior periods. Any of these adverse events could affect project economics, resulting in the same potential choice between additional infusion of capital and write-off of project investment.

Risk mitigation

At the development stage, the critical risk mitigation technique essentially relates to a combination of project selection and 'off-ramp' techniques and cost controls on project expenditures. Even the best-conceived project concept may be rendered unattractive by changes in power or thermal energy markets or regulatory climate. For that reason, companies have developed exit strategies that allow for cessation of development funding in the event of project failure or unexpected project delays.

Cost controls are also a critical risk-management strategy during the development period. Among other techniques, a company may require potential equipment suppliers or constructors to bear their own preliminary design costs. In addition, a joint-venture partner may be required to share development costs. Because of the potential lengthy period for project development, these cost-control strategies are common.

Risk mitigation techniques during the construction and operating periods essentially relate to the choice of finance strategy. As described above, project financing, as well as the related contracting approaches, have become an essential feature of energy industry risk management strategies.

Certain unique risks presented by energy project investment are mitigated through a combination of agreements with the energy purchaser, foreign government, insurance providers and lenders. For example, change-of-law risks are mitigated through a combination of insurance packages, contract price adjustment provisions and government obligations. Below is a discussion of four main blocks of risk mitigation agreements: currency arrangements, governmental support obligations, power or thermal energy sales agreement provisions, and insurance packages.

Currency arrangements

The available methods for managing currency risks can be characterized by the extent of necessary government involvement. If the government is willing to allocate sufficient funds in advance, one approach is for the foreign government to obtain an irrevocable standby letter of credit from a money centre bank that will ensure payment of utility contractual obligations over the term of the financing. Similarly, a government can simply provide a sovereign guarantee of project debt. In most cases, however, governments are either unwilling or unable to dedicate hard currency reserves, or to provide direct-debt guarantees. For example, arrangements with the IMF, World Bank or other lenders could be jeopardized by the conventional approaches described above.

Another approach historically used involves currency 'swap' arrangements and block fund transfers. These vehicles facilitate the availability of hard currency without the necessity for government involvement. They are most often used when companies in need of local currency or registered capital are matched, through traders, with companies needing dollars or other hard currency. This approach often amounts to trading currency in the 'spot' market; however, this may be of limited assistance in a project financing.

Finally, a combination of barter and counter-trade vehicles may be utilized where the government wishes to facilitate project financing while minimizing its credit exposure. Under this approach, the government agrees to make available sufficient quantities of commodity stocks that it owns or controls (for instance, sugar or cocoa) for the purpose of generating hard currency in the international markets. The project sponsor arranges for 'forward sales' of such commodities to 'counter-parties' who will pay for the commodities in hard currency. The payments of the counter-party are deposited in a trust and then allocated to debt service, project expense distributions, payments to project sponsors and other reimbursements as provided for in the basic contract arrangements for the project.

The need for government assurances may vary from country to country, and depend on the economic cycle. Examples such as Argentina's economic crisis in 2002 highlight the potential complexity of problems that may arise.

Government support obligations

Government support arrangements take three forms. First, as discussed in Section 9.3.1 (Institutional framework), economic and structural reforms are put in place to facilitate private investment in the energy sector. Similarly, as discussed below, steps are taken to enhance the ability of the utility to pay for power or thermal energy deliveries. Finally, agreements are executed that delineate the extent of governmental assistance and define obligations in the event of change of law or *force majeure*.

Government support may be needed if the purchaser is not creditworthy in its own right. For electricity generation projects, an essential feature in the success of those projects that have closed financing arrangements has been the creation of enforceable contractual commitments by the utility to provide liquidity instruments (such as letters of credit) and funding of payment reserve contingencies, all designed to assure lenders that funds will be available to the developer to pay debt service in all but the most catastrophic of situations. Often, standby payment commitments from regional or national government banks have backstopped these arrangements.

For power projects, a related series of approaches to facilitate purchaser creditworthiness is based on the fact that one of the reasons some foreign utilities may be deemed to have less than investment-grade credit is that prior government policies have caused these utilities to sell electric energy at below the marginal cost of generation and transmission. However, the decision of a government to subsidize service to one segment of the customer base of the utility does not mean the entire customer base is incapable of paying for services priced at or above production costs. In other words, if part of the customer base of the utility is comprised of large, reliable, creditworthy customers, the electric energy sales contract can be structured to channel payments from such customers to private developers. A variation on this approach is for electricity to be sold directly to end-users who are creditworthy purchasers.

The contract entered into between the host government and the project sponsor is typically referred to as the 'Implementation', 'Co-operation', or 'Co-ordination' agreement. Traditionally, this document has been used to memorialize transfer arrangements (after lenders and equity have received their anticipated return on investment) for BOT projects, to confirm that the plant, equipment and spare parts of the project will not be subject to governmental import controls or excessive duties or tariffs, to ensure the availability of foreign exchange and provide protection against devaluation, to specify tax treatment, and to acknowledge rules governing employment of domestic and foreign nationals.

In the context of currency risk mitigation, government participation has taken at least two forms. In the case of countries that lack readily convertible currencies, the Implementation Agreement has been used to memorialize the commitment of the government to make forward sales of state-owned or controlled commodities (for instance, sugar cane) or to pledge revenues from utility sales to industrial customers for payment pursuant to the energy sales agreement. Where the host country is willing to provide credit support in the form of a liquidity commitment (for example, letter of credit), the Implementation Agreement may also serve to establish the government's obligation to issue such instruments through the central bank or other financial institution.

Other provisions of the Implementation Agreement accomplish the following purposes:

- Grant a concession to use natural resources (unless covered by a separate concession agreement).
- Clarify and confirm legal authority to engage in the generation, transmission, or distribution of electricity or thermal energy.
- Acknowledge the authority to do business (including confirmation of legality of ownership vehicle and restrictions, if any, on change of ownership or control).
- Memorialize the authority to sell electricity or thermal energy (including right to obtain water supplies for power generation) and to use energy resources (if the energy input is owned or controlled by the government).
- Guarantee the performance of government-owned water and back-up energy suppliers, if any.
- Guarantee the payment obligations of government-owned purchasing utilities (or acknowledge back-up support arrangements such as escrowed funds and letters of credit).
- Identify required governmental approvals and mandate governmental support in expeditious processing of permits.
- Specify the consequences of a change in law and sustained *force majeure* periods (for example, tariff adjustments or government obligation to buy the facilities at a price that retires debt and provides return of equity).

- Provide protection against uninsurable *force majeure* events, such as war, insurrection, terrorism, general strikes and natural events to which a country may be especially vulnerable (for instance, floods and cyclones).
- Delineate the tax and customs status (with acknowledgement of specific effects of change in law).
- Specify the procedures for resolution of disputes when government agencies have competing interests (an occasional problem when the interests of a government-owned energy purchaser differ from those of the government entity that owns or controls the geothermal resources).
- Establish procedures governing the availability of foreign exchange, convertibility protection, and exchange risk coverage.
- Clarify immigration procedures.
- Provide a covenant against expropriation and assurance of no discriminatory action.
- The removal of foreign ownership limits.

These covenants are often requested by sponsored power projects. Although not every project will have an Implementation Agreement, project sponsors, investors and lenders will seek assurances that these matters are covered in binding agreements or applicable laws.

Energy sales agreement provision

In power-project financings of energy facilities, the power sales or thermal-energy sales agreement represents the primary source of revenue for the project. Unless the sponsor and its funding sources are willing to bear market risks, the agreement must obligate the energy purchaser to purchase enough power or thermal energy from the project to support the project debt. A firm commitment at a specified revenue level or levels, not subject to material changes or offsets, is the classic technique to mitigate market risks. A commitment to take if delivered, subject at most to very narrow maintenance and emergency exceptions, is a preferable approach.

The key contract terms and mitigation approaches are identified below:

- *Term*: It is important that the contract have a term that is sufficient to support the financing. The term also must fit with the schedule of projected construction and operation.
- *Prices*: The basic goal of the energy sales agreement is to establish a firm commitment to purchase power or thermal

energy, at a specified revenue level or levels, and without the possibility of material changes or offsets. The absence of a firm price for the output of the project could lead to instability of the revenue flow. Pricing may take the form of a capacity charge (sufficient to cover fixed costs) and an energy component (to cover variable costs).

- *Conditions*: Conditions precedent to the obligation to the purchaser to purchase the energy output should be minimized.
- *Termination*: The grounds for termination should be narrow in scope. The energy sales agreement should give a successor party the right to continue performance in the event of a bankruptcy.
- *Force majeure*: The ability of the purchaser to terminate or modify the energy sales agreement in the event of governmental action should be eliminated, and the obligation to continue capacity payments (at a level sufficient to cover debt service) should be clarified during *force majeure* periods.
- *Government jurisdiction*: The energy sales contract should not contemplate continuing governmental review of the contract.
- *Operations*: The energy sales contract should contemplate the details of interconnection requirements and charges, design and construction requirements, equipment and metering, and continuation of service if an outage occurs.
- *Additional capital or operating costs*: Energy sales agreements must be examined for potential imposition of additional capital or operating costs. Of course, other contract provisions may also provide risk-mitigation objectives. In the international market, a preferred, but less common, approach is to obtain such clauses as price adjustment provisions that adjust the energy sales rate to accommodate changes in project economics as a result of changes in environmental laws.
- *Currency mix*: Rates may be payable in a basket of currency that reflects the mix of currencies in which the project's obligations (for example, operations, debt service and profit) are payable.
- *Currency fluctuation*: The rate may be subject to change to reflect changes in exchange rates.
- *Change of law*: To the extent changes in law (for instance, tax laws or environmental rules) require construction of additional facilities or incurrence of increased expenses, the rate may be subject to adjustment.

- *Disputes*: The experiences of corporate investors caught up in Argentina's economic crisis in 2002 demonstrate that it is important for investors to nominate an appropriate forum for international arbitration. Project sponsors will need to weigh the comparative benefits of alternative fora for dispute resolution.

As discussed above, related matters may be covered in the Implementation Agreement.

Insurance packages

Many techniques exist to deal with overall country risks. The underlying concerns include political instability, war, terrorism, unknown legal requirements, a change of law, currency convertibility and unknown conditions for performing the work.

As discussed above, certain of these risks can be dealt with contractually. In many foreign projects, the Implementation Agreement incorporates the government's granting of certain protection and assurances. Certain risks, however, cannot be adequately covered by contracts with other parties. A primary example is casualty losses. The primary means of addressing these risks is through insurance. For example, all-risks insurance and business-interruption insurance are the principal protection against casualty losses and attendant loss of revenues.

In the international context, political risk insurance is available from a limited number of sources to offset certain risks. The basic risks covered are expropriation, currency inconvertibility and breach of contract by governmental bodies.

In the wake of the 11 September 2001 terrorist attacks in New York City and Washington, D.C., terrorism insurance has become a subject of renewed importance, not just in relation to projects located in the United States, but on a worldwide basis. Whereas terrorism was once included in general casualty insurance policies, the heavy losses suffered by the insurance industry as a result of the 11 September attacks have led to the creation of terrorism insurance as a separate, and often expensive, policy which financiers frequently require project sponsors to obtain. The cost and availability of terrorism insurance has therefore become an additional factor to be considered in relation to project budgets and financing.

Other project agreements

Consummation of the project financing of a power plant or thermal energy system requires that the developer obtain all permits necessary for the construction of the project, and that all of the contractual arrangements relating to development of the project be negotiated and executed. The purpose of obtaining the permits and entering into the agreements is to assure the project lender that it may advance funds to the project with a high degree of confidence that the project will be successfully developed and implemented. There follows a brief description of site selection, permissions and construction issues.

Site selection and permissions

Among the most fundamental aspects of any project financing are those involving site selection and permissions. Lenders typically are unwilling to fund project construction unless they are assured that the developer has definitive control of the project site and has in hand all permits necessary to initiate construction of the project. A key permissions consideration for power supply projects, of course, is receipt of sufficient rights to obtain, transport and use fuel for power generation. The developer may establish its control over the site by either acquiring the site directly, or entering into a long-term lease. Whichever alternative is selected, the lender will require appropriate evidence of ownership, title insurance, and appropriate mortgages or other security arrangements granting the lender a first lien security position with respect to all real property comprising the project.

When power facilities or thermal energy systems are built in industrial areas, environmentally related issues frequently arise in connection with the site-acquisition process. Lenders will require appropriate assurances that by lending to the project, they will not incur any liability under applicable environmental laws, either by virtue of their security interest in the real property comprising the facility or as a result of any foreclosure on that interest. It is now routine for lenders to make their financing conditional on the availability of a satisfactory environmental audit of the project site. Firms specializing in environmental analysis commonly provide such audits.

Lenders also will not finance project construction unless all permits necessary to begin operation of the project have been obtained, or evidence exists that they can be obtained prior to operation. The permits required include air quality permits, waste disposal permits and water discharge permits. In addition, all required regulatory filings for power or thermal energy sales must be in place.

Construction contract

Certain provisions of the construction contract are important in energy project development. Key provisions of construction contracts that need to mesh with the energy sales contract include the following:

- *Fixed price*: The contractor may agree to perform to construct the facility for a fixed price, which must apply regardless of the actual price of construction.
- *Completion date*: The contract may establish the date by which construction of the project is to be completed. The contractor is typically subject to penalties in respect of a failure to complete the project in a timely fashion, in amounts sufficient to pay the additional interest on the construction debt resulting from the failure to provide timely completion. In addition, contractors may be entitled to bonus payments for early completion of the project.
- *Project acceptance*: The construction contract normally establishes a set of operating standards that must be satisfied by the project before it will be subject to acceptance by the developer. If the project does not meet the contractually established performance standards, the contractor is usually required to pay damages to the developer and lender in amounts intended to compensate them for the increased construction cost and/or lost energy-sales revenues resulting from the failure of the facility to operate at the specified performance levels.
- *Warranties*: The contract should provide performance warranties on workmanship, engineering and mechanical parts, providing for replacement of defective parts and repair of defective work for a specified time (typically, two years) at no cost to the developer.
- *Bonding*: Bonding may not be required of very large and financially strong construction companies. All other construction companies, however, are typically required to post payment and performance bonds for the full amount of the construction contract, which are payable to the developer and lender in the event of default by the contractor.
- *Insurance*: Risk of loss pending completion of construction typically is placed on the contractor, which must insure the project in amounts equal to the full replacement cost of the facility. The contractor also must provide appropriate liability insurance for the construction.
- *Retainage*: Construction contracts generally provide for payment of the contractor in a series of instalments that become due as construction milestones are achieved. Ordinarily, however, the contractor will not be paid a

certain portion of each milestone payment until final satisfactory completion of the project. The amount of the retention is typically from 5 to 10 per cent of each construction draw.

Enforceability

In the end, contracts are only as good as their enforceability. It cannot be taken for granted that the laws of the host government will assure that contractual provisions are enforceable. Some questions relate to formal legal rights, such as the ability to perfect security in revenues in accounts or revenues subsequently realized. Other legal issues may relate to issues of when ownership interests vest. There may also be issues as to whether a 'foreign' lender may enjoy security interests in a collateral package.

Legal steps can be taken to improve the position of a developer. For example, choice of law clauses, international arbitration procedures, and waivers of sovereign immunity are critical. But the ultimate protection is a well-conceived project, one that will make economic sense over time, and one with the proper mix of local support (including local partners) to best ensure long-term viability.

For Section 9.3, see the following: Augenblick and Scott Custer (1990); Benoit (1996); Bond and Carter (1994); Cook (1996); Sifford et al., (1985); UNCITRAL (1998); World Bank Industry and Energy Department (1994).

REFERENCES

AUGENBLICK, M.; SCOTT CUSTER, B., JR. 1990. *The Build, Operate and Transfer ('BOT') approach to infrastructure projects in developing countries.* World Bank Working Paper No. 498, Washington D.C., World Bank.

BALDI, P. 1990. General development of a geothermal project. In: M. H. Dickson and M. Fanelli (eds.), *Small Geothermal Resources: A guide to development and utilization*, pp. 203–15. New York, UNITAR.

BENOIT, P. 1996. *Project Finance at the World Bank: An overview of policies and instruments.* World Bank Technical Paper No. 312, Washington D.C., World Bank.

BLOOMQUIST, R. G. 1990. *Load Following and Dispatchability in the Larderello Geothermal Field Italy.* US Department of Energy, Contract C90102925.

BLOOMQUIST, R. G.; SIFFORD, B. A. 1995. *Evolving Power Plant Designs Prepare American Geothermal Industry for the 21st Century.* Proc. World Geothermal Conference '95. Florence, pp. 639–44.

BOND, G.; CARTER, L. 1994. *Financing Private Infrastructure Projects: Emerging trends from IFC's experience.* International Finance Corporation Discussion Paper, 23, Washington D.C.

BORCIER, W.; MARTIN, S.; VIANI, B.; BRUTEN, C. 2001. Developing a process for commercial silica production from geothermal brines, *Geothermal Resource Council Transactions*, Vol. 25, pp. 487–91.

BORCIER, W. 2002. *personal communication.*

CAMPBELL, R. G. 1995. The power plant. In: R.G. Bloomquist (ed.), *Drafting a Geothermal Project for Funding*, pp. 73–94. Pisa, WGC '95 Pre-Congress Courses.

CLUTTER, T. J. 2000. Mining economic benefits from geothermal brine. *Geo-Heat Center Quarterly Bulletin*, Vol. 21, No. 2, p. 123.

COOK, J. 1996. Infrastructure project finance in Latin America. *24 Int'l Business Law.*

ELIASSON, E. T.; ARMANNSSON, H.; FRIDLEIFSSON, I. B.; GUNNARSDOTTIR, M. J.; BJORNSSON, O.; THORHALLSSON, S.; KARLSSON, T. 1990. Space and district heating. In: M. H. Dickson and M. Fanelli (eds.), *Small Geothermal Resources: A guide to development and utilization*, pp. 101–28. New York, UNITAR.

FESMIRE, V. R. 1985. Reliability through design, the Santa Fe geothermal plant no. 1. *Trans. Geothermal Resource Council*, Vol. 9, pp. 133–7.

INDEPENDENT POWER. 1989. February and June editions. New York, McGraw-Hill, Inc.

KOENIG, J. B. 1995. The geothermal resource: data requirements of the fuel supply. In: R.G. Bloomquist (ed.), *Drafting a Geothermal Project for Funding*, pp. 43–71. Pisa, WGC '95 Pre-Congress Courses.

KLEINHAUS, P. V.; PRIDEAU, D. L. 1985. *Design, Start-Up, and Operation of SMUDGEO #1*. PCEA Annual Engineering and Operating Conference, TP 85-86.

MCKAY, G. D.; TUCKER, R. E. 1985. *Integrating Geothermal Power Plant Design with Steam Field Operation for an 80 MW Plant in The Geysers*. San Diego, PCEA EPRI/IIE Geothermal Conference and Workshop.

PARKER, D. S.; SIFFORD, A. W.; BLOOMQUIST, R. G. 1985. Estimating the levelized cost of geothermal electricity. *Trans. Geothermal Resources Council*, Vol. 9, pp. 189–93.

SIFFORD, A. W.; BLOOMQUIST, R. G.; PARKER, D. S. 1985. Estimating development costs of geothermal electricity. *Trans. Geothermal Resources Council*, Vol. 9, pp. 195–99.

STEFANSSON, V. 1999. *Economic Aspects of Geothermal Development*. Ljubljana, Slovenia, International Workshop on Direct Use of Geothermal Energy.

SULLIVAN, R. 2001. *Personal communication.*

UNCITRAL. 1998. *Legal Guide to Drawing Up International Contracts for Construction of Industrial Works*, New York, UN Publications A/CN/9/SER.B/2.

WORLD BANK INDUSTRY AND ENERGY DEPARTMENT. 1994. *Submission and Evaluation of Proposal for Private Power General Projects in Developing Countries*. Occasional Paper No. 2, Washington D.C., World Bank.

SELF-ASSESSMENT QUESTIONS

1. What are the primary factors affecting project economics common to all forms of geothermal development, whether power production or direct-use applications?

2. The fuel for a geothermal facility (that is, the geothermal resource) is similar to oil, gas and biomass in terms of its availability and cost: true or false?

3. Explain the similarities or differences between the fuel for a geothermal project and fuels such as coal, gas, oil and biomass.

4. Why are productivity and longevity of the geothermal reservoir so critical to obtaining project financing?

5. Explain why detailed data collection and accurate data analysis and interpretation are so important.

6. What are the primary factors that result in significantly higher capital cost for a binary plant than a steam cycle plant?

7. Why do binary plants have significantly higher parasitic loads than steam cycle plants?

8. Why does the power sales contract play such an important role in influencing power plant design?

9. Name one mechanism that can be used in the equipment procurement process to ensure steam use efficiency.

10. What is partial-arc admission and how does it contribute to plant operating efficiency?

11. The design of the cooling tower is simply a matter of personal preference and has no impact on cost or operation: true or false?

12. In a steam cycle plant, a turbine bypass plays a significant role in environmental compliance but plays no role in plant availability: true or false? Explain your answer.

13. Many direct-use geothermal projects depend totally on geothermal energy to meet thermal demand. What effect on economics would meeting peak demand with a fossil-fuel-fired boiler have, and what other reasons can you give for considering this approach?

14. What technology allows geothermal resources to be used to provide cooling?

15. Which sources provide revenue generation in (1) electricity generation, and (2) direct-use applications?

16. What factors are critical to ensuring that a generating plant produces maximum revenue?

17. Why are operation and maintenance important not only to revenue generation, but also to the obtaining of financing?

18. Geothermal power plants are generally considered to be base-load units. However, certain factors may make load-following or even operating in a dispatchable mode necessary or economically attractive. What are these factors, and what design consideration should be taken into consideration to ensure maximum operation flexibility and fuel-use efficiency?

19. Why is revenue potential for such direct-use applications as greenhouses or industrial processing almost entirely independent of the viability of the geothermal system?

20. Thermal sales from a district energy application are very weather-dependent. What contractual provision ensures that revenues will equal or exceed the cost of providing service?

21. When should cascading be considered and what impact can it have upon project economics?

22. What is co-production? Give three ways in which co-production can improve power plant economics.

23. Explain why policy reform may be critical to the success of governmental efforts to encourage private sector development of geothermal facilities.

24. What is a BOT project?

25. Under what circumstances may 'payment backstops' be necessary?

26. How does 'project financing' differ from conventional means used to finance energy projects?

27. What are examples of market risk, and at what stages of project development can they arise?

28. What is an Implementation Agreement?

29. How do Energy Sales Agreements serve to mitigate market risks?

ANSWERS

1. Primary factors include: (a) fuel supply, (b) design and construction, (c) revenue generation, and (d) financing.

2. False.

3. Geothermal 'fuel' is available only after extensive exploration and development drilling. It cannot be transported over long distances, cannot be purchased on the open market or legislated into existence.

4. The ability to generate revenue to cover financial obligations is totally dependent upon an adequate fuel supply over the economic life of the project.

5. Detailed and accurate data are critical to making decisions concerning project design, operation and maintenance, as is provision of documentation to potential financiers that the project will be able to meet its financial obligations over the economic life of the project.

6. The heat exchangers, air-cooled condensers, well pumps and circulating pumps.

7. Because of the need to pump the geothermal fluids under pressure through the heat exchanger, and the need to pump the working fluid through the heat exchanger and to the turbine.

8. The power sales contract establishes terms of payment including all incentives and/or penalties.

9. The use of procurement evaluation criteria that penalize inefficient use of steam.

10. Partial-arc admission refers to a turbine design that allows for steam to enter the turbine through differing portions of the turbine blade sets, and allows for the turbine to be operated at various steam-pressure levels and at different outputs.

11. False.

12. False, the turbine bypass also allows the plant to come back on-line more quickly after a forced outage.

13. The use of a fossil-fuel-fired boiler to meet peak demand could dramatically reduce the number of wells that would be required, as well as allow for the use of smaller diameter piping. Other benefits include a secure back-up and the ability to meet temperature requirements in excess of that available from geothermal.

14. Absorption chiller.

15. (1) Sale of electricity and, possibly, thermal energy from cascade users. (2) Thermal sales, product sales (for example, flowers or plants from a greenhouse) and value-added service (for example, vegetable dehydration).

16. Ability to meet the conditions of the power sales contract including the flexibility of operation, a high level of availability, and sound operation and maintenance procedures and practices.

17. Financiers look very closely at the operation and maintenance team because of the critical impact sound operation and maintenance practices have upon the capacity and availability of a facility, and hence the potential of a facility to generate revenue over the economic life of the project.

18. Both contractual provisions of the power sales contract and/or reservoir concerns can provide incentives for operating in a load-following or dispatchable mode. Design consideration includes use of redundant equipment, use of a turbine bypass, partial-arc admission, computerized well-field operation, and remote operation.

19. Revenue generation for these projects is dependent upon the sale of a product (for instance, vegetables) or providing a value-added service (for example, lumber drying).

20. Thermal sales contracts should contain both capacity and variable-cost provisions. The capacity portion of the bill should be sufficient to cover debt service where other costs such as O&M and fuel make up the variable-cost portion of the bill.

21. Cascading should always be considered, and can often serve as a secondary source of revenue generation.

22. Co-production is the production of silica and other marketable products from geothermal brines. Co-production materials can be a source of additional revenue. Removal of silica minimizes re-injection problems; the removal of silica also may allow for additional geothermal energy extraction in bottoming cycles.

23. The country may not yet have in place laws and regulations that are conducive to long-term investment in the energy sector.

24. A build–operate–transfer project.

25. A payment backstop, such as a letter of credit or central bank guarantee, may be necessary if the energy purchaser cannot provide meaningful assurances of its practical ability to make timely payments for delivered electricity.

26. Unlike sovereign-guaranteed financing (where the government guarantees debt payment) and corporate finance (where the sponsor pledges its corporate assets as security for the loan), the lender in project financing looks solely to the revenues to be produced from the project being financed, and the underlying contracts and physical assets, as security for repayment.

27. Market risk can exist as a consequence of changes in the near-term and long-term need for energy of the energy purchaser. Market risk can also arise from changes in the prices a purchaser is willing (or able) to offer to suppliers from new facilities. Market risk can exist at the development, construction and operation stages.

28. An Implementation Agreement is a contract by which the government makes certain commitments to a project sponsor in order to facilitate project development, financing, construction and operation.

29. An Energy Sales Agreement serves to mitigate market risks by establishing the contract term, purchase price, and sales quantity.